20th Conference on Power System Engineering 2021

MATEC Web of Conferences Volume 345 (2021)

Pilsen, Czech Republic
7 - 9 September 2021

Editors:

Lukas Richter
Martin Pelikan
Michal Volf
Michaela Vackova

ISBN: 978-1-7138-3801-2

Printed from e-media with permission by:

Curran Associates, Inc.
57 Morehouse Lane
Red Hook, NY 12571

Some format issues inherent in the e-media version may also appear in this print version.

For additional information, please contact EDP Sciences – Web of Conferences
at the address below.

EDP Sciences – Web of Conferences
17, Avenue du Hoggar
Parc d'Activité de Courtabœuf
BP 112
F-91944 Les Ulis Cedex A
France

Phone: +33 (0) 1 69 18 75 75
Fax: +33 (0) 1 69 28 84 91

contact-edps@webofconferences.org

Additional copies of this publication are available from:

Curran Associates, Inc.
57 Morehouse Lane
Red Hook, NY 12571 USA
Phone: 845-758-0400
Fax: 845-758-2633
Email: curran@proceedings.com
Web: www.proceedings.com

TABLE OF CONTENTS

MATEC Web of Conferences **345**, 00001 (2021) https://doi.org/10.1051/matecconf/202134500001
Power System Engineering 2021

20th Conference on Power System Engineering

Lukas Richter[1*], *Martin* Pelikan[1], *Michal* Volf[1] and *Michaela* Vackova[1]

[1]University of West Bohemia in Pilsen, Univerzitni 22, 306 14 Pilsen, Czech Republic.

Abstract. The Conference on Power System Engineering has a long-term tradition. It is an annual meeting of professors and experts from departments and universities where power system engineering is taught. The 20th year of this event was held in Pilsen (Czech Republic) between 7th and 9th September 2021 and was hosted by Department of Power System Engineering of Faculty of Mechanical Engineering, University of West Bohemia in Pilsen.

1 Preface

There were 74 delegates from academic and industrial field who came from the Czech Republic, the Slovak Republic, Germany, Hungary, Poland, Ukraine and Russia. The conference was launched by the rector of the University of West Bohemia. The organizers received 35 contributions. 34 contributions have gone through a professional review for the publication. All presentations brought contributions to the fields of equipment for classic and nuclear power plants including their operation, fluid mechanics and flow simulations, thermodynamics, heat transfer, diagnostics of power machines and equipment etc.

The social evening with a programme gave the delegates an opportunity to meet their colleagues, share and discuss their experience and results of their research projects in less formal environment. The participants could also visit the world-famous brewery Pilsner Urquell. The programme of the conference included also a student section, in which especially Ph.D. students have an opportunity to present their contributions or posters.

The conference was financially supported by a Visegrad grant. The main goal of the grant is to build an international network of scientists and experts in the V4 region with the focus on energy savings, efficiency and reliability increase in various industrial processes, development and implementation of emission-free technologies. At a time of transformation of the energy industry caused by the ever-increasing demand not only for electricity, but also for other forms of energy, decarbonization and the reduction of fossil fuel reserves, it is necessary to work especially in academia to develop new technologies and methods to implement energy savings. The issue of energy savings is currently addressed to some extent by all the grant partners. However, since each workplace has a certain area of focus, the results of such research have limited scope. Only the international cooperation of these workplaces offers a comprehensive solution, because only then is it possible to communicate experience and knowledge in specific areas in order to harmonize the activities of the individual workplaces.

---

* Corresponding editor: richterl@kke.zcu.cz

Fig. 1. Visegrad fund – Project ID/Title: 22110166/ Conference on Power System Engineering 2021.

The conference organizers would like to thank all industrial and academic partners of the conference for their financial support. The Institute of Thermomechanics of the Academy of Sciences of the Czech Republic contributed financially to the organization of the conference within the framework of the Strategy 21. Other significant financial support came from Doosan Škoda Power, AFRY, NUM solution, ATMOS Chrast, Helago, Techsoft Engineering, TechSim Engineering and Czech Society for Mechanics.

Fig. 2. Logos of sponsoring organizations.

Statement of Peer review

In submitting conference proceedings to *Web of Conferences*, the editors of the proceedings certify to the Publisher that

1. They adhere to its **Policy on Publishing Integrity** in order to safeguard good scientific practice in publishing.
2. All articles have been subjected to peer review administered by the proceedings editors.
3. Reviews have been conducted by expert referees, who have been requested to provide unbiased and constructive comments aimed, whenever possible, at improving the work.
4. Proceedings editors have taken all reasonable steps to ensure the quality of the materials they publish and their decision to accept or reject a paper for publication has been based only on the merits of the work and the relevance to the journal.

Title, date and place of the conference

20th Conference on Power System Engineering
September 7 – 9, 2021 in Pilsen, Czech Republic

Proceedings editor(s): Lukas Richter, Martin Pelikan, Michal Volf, Michaela Vackova

Date and editor's signature 30. 9. 2021

17, avenue du Hoggar - PA de Courtabœuf – BP 112 - 91944 Les Ulis Cedex A (France)
Tél. : 33 (0)1 69 18 75 75 - Fax : 33(0)1 69 07 45 17 – www.edpsciences.org

Machine learning model designed to predict the amount of CO_2 produced by a small pellet boiler

Alexander Backa[1*], *Juraj* Drga[1], *Lucia* Martvoňová[1], and *Mária* Polačiková[1]

[1]KET, SjF, Žilinská Univerzita v Žiline, Univerzitná 1, 010 26 Žilina.

Abstract. Emissions, including CO_2 emissions, are generated during the combustion process. Perfect combustion of biomass should not lead to the formation of CO, but all carbon should burn perfectly and change to CO_2 by the oxidation process. Under real conditions, complete combustion never occurs and part of the carbon is not burned at all or only imperfectly to form CO. The aim of the work was to create a prediction model of machine learning, which allows to predict in advance the amount of CO_2 generated during the combustion of wood pellets. This model uses machine learning regression methods. The most accurate model (Gaussian process) showed a root-mean-square error, RMSE = 0.55. The resulting mathematical model was subsequently verified on independent measurements, where the ability of the model to correctly predict the amount of CO_2 generated in % was demonstrated. The average deviation of the measured and predicted amount of CO_2 represented a difference of 0.53 %, which is 8.8 % of the total measured range (3.08 - 9.2). Such a model can be modified and used in the prediction of other combustion parameters.

1 Introduction

A number of households also include a heat source for heating, in which energy is converted into heat, which is used to ensure thermal comfort. Mankind used mainly biomass, coal and later also natural gas as a fuel source. In recent decades, innovations have also taken place in the field of thermal technology, and less traditional sources have been used, which we also call alternatives. Today's users often demand comfort, which includes the operation of the combustion device itself, with those that are automated being preferred or without the need for constant control of the combustion mode, if at all. Such automated heat sources also include pellet boilers, where the operator provides refueling in the form of pellets, cleaning of the burner and ashtray and basic maintenance. The most used pellets are wood pellets, where cleanliness and low humidity are important. However, other types of pellets are increasingly being produced (e.g. from phytomass) or are enriched with additives such as bark, paper sludge, coffee and others.

When burning pellets in order to obtain energy, it is often necessary to maintain some combustion parameters in the required range of values so that undesirable conditions do not occur or, conversely, in order to achieve the necessary combustion outputs. These values can

* Corresponding author: alexander.backa@fstroj.uniza.sk

be influenced in time by setting the combustion device, the choice of fuel and also by changing the physic-chemical properties of the media that enter the combustion process [1]. Thanks to advanced diagnostic measurement methods, the combustion process can be extensively monitored [2]. In some applications it is necessary not only to monitor, but also to evaluate changes in quantities using prediction. Due to the complex nature of combustion, it is difficult to accurately predict the state of the investigated quantity, so creating a quality model is often difficult [3]. In most existing models, differential equations describing the combustion process are used in the calculation [4]. The disadvantage of most existing models is their complex computational technique and complex transformation with the necessary changes to the model [5]. An alternative to the existing mathematical models used is the use of artificial intelligence, for the purposes of prediction, which has been increasingly used in recent years in the research of combustion and combustion plants [6].

The study by Baruah et al. [7] describes the development of a model of biomass gasification based on an artificial neural network (ANN) in gasification plate gasifiers with a fixed bed. The prediction of the concentration of CH_4, CO, CO_2 and H_2 was calculated using a model based on ANN and the results show a good agreement with the experimental outputs by achieving a coefficient of determination - $R^2 > 0.98$ for CH_4, CO, CO_2 and H_2 [8].

One of the sub-areas of artificial intelligence is machine learning, in which it is possible to use classification or regression, where the created model is learned from a set of training data. In classification, the output is the classification of the training example into a selected set, and in regression, the output is a specific value corresponding to the training example [9]. To predicting some combustion parameters, which are based on other measured quantities of combustion, it is therefore appropriate to use regression techniques to obtain specific values. Using this technique, a model is also created in this article, where the amount of CO_2 emissions generated during the pellet combustion process is determined.

2 Procedures and methods

Input data for the regression model calculation were obtained from available measurements for 100% wood pellets (FIG. 1).

Fig. 1. 100% wood pellets

Wood pellets were made from spruce sawdust, which underwent a crushing process and subsequent pelletization. Basic parameters - physical and chemical composition, deformation temperature, moisture content and ash content of used wood pellets are given in Table 1. Fuel properties were determined by thermogravimetric analyzer LECO TGA701, LECO CHN628 device for chemical analysis and also LECO AC 500 device for determination of higher heating value (FIG. 2).

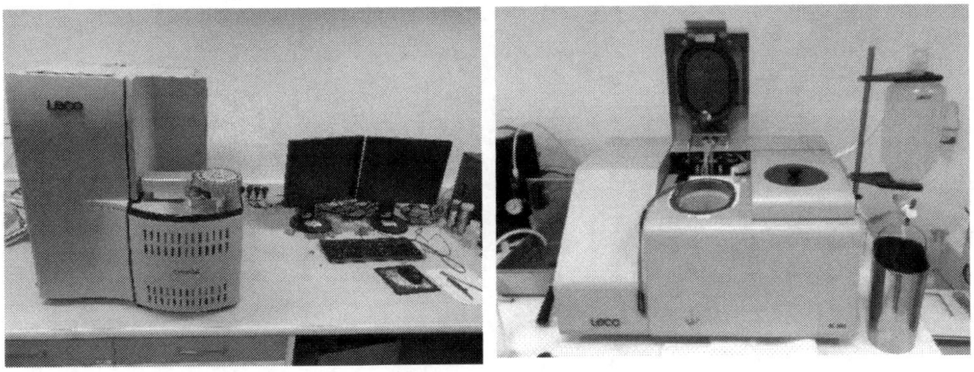

Fig. 2. LECO CHN628 (on the left) and LECO AC 500 (on the right)

The performed measurements were realized on the LOKCA ÚSPOR boiler (FIG. 3), where a continuous measurement of parameters was performed.

To obtain the results, a heat source with a nominal heat output of 18 kW was used, which is primarily intended for burning wood pellets. This combustion device was equipped with an additionally mounted rotary burner the original. The retort burner was removed and replaced by a rotary burner with a nominal output of 25 kW.

The boiler contains a water exchanger with a volume of 55 l and a 200 l fuel tank in which a screw conveyor is located. The pellets are fed from above and transported to the rotary burner, where they are mixed with the combustion air, and by gradually burning and shifting, the burnt fuel is forced into the ashtray.

The control system was selected for a preset maximum boiler temperature of 70 °C and the fuel dosing to the rotary burner was adjusted accordingly. The heat produced from the heating system was dissipated by means of a heat exchanger to the cooling tower. The regulation was taken care of by the MAGA S. CONTROL device, which automatically controlled the combustion air supply, fuel dosing and maintained the maximum boiler temperature using the control electronics and our preset values.

Table 1 Basic characteristics of burned pellet samples

INPUT MEASUREMENT		
Proximate analysis (from TGA)	**Unit**	**100% wood pellets**
Moisture	%wt	3,86
Ash content	%wt	0,49
Volatile combustible matter	%wt	74,99
Fixed carbon	%wt	16,31
Ultimate analysis		
Carbon	%wt	48,78
Hydrogen	%wt	6,31
Nitrogen	%wt	1,25
Calorimetric analysis		
Higher heating value	MJ.kg-1	18,29
Lower heating value	MJ.kg-1	16,82

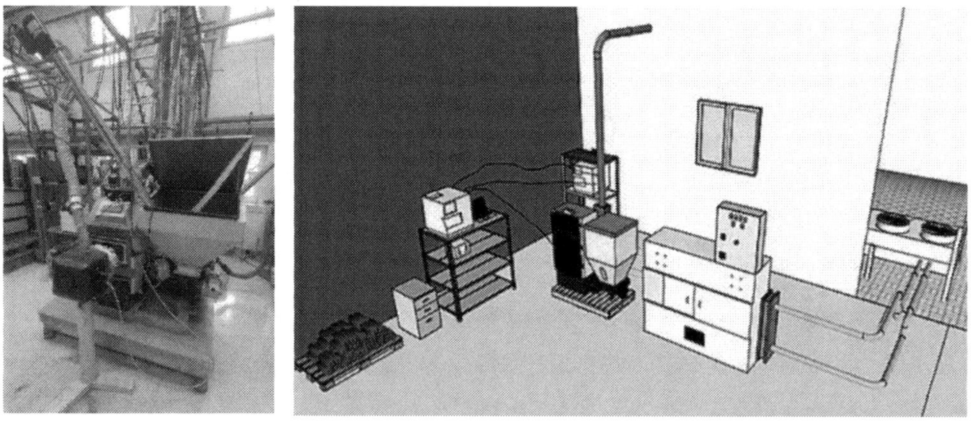

Fig. 3. boiler LOKCA ÚSPOR 18 with a rotary burner with an output of 25 kW (on the left) and boiler room scheme (on the right)

A total of 371 measurements were recorded in this way, which were used as inputs to the proposed model. The selected set of parameters - Table 2, which was selected for this model, can be changed depending on the need to improve the predictive quality of the model.

Table 2 Measured parameters during the combustion of 100% wood pellets

INPUT MEASUREMENT	
Quantity	**Unit**
Ambient air temperature	°C
Chimney temperature	°C
Chimney draft	Pa
O_2 in the flue gas	%
CO in the flue gas	ppm
Heat output	kW
CO_2 in the flue gas	%

To predict the amount of CO_2 produced, it would be appropriate to add parameters such as humidity and speed of the supplied air, abrasion resistance of the pellets and the amount of fuel supplied. When creating a comprehensive model for different types of fuels and combustion samples, such a model would also include characteristics that differ for different samples. It would be appropriate to add quantities such as chemical composition, moisture content and amount of ash to such a model. The selection of specific quantities was based on knowledge of the impact on CO_2 production and also on the basis of performed measurements and the availability of measuring devices for the combustion equipment and the combustion process. If significant influence quantities on the investigated parameter are neglected, it is necessary to keep them constant in time in new predictions or as changing as during the measurements from which the mathematical model was created. Only thanks to this can the prediction be correct and sufficiently accurate.

Subsequently, the parameters were entered into the Matlab program, where using the regression module - Regression Learner App, a model was built with the input parameters listed in Table 1, including the amount of CO_2 emissions, which were considered as the output (determined) quantity. The prediction model was developed to determine a specific value depending on the selected parameters, while the evaluation of success depends on the performance of tests in k-fold cross-validation using calculated parameters - coefficient of determination (R^2), mean square deviation (RMSE) and also root mean square error (MSE) [10]. The input data set is divided into subsets. One subset serves as a test set, the remaining subsets serve as training sets. The classifier trains the model on a trained set and uses the test set to test the accuracy and performance of this model. This process is repeated several times with different subsets. In multiple cross-validation, a value of 10 was chosen, which is most often used because it is the most objective [11].

$$R^2 = 1 - \sum_{i}^{m} \frac{(Mi - Pi)^2}{(Mi - Si)^2} \ (-) \tag{1}$$

$$RMSE = \sqrt{\frac{1}{m} \sum_{i}^{m} (Mi - Pi)^2} \ (-) \tag{2}$$

$$MSE = \frac{1}{m} \sum_{i}^{m} (Mi - Pi)^2 \ (-) \tag{3}$$

Where M_i is the measured value, P_i is the predicted value, S_i is the mean value and m is the number of data points. These parameters are used to compare individual models, but they also evaluate the success of the descriptive function, which is based on a specific data set. RMSE is used for the purpose of comparing the observed values with the values predicted by the model. A value of 0 would mean a perfect match of the data, so a lower RMSE value is generally better. The coefficient of determination R^2 shows how the regression prediction approaches the actual data. The value of R^2 normally ranges from 0 to 1, and if this value is closer to the number 1, the regression predictions are better expressed using the proposed regression function. The value of R^2 is a measure of how well the measured results are replicated by the model [12].

For control purposes, it is appropriate to predict the new data on an independent measurement using the created model. The model created in this way can then be verified and used in predicting the selected quantity in independent measurements, thanks to which it is possible to determine the approximate value of the sought quantity without the need for its measurement.

The regression models used in the calculation in this research were linear regression, regression decision trees, supporting vector machines, Gaussian process regression models, regression tree sets, and neural networks. After testing all types of machine learning, the one with the lowest RMSE was chosen.

3 Regression model of machine learning CO_2 emissions

After reading the measured data, it was necessary to choose a validation method, while for a specific application, cross-validation with 10 subsets was selected. This type of validation appeared to be the most appropriate when performing various experiments with changing validation settings.

Subsequently, testing of individual types of regression models was started, while the lowest value of the mean square deviation, RMSE = 0.035, was shown by the regression model of the Gaussian process. Figure 4 shows the selected regression model, where the measured and prediction data are shown. The "x" axis represents the measurement sequence and the "y" axis the value of CO_2 emissions in %, while it is possible to graphically observe an excellent agreement, which is also confirmed by the coefficient of determination R^2 in this case with 1.00 and the root mean square error with MSE = 0.001. Figure 5 shows the deviations of the predicted values from the actual values entering the model using the prediction function. Matlab code was subsequently generated for this model. Using this code, it is possible to further predict (also) states other than those that were inserted into the model by entering all input quantities with the required values. In this way, it is possible to predict new model situations.

Fig. 4. Comparison of CO_2 emissions (%) of predicted data (yellow) from measured data (blue)

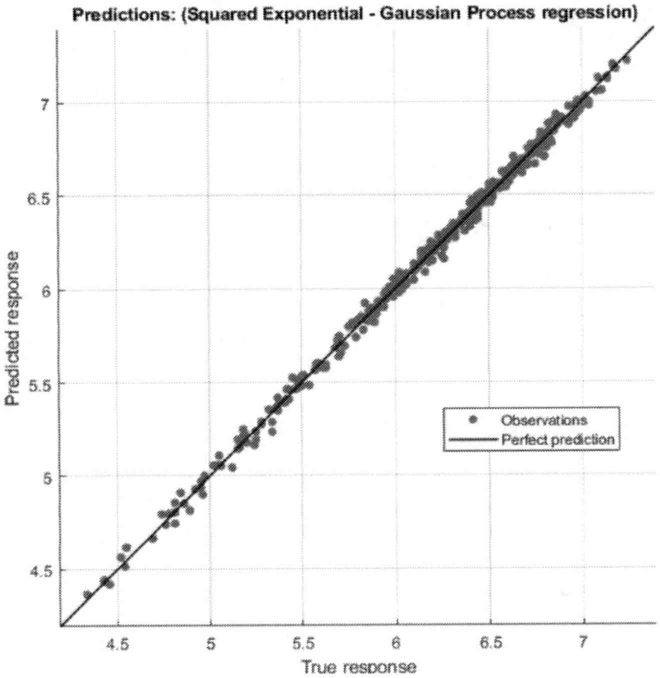

Fig. 5. Prediction accuracy indicator

To verify the functionality of this model, new special measurements (180 special measurements) were performed using the same type of fuel and also the combustion equipment. These new data were not the subject of the originally created model.

When predicting new data, it is necessary to keep all settings and conditions that entered the model the same as in the measurements for this model, where only some specific parameters were selected. A new 180 measured records were inserted into the created model for the prediction of CO_2 emissions and a value in % was determined for each measurement. This value was then compared with that predicted by the proposed prediction model, compiled from the originally measured data. In this way, the new error RMSE = 0.55 with the value MSE = 0.31 was determined, as well as the average deviation of the predicted data from the measured values, namely 0.53 %. Such a result shows that the proposed model is functional and can be used to predict the amount of CO_2 generated in the flue gas. A graphical comparison of the predicted and measured data is shown in FIG. 6.

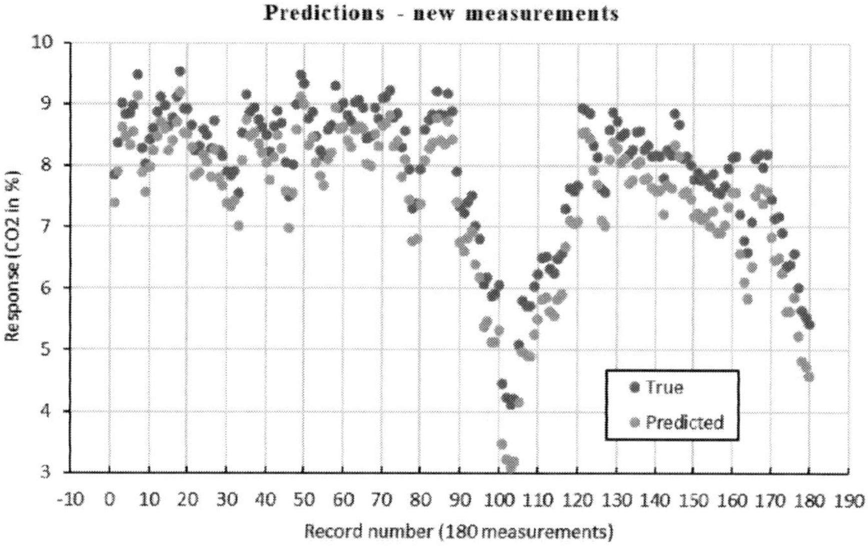

Fig. 6. Prediction of a new model situation for 180 measured records using the proposed model

When using such a model in practice, it is necessary to realize that the model will always show a certain deviation from the real measured quantity. However, this deviation can be eliminated by adding other quantities that have a significant effect on the parameter sought. It is also possible to maintain the "learning" of the regression model by adding new measurements.

4 Conclusion

An interesting concept for further direction would be to link the information obtained from a specific model with specific combustion phenomena and their application to create a comprehensive combustion model using artificial intelligence tools. It would also be possible to implement such a model directly in the combustion plant, which would be improved through further measurements and based on the model could correctly select the necessary parameters to achieve the optimal value of power or other parameters (e. g. emissions). From an environmental point of view, the prediction models could also be used for monitoring and evaluation of produced emissions at a specific time, using combustion plants connected via a network to a central register, which would obtain predicted values of discharges of certain emission values based on sent settings of individual combustion plants.

In this paper, a regression model was built, which allows the calculation of the selected parameter - CO_2 emissions, using the input measured parameters. The usability of such an application lies in the possibility of the correct setting of the combustion device, as well as the adoption of appropriate measures for efficient and high-quality combustion in order to achieve the required amount of CO_2 production. It is also possible to predict the values of a particular quantity using the model without the need for its continuous measurement. In a similar way, it is possible to build a model that would allow the prediction of other parameters.

The success of the proposed model was later determined by independent measurements on the same boiler using also wood biomass pellets. It was found that the predicted values were different on average by 0.53 % compared to those measured. This value represents an error of 8.8 % of the total measured interval (3.08 - 9.19).

This research was funded by VEGA 1/0479/19 "Influence of combustion conditions on production of solid pollutants in small heat sources."

References

1. JANDAČKA, J. – MIČIETA, M. – HOLUBČÍK. M. – NOSEK, R., (2016): *Inovácie na zefektívnenie procesu spaľovania biomasy*. Žilina: EDIS - vydavateľské centrum ŽU, 265 s. ISBN 978-80-554-1236-8.
2. HERNÁNDEZ, R. – BALLESTER, J. Flame imaging as a diagnostic tool for industrial combustion. Combustion and Flame. (2008). 155. s. 509-528.
3. HAN, Z. – LI, J. – HOSSAIN, M. – XU, CH. Prediction of combustion state through a semi-supervised learning model and flame imaging. Fuel. (2021). 289 s.
4. ZHENG, Z. et. al. Progress in the Application of Machine Learning in Combustion Studies. ES Energy & Environment. (2020).
5. MORARI, M. – LEE, J. H. Model predictive control: past, present and future. Computers & Chemical Engineering. (1999). 23(4-5). s. 667-682.
6. ELMAZ, F. – YÜCEL, Ö. – MUTLU, A. I. Predictive modeling of biomass gasification with machine learning-based regression methods. Energy. (2020). 191, ISSN: 0360-544.
7. BARUAH, D., BARUAH, D.C., HAZARIKA, M.K., (2017). Artificial neural network based modeling of biomass gasification in fixed bed downdraft gasifiers. Biomass and Bioenergy 98, s. 264–271.
8. MUTLU, A.Y., YUCEL, O., (2018). An artificial intelligence based approach to predicting syngas composition for downdraft biomass gasification. Energy 165, s. 895–901.
9. ZHOU, Z.-H. A brief introduction to weakly supervised learning. National Science Review. (2018). 5(1). s. 44-53.
10. BINOD, J. R. - RAVI, M. R. Development of numerical model for analysis of biomass based pottery furnace. [autorcafé], (2019), s.l.: Summer Research Fellowship Programme of India's Science Academies.
11. AMBROISE, C. - MCLACHLAN G.J. Selection bias in gene extraction on the basis of microarray gene-expression data. (2002). Proc Natl Acad Sci, 99 (10), pp. 6562-6566
12. DRAPER, N. R. - SMITH, H. Applied Regression Analysis. (1998). Wiley-Interscience. ISBN 978-0-471-17082-2

MATEC Web of Conferences **345**, 00003 (2021) https://doi.org/10.1051/matecconf/202134500003
Power System Engineering 2021

A study of the electrostatic charge of the coarse water droplets

Ondřej Bartoš[1*], and *Václav* Peter[1]

[1]CTU in Prague, Energy Engineering Department, Technická 4, Prague 6, 166 07, Czechia

Abstract. The aim of this paper is introduce the results of a measurement of the electrostatic charge of water droplets during wet steam expansion. The analysis of the charge origin was done. The achieved result shows that there is not a significant charging mechanism for the fine droplets. The dependence between the specific electrical current and the amount of liquid phase collected by the probe was found. Some models were studied to determine the origin of the electrical current. Finally a model was proposed, where the el. current is generated by the disruption of thin liquid film on the surface of the probe. The charging model was later applied for the measurement of the liquid film disruption in the new wind tunnel designed for the simulation of the flow field in the steam turbines. The measurement was performed with the air as a flowing medium and two modal droplets size distribution was found.

1 Introduction

The presence of the liquid phase in the flowing wet steam has a very unfavourable effect on the operating characteristics and reliability of the steam turbines. Electrostatic charge could be involved in the formation of a liquid phase in the flow section of steam turbines. Knowledge of the electrostatic charge phenomena in steam turbines is therefore essential, especially on these days, this knowledge has a potential to enhance the steam turbines. The presence of electrostatic charge in the flowing steam has been proven by measurement on steam turbines [1,2,3 &5], which showed that the charge distribution in the flow was not homogeneous and was probably influenced by the operation and chemical mode of the thermal cycle. For these reasons, the existing wind tunnel, located at the Department of Energy Engineering, Faculty of Mechanical Engineering of the Czech Technical University in Prague, was modified for research of electrostatic charge in the flowing wet steam. The measurement was made by electrostatic probe in Laval nozzle which is the part of the tunnel, where due to rapid expansion of steam, a phase change occurs, similarly as in a steam turbine. The advantage of this procedure is in a better knowledge of the steam flow field and inflow parameters in the nozzle than in a turbine measurement.

* Corresponding author: ondrej.bartos@fs.cvut.cz

2 Measurement

The electric current ***I*** generated by the water droplets carried in the steam flow upon impact on the probe can be described by the formula:

$$I = A.c.N_V. \int_{rk,\min}^{rk,\max} \eta_z(r_k).\alpha(r_k).q(r_k).\varphi(r_k).dr_k \qquad (1)$$

Where r_k is the radius of the droplet, A is the projection of the probe surface perpendicular to the flow axis, c is the steam velocity, N_V is the number density of droplets in $1m^3$, η_z (r_k) is the drop impact efficiency on the probe, α (r_k) is the probability of the charge transfer from the droplet to the probe, $\varphi(r_k)$ distributing the droplets in the stream by the radius.

The measurement stand is shown schematically on fig. 1 as a part of the steam wind tunnel. To measure the necessary values along the nozzle axis, the nozzle is equipped with a sliding meridian. The features of the steam line and the experimental methods used for the experiment made it possible to take advantage of some simplifications in the relationship (1). It can be assumed that due to the size of the steam nozzle, the flow is one-dimensional and the distribution of droplets in the stream of wet steam is uniform. The steam flow rate was derived from the stagnation input parameters of steam p_0, T_0 and the static pressure at the measurement site p_s. To determine the number and size of the water droplets in a stream of wet steam, the two wavelength method was used. It is based on the measurement of the two different wavelengths attenuation of monochromatic light when passing through a layer of wet steam. The use of this method makes it possible to assume that the liquid phase is formed in a small confined area and therefore has similar properties. It is therefore possible to introduce the assumption that the flowing mixture is monodisperse, $\varphi(r_k) = 1$. The current on the electrostatic probe was measured with a micro-ammeter. The charge transfer coefficient was assumed $\alpha(r_k) = 1$. Creating a capture efficiency model and a detailed description of the measurement in a steam nozzle is given in [4].

By applying relation (1) to an electrostatic probe placed in a stream of wet steam, it is possible to write a balance according to the charge conservation law (2) as shown in fig. 2.

The index p denotes fine droplets formed by nucleation and subsequent condensation in the steam stream and the index s with coarse droplets leaving the probe surface, m is the mass flow of the fine droplets onto the probe surface.

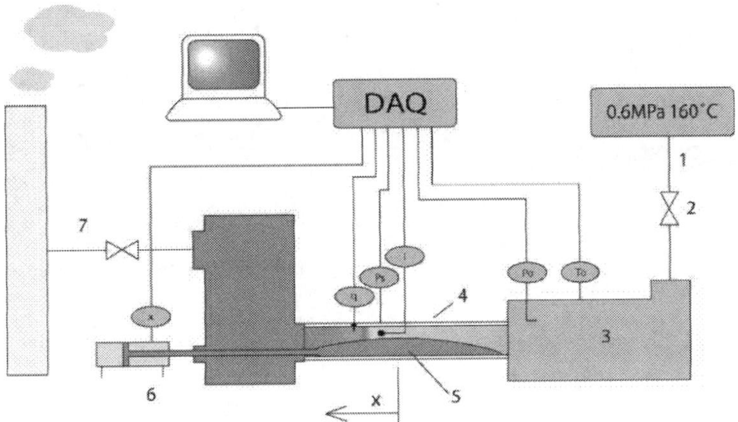

Fig. 1. Schema of the measurement in steam nozzle. 1-steam inlet, 2-gate valve and control valve, 3-settling chamber, 4-nozzle, 5- meridian, 6-hydraulic cylinder, 7-outlet to atmosphere.

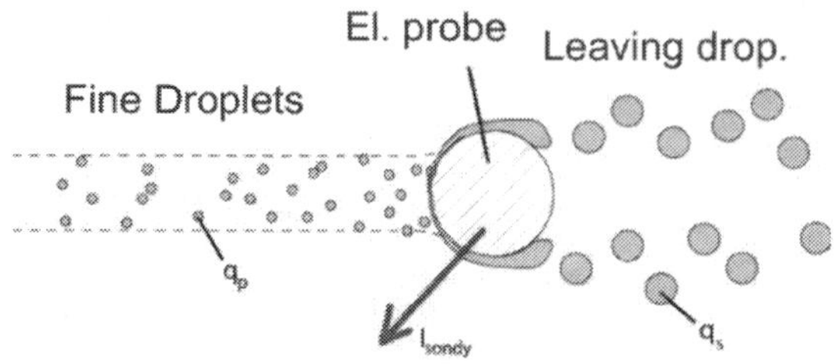

Fig. 2. The idea of distribution of electric charge between drops and probe.

$$A\, cN_V \eta_{z(r,Ma)} \overline{q}_p = I_{sondy} + \frac{\dot{m}}{\frac{4}{3}\pi r_s^3 \rho'} \overline{q}_s \tag{2}$$

3 Electrostatic probe

Fig. 3 schematically shows the overall arrangement of the electrostatic probe. The recorded electric current is the result of the interaction between the droplets and the active part of the probe. The projection area of the active part of the probe is 1.4 x 2 mm. This probe can measure across the nozzle channel. The flow field influence for different positions of the active part can be considered constant for all positions of the active surface of the probe. The advantage of this probe arrangement is that it is possible to measure across the channel.

4 Achieved data

Measurement with the el. probe was divided into four sets of measurements, L1-L4, divided according to the stagnation pressure and temperature (Tab.1). The moving speed of the meridian was in the range of 2-3 mm / s, this velocity was determined as an optimum for eliminating the influence of the meridian leakages. The measurements were made as the meridian moved back and forth, and the values were averaged.

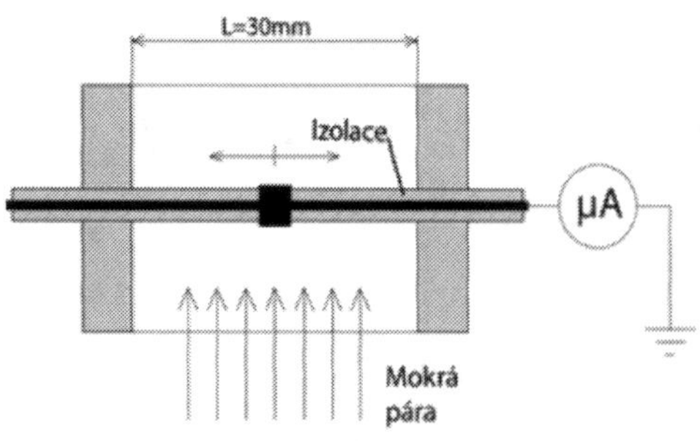

Fig. 3. Measurement schema of the electrostatic probe

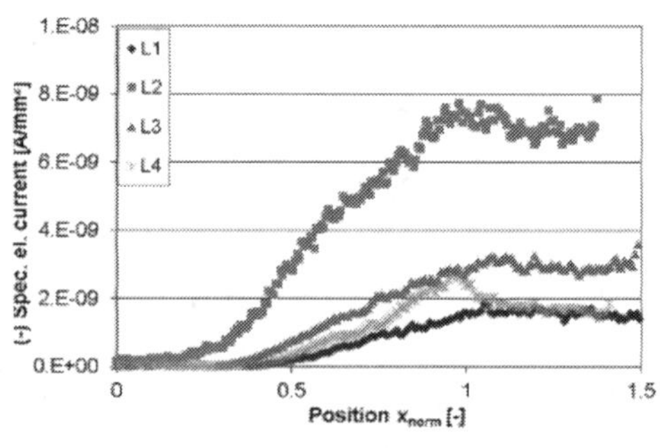

Fig. 4. current profile at the standardized coordinate

The position coordinate was normalized so that $x_{norm} = 0$ corresponds to a light attenuation value of 5% and $x_{norm} = 1$ corresponds to the end of expansion. This method of normalizing the position coordinate is necessary because of the different steam input parameters and thus the different expansion patterns.

The specific electrical current through the probe was referenced to the unit area of the probe. This evaluation was performed for an average value across the channel, ± 10 mm from the centre of the longitudinal axis of the nozzle.

Table 1. Table of stagnation parameters

	p_0 [kPa]	T_0 [°C]
L1	248	155
L2	249	148
L3	238	149
L4	241	153

The measurement results are shown on fig. 4, where is visible an increase of the specific electrical current in the nozzle. The el. probe measurements show that the electrical current reaches maximum values in the range of $(-1 \div -7)\cdot 10^{-9}$ A / mm^2.

5 Data processing

An important question, which had to be clarified, was whether there is an increase of the charge on the individual fine droplet during expansion or whether the measured increase in el. current (fig. 4) is caused by other influences. This idea is based on the notion that the electric charge is already carried by the droplets that arise during spontaneous condensation. However, it is also possible to generalize this problem and observe what parameters affect the value of the measured electric current.

All measured quantities were checked and their conformity with the profile of specific electric current measured by electrostatic probe was monitored. For comparison, it was necessary to standardize all quantities. The average value of the relevant variable around $x_{norm} = 1$ was chosen for standardization.

The best match of the measured specific electrical current was achieved when it was compared to the mass flow of the fine droplets on the probe surface according to the following formula:

$$\dot{m} = A\ c N_V \eta_z \rho_k \frac{4}{3} \pi r_k^3 \tag{3}$$

The comparison was made on the basis of the standardization relation:

$$\frac{i}{i_0} \approx \frac{\dot{m}}{\dot{m}_0} = \frac{c}{c_0} \frac{N_V}{N_{V0}} \frac{\eta_z}{\eta_{z0}} \frac{r_k^3}{r_{k0}^3} \tag{4}$$

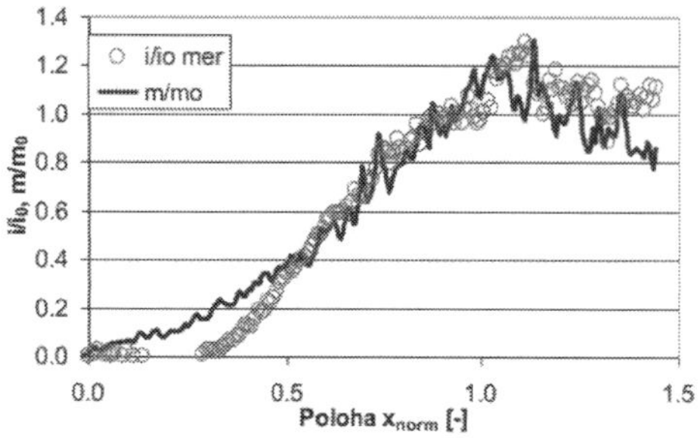

Fig. 6. Course of specific el. current and normalized mass of the liquid incident on the probe surface.

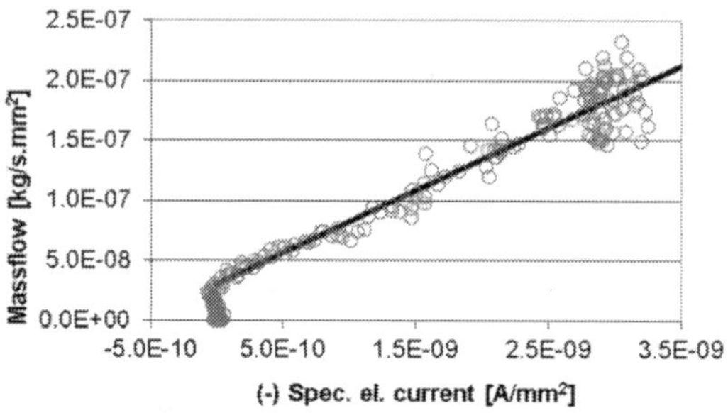

Fig. 5. Dependence between measured current and mass flow of water on the probe surface.

Figure 5 shows the waveforms of the measured normalized current and the normalized mass flow of water falling on the probe surface.

It can be seen on fig. 6 that after an initial steep increase in the zero electric current region, the linear dependence of the mass flow of the liquid phase on the surface of the electrostatic probe continues as a function of the specific electric current.

This is similar for all measurements with the electrostatic probe in a steam nozzle. The dependence was determined from all measurements by the electrostatic probe (5). The coefficients were determined as arithmetic averages and the uncertainty was determined as a standard deviation. All these values are affected by a relatively small number of data.

$$\dot{m} = -57{,}21(\pm\, 15{,}48)\dot{\imath} + 3{,}7.10^{-8}\left(\pm\, 2{,}0.10^{-8}\right) \tag{5}$$

6 Droplets charging

Several known methods of charging aerosols have been investigated (diffuse bipolar and unipolar, electrostatic field charging, water droplet charge dropping on the probe surface). The measured reality best described the model based on the disintegration of the water film that forms on the surface of the probe. This proposal is based on work on an electric double-layer [6,7&10] for monovalent ion concentrations above $\boldsymbol{C} = 1 \cdot 10^{-4}$ mol/l. The water that collects on the electrostatic probe is created directly during spontaneous condensation and therefore we can assume higher concentrations of chemical impurities [8] than in the bulk condensate. According to the derivation at work [9], the charge magnitude of the resulting drop is more dependent on geometric characteristics than on electrochemical properties and can be estimated approximately $\boldsymbol{q_s} = 3{,}3 \cdot 10^{-15}$C. Since it is obvious that the radius of entrained droplets is very important, it was determined from the Weber number:

$$\mathrm{We} = \frac{2\rho_g c^2 r_k}{\sigma} \tag{6}$$

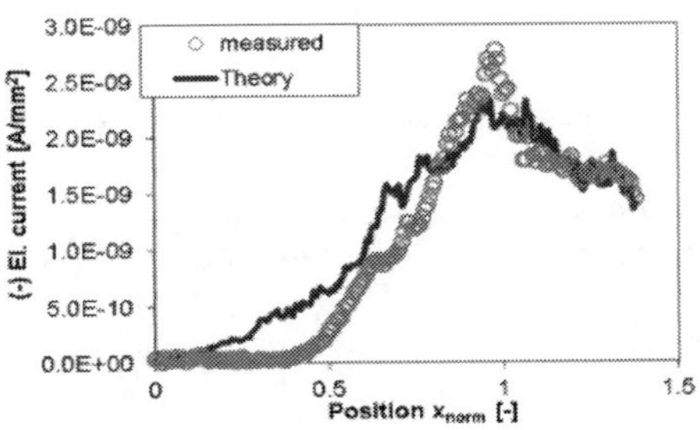

Fig. 8. Course of measured specific current and calculation for We = 18.

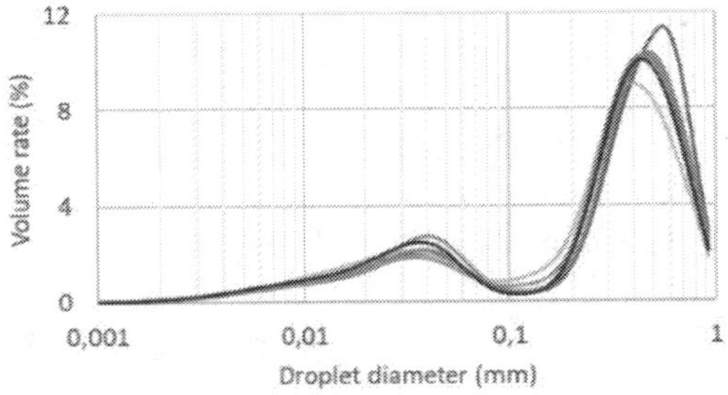

Fig. 7. Course water droplets distribution.

Where ρ_g is the steam density, σ is the surface tension of the liquid. The critical Weber number was chosen according to [8,10] We_{kr} = 18. Figure 7. shows the profile of the measured current and the electric current calculated from the theory described above.

7 New results

For the verification of the above mentioned results and theories a new wind tunnel was build. This tunnel is equipped with more advance measurement technic for the evaluation of the size distribution function of coarse droplets, light scattering method (Spraytec – Malvern Panalytical Ltd.) and photogrammetric method. The tunnel is working with the air and water is pumped to the flow through the aerofoil. The results of the size distribution of the droplets are on the fig. 8. The bi-modal distribution is clearly visible. The first modus is approx. 40 μm and the second is 400 μm. The Weber number for the first modus is around We=10 and for the second one is more than We=100 The Weber number was determined with respect to the velocity profile in the nozzle [11,12].

The measurement of the electrical current on the electrostatic probe inserted to the flow of the drifted droplets is another parameter which could be used for the droplets identification. This measurement is following the previous work done in the steam nozzles. The charge measurement is easy, but the interpretation of the results is very complex. From this reason the profile of the measured current is presented on the fig. 9. After many simplifications one can expect the mean charge of the unique droplet app. 10^3e.

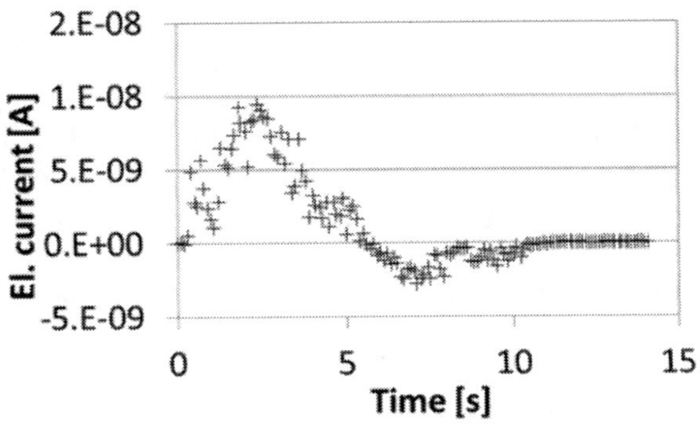

Fig. 9. El. current dep. on time

8 Summary

The formation of electrostatic charge on the droplets during the expansion of steam is a very complex and still unresolved phenomenon. An important result of this work is to find a connection between the amount of water falling on the probe surface and the measured electric current. The similar knowledge was yielded from the measurement in the new wind tunnel with air. The importance of the understanding of the liquid film break up and distribution of the el. charge within these process is apparent.

We gratefully acknowledge the support by the Grant Agency of the Czech Technical University in Prague, Grant No. SGS 21/154 and National Centre for Energy TN01000007.

References

1. Cross, J.A., Electrostatics, principles, problems and applications, Adam Hilger, Bristol,1987.
2. Dooley R.B., Rieger N.F., Bakhtar F.: Studie sof elektrostatic charge effects relating to power output from steam turbine, PowerPlant Chemistry, 2005, 7(2).
3. Electrostatic charge and its influence on the condensation of steam in a turbine, EPRI, Palo Alto. CA: 2001.1001332.
4. Hinds W.C.: Aerosol technology. JohnWiley,New York,1999.
5. Iribarne, J.V., Mason, B.J., Electrification accompanying the bursting of bubbles in water and dilute aqueous solution, Transaction of Faraday society, Vol. 63, pp.2234-2245, 1967.
6. Lindsay, W.T., Behavior of impurities in steam turbines, Power engeneering, Vol. 5, pp. 68-72, 1975.
7. Moore, M.J., Sieverding, C.H.: Two-phase flow in turbines and separators, Theory,Instrumentation, Engineering, Hemisphere Publishing corporation, 1976.
8. O. Bartoš, X. Cai, M. Kolovratník, EPJ Web of Conferences. 2014, vol. 67, ISSN 2100-014X
9. H. Liu, Science and engineering of droplets: fundamentals and applications. Norwich, NY: Noyes Publications, c2000. ISBN 0-8155-1436-0.
10. J. Valha: Proudění mokré páry a její erozivní účinky v průtočné části parní turbíny. SVUSS, Sborník referátů, 1978
11. O. Bartoš, X. Cai and M. Kolovratník. Experimental investigation of coarse water droplets in steam turbines by the adapted photogrammetry method. In The Baumann Centenary Conference Cambridge: ISBN 978-0-903428-35-4.
12. L. Měšťanová and O. Bartoš. A study of the coarse water droplets formation in the nozzle. In: EFM 2018. Experimental Fluid Mechanics 2018, Praha, 2018-11-13/2018-11-16. Liberec: TUL, Fakulta mechatroniky a mezioborových inženýrských studií, 2018. 381-385.

MATEC Web of Conferences **345**, 00004 (2021) https://doi.org/10.1051/matecconf/202134500004
Power System Engineering 2021

Heat pumps in the future

Tímea Gresz[1], *Dávid* Faragó[1], and *Péter* Bencs[1*]

[1]Institute of Energy Engineering and Chemical Machinery, Department of Fluid and Heat Engineering, University of Miskolc, H-3515 Miskolc, Hungary

Abstract. It has been discovered long ago that geothermal energy can be used to substitute various types of fossil fuels. Heat pumps are devices that can transfer the earth's thermal energy from a cooler to a warmer space. Their deployment would not only benefit large companies but could even reduce the everyday costs of individuals. Among their many advantages, possibly the most important one is that it is environmentally friendly. Their power demand can ideally be satisfied using other renewable energy sources. In this article, the current situation of heat pumps and geothermal energy in Hungary, Sweden and Germany are examined including incentives and grants. Afterwards, the current situation and operation of hybrid systems in Sweden and Egypt are studied.

1 Introduction

It is a well-known fact that fossil fuels – such as oil, natural gas, etc. – are finite and are becoming scarce, but the most serious issue is the inherent CO_2 emissions involved with them. Previously undiscovered oil and gas deposits are being found every so often, and there are also cases where extraction used to not be economically viable, but with increasing prices and the availability of more advanced technology, nowadays it is.

Heat pumps can be used to provide an alternative to fossil fuels for heating and providing hot water. They are environmentally friendly and can produce heat relatively quickly and cheaply. Heat pumps use electricity to produce hot water and are also suitable for either heating or cooling. Their main advantages are that they do not require a lot of space and are suitable to be operated at high performance [1].

The principle of the heat pump was first conceived by James Joule and William Thomson in 1852. Peter Ritter von Rittinger implemented the world's first industrial heat pump based on Carnot's thermodynamic writings. The first building that was continuously heated solely by a heat pump was instituted in 1938 in Zürich, the heat source being the water of the Limmat river.

In 1948, László Heller achieved a technological breakthrough with the development of the compressor heat pump. For example, the heat pump of the Bundestag building was designed in such a way that it could provide heating during the heating season, and cooling in summer.

* Corresponding author: arambp@uni-miskolc.hu

The energy demand of heat pump systems can be provided by renewable energy sources such as solar, wind, hydroelectric, biomass and geothermal [3].

Table 1. Average soil temperatures in Hungary

Depth	Temperature
500 m	35-40 °C
1000 m	55-60 °C
2000 m	100-110 °C

Even though most of the major geothermal energy sources are usually found around volcanically active areas, the geothermal potential of Hungary is significant. On average, soil temperatures increase by about 45 °C/km with depth, detailed in Table 1.

In warmer areas, this gradient can be of much higher values (as high as 120-130 °C/km). More than 70% of the territory of Hungary has thermal water deposits with a minimum temperature of 30 °C [4]. Soil temperature distributions at different depths are shown in Figure 1. Dark blue indicates 40 °C, yellow indicates 110 °C, and brown indicates the soil temperature of 150 °C.

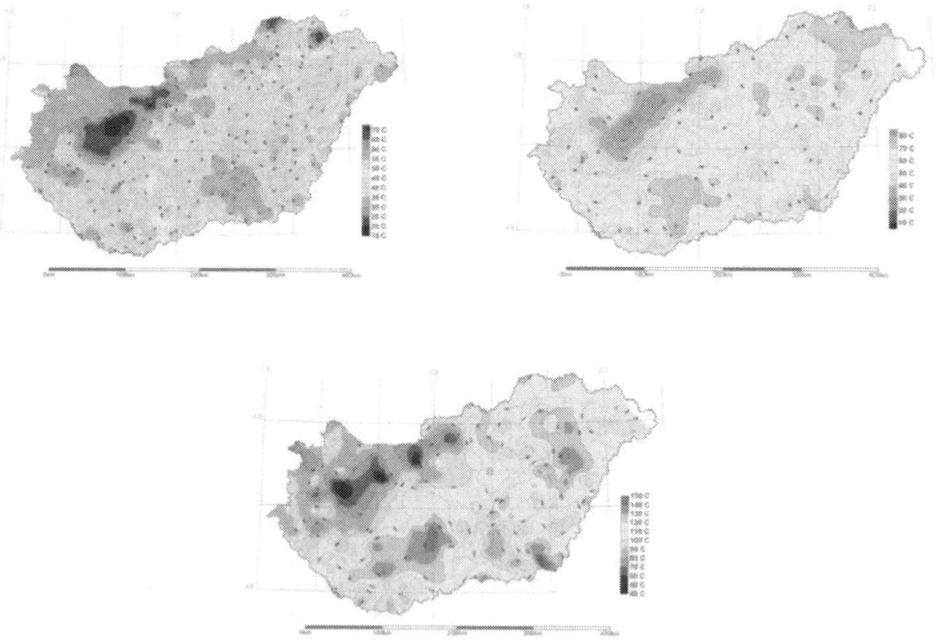

Fig. 1. Temperature distribution of Hungary at 500, 1000, 2000 m depth [4]

2 Heat pumps and their applications

The primary heat source of modern heat pumps is the geothermal heat of the ground. From there, heat is extracted by a horizontal closed loop field of pipes or vertical probes in boreholes. The principle of operation is practically the same as that of a refrigerator but in reverse. The refrigerant is heated in a piping system, which is located 1.5-2 meters below the ground to avoid any major temperature change due to external effects. This heat is absorbed by the circulating refrigerant through the heat exchanger on one side. The temperature of the fluid further increases while it flows up the narrow-section tube, then the heat is released on the other side [2].

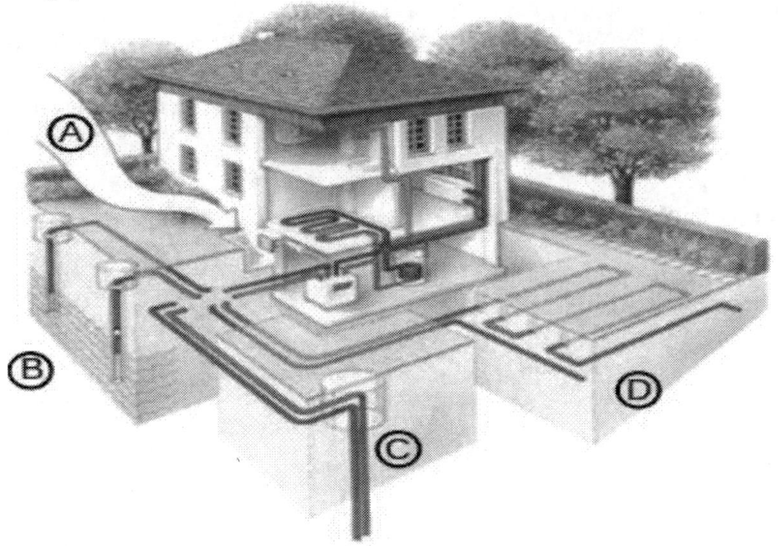

Fig. 2. Types of heat pumps [7]

The main types of heat pumps, illustrated in Figure 2: [2]

- air-water type (A),
- water-water type (B),
- ground-water probe (C),
- ground-water collector (D).

One of the most popular types is the vertical arrangement of the ground source heat pumps for several reasons. It has the best efficiency, it is environmentally friendly, there is no combustion product and no need for people to take care of ash and flue gas disposal, no need to buy fuel, no need to carry it from the storage to the boiler.

"COP (Coefficient of Performance) is the ratio of the heating capacity of a heat pump to its effective power input." [3].

Water-water heat pumps possess the highest COP of which the large body water is the highest (COP: 4-6), closely followed by groundwater-based heat pumps (COP: 3-5). Air-water type heat pumps have a COP of 2-3 if the air temperature is above 0 °C, and 1-2 if subzero. The coefficient of performance of an air-air type heat exchanger falls between 1 and 3, but this largely depends on the ambient temperature [2].

Heat pumps, besides heating and cooling, can also be used to power household appliances, such as washers and dryers, where a heat pump can perform the drying function. Implementing such methods, the energy efficiency of the household can be greatly improved, as the most energy-intensive household appliances are those that use water during their operation [5].

They can be used for both heating and heat retention, not only in households but also in fish farming systems. An air-water heat pump was used to heat and hold the water temperature in a fish tank, with a COP not falling below 4.31. This solution is not only environmentally friendly but also cost-effective as it leads to a much higher reproductive rate of fish [6].

The possible functions of heat pumps are not limited to heating or hot water production, they can be used for cooling purposes as well. This can be achieved in two different ways, active or passive cooling. Passive cooling is one of the cheapest ways to lower the temperature inside a building. However, when the summer heatwave is raging endlessly, passive cooling is no longer sufficient. In active cooling systems, a combined buffer tank cannot be used, so a separate tank is needed for cooling-heating, and a separate tank for hot water production. Cooling with a heat pump is 67-75% cheaper compared to cooling with an air conditioner. On contrary, cooling with a heat pump requires three times as much energy as heating [7].

3 Geothermal energy in Europe

Europe aims to become the world's first climate-neutral continent by 2050. In 2018, 21.1% of the energy used for cooling and heating was provided by renewable sources (Table 2), an increase of almost 10% compared to 2004 [8].

Table 2. Rate of energy produced from renewable energy sources for cooling and heating [%] [8]

	2004	2006	2008	2010	2012	2014	2016	2018
Germany	7.2	8.4	10.3	12.1	13.4	13.4	13.0	13.6
Sweden	46.6	54.2	57.6	59.1	62.7	64.6	65.5	65.4
Hungary	6.4	11.4	12.0	18.1	23.3	21.3	21.0	18.1

In Sweden, the temperature gradient is 15-20 °C/km, clearly much lower than in Hungary, yet heat pumps are much more common. Sweden is at the forefront of geothermal energy usage within Europe, in fact even surpasses the USA in this regard. The proliferation of heat pumps, with 30% of single-family homes using them from as early as 2008, is largely due to state subsidies and a system of regional, interest-free loans [4].

Figure 3 shows the distribution of the geothermal heat map of Europe. It can be seen that the model temperature is much lower in Sweden. Hungary has extraordinarily high values, but Germany and France also have values of around 120-150 °C at a depth of 5 km.

Geothermal heat pumps are becoming extremely widespread in Germany as well. In this case, it is mainly due to the Renewable Energy Act, which set a very favorable price for electricity from renewable energy sources [4]. In Germany, the BAFA and MAP deal with energy upgrading of existing buildings. For new buildings, only innovative and efficient heat pump systems are financed. From 1. January 2020, 35 % of the costs for the construction of heat pumps and related environmental measures will be covered, which can be further increased to up to 45% if oil heaters are replaced [12]. Financial support will also be provided if the heat pump is only installed for heating and the hot water is produced from non-renewable energy sources. The funding has also been extended to hybrid heating systems. Applications are open to individuals, municipalities, corporations, non-profit organizations, etc. Air-air heat pumps, self-built or second-hand systems are not eligible for financial support [13].

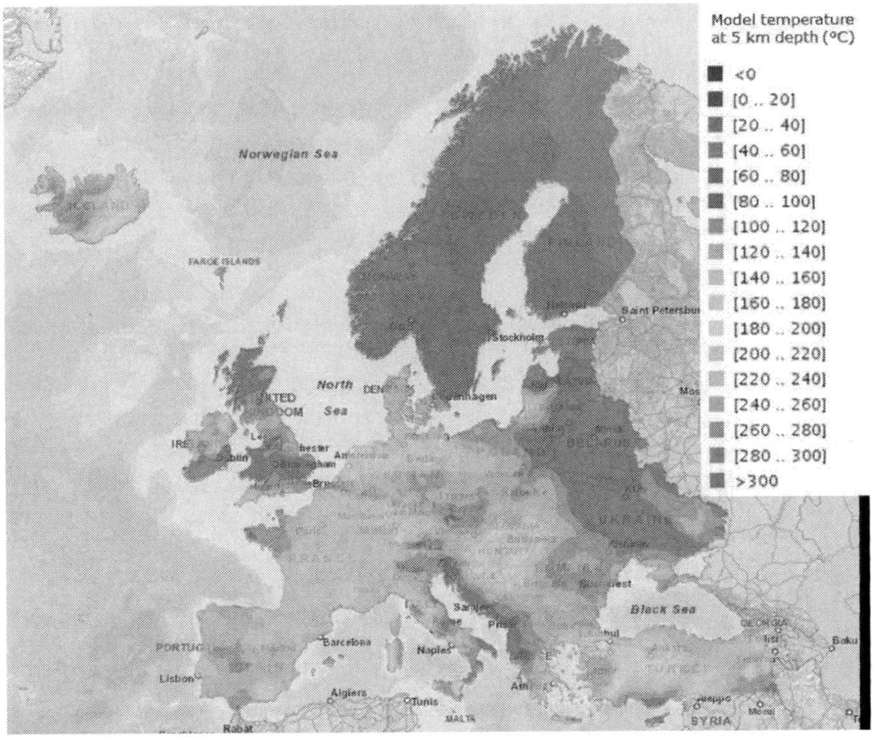

Fig. 3. Geothermal map of Europe

In Hungary, the first geothermal power plant to produce electricity was established in Tura (Figure 4). The plant runs Organic Rankine Cycle (ORC) technology to generate 2.7 MW of electricity. Construction of the plant started in 2017 and the first part was commissioned in 2018 [15].

Fig. 4. Geothermal plant in Tura, Hungary [15]

One of the most important reasons that heat pumps are not as widespread in Hungary as in other countries is probably that tenders are often available for only a month. A more predictable and more general incentive, offering at least 30% subsidies, would be recommended to encourage the public to invest in heat pipe systems. Experiences show that such high grants are only necessary for about 5-10 years, when heat pump systems are introduced to the market to help their early spread [9]. After this period, the subsidy rate could be decreased to around 10%. In addition to domestic or EU subsidies, it would also be important to reduce the price of electric power for heat pumps during the heating season. This approach has already been implemented by Germany, generally making the payback period of heat pump investments significantly shorter [10].

There used to be a well-functioning tendering system between 2000-2005; it was designed to promote the comprehensive use of renewable energy sources. After that, only the Environmental Protection and Infrastructure Operational Programme offered tenders, but these were not open to the public as they were available only for large-scale projects. This was followed by the Environment and Energy Operational Programme, which ran between 2009-2015, but once again, tender intervals were not predictable [11]. A major drawback was that the region of Central Hungary was not eligible for the Environmental Protection and Infrastructure Operational Programme fundings.

To install a heat pump system, a water permit is required, which must be obtained from the General Directorate of Water Management, where an application must first be submitted. There are, of course, costs involved in the procedure, depending on:

- water usage, water quantity,
- investment cost, etc.

The costs of the authorities to be paid by the customer:

- Directorate of Plant Protection and Soil Conservation
- Forestry Directorate
- Ministry of Agriculture, Department of Land Administration and Geoinformation
- National Directorate General for Disaster Management
- National Transport Authority

The payback period for ground source heat pumps in Hungary are [14]:

a) For an existing building with a heated volume of 682 m^3:
    - without tender: 8.1 years payback period,
    - with tender: reduced to 5.8 years.

b) For a new building where it could be used for both heating and cooling:
    - without tender: 4.2 years,
    - with tender: reduced to 2.2 years.

There are similar payback periods in Germany, without tender it takes 7-10 years, but with the BAFA tender, the payback period is much shorter.

Today's new hybrid heat pump systems use renewable energy sources (such as solar or geothermal energy) to produce heating or domestic hot water. This new system has been developed and pioneered in Sweden, and a test plant was completed by 2015. The new system consists of a hybrid solar panel, a ground source heat pump, and a borehole thermal energy storage. The hybrid solar panel differs from conventional solar panels in that the PV panels circulate a cooling medium, such as water or air. During the operation of the system, the water is heated to about 10 °C on the PV panel and then fed to the cold-side ground-source heat pump and used as a heat source. If excess heat is generated, it is fed into boreholes where the ground absorbs the unused heat energy. Boreholes contribute greatly to increasing the efficiency of heat pumps. In boreholes, we can store the excess energy produced during hot summer months and use it in the colder seasons as required [16].

In recent years, a new cooling system for photovoltaic panels has been tested in the Port Said area of Egypt. A ground-air heat exchanger is used for cooling. With this method, the cooling of the solar panel is solved, and its performance is improved at the same time. The PV_R notation used in Table 3 can be taken as a reference, while the PV_C shows the details of the hybrid system. The costs are given in US dollars. It can be observed that when coupled with a heat pump, the system does not represent a very large additional cost, namely an extra \$150 [17].

For a conventional solar PV system, the levelized cost of energy (LCOE) is 0.87 (US \$/kWh), while for a hybrid system the LCOE is 0.76 (US \$/kWh). Additionally, the relative energy cost is reduced by 12% and allows the generation of 36 kWh of electricity and the avoidance of 13986 g of CO_2 emissions during the summer season [17].

Table 3. Cost distribution of the hybrid system and solar panel [17]

Spare parts	PVR (\$)	PVC (\$)
PV panels	300	300
Galvanized steel rack and structures	40	40
Battery	225	225
Solar charge controller (MPPT)	112	112
Insulation	-	10
Piping and regulators	-	15
Air blower	-	35
Ground-air heat exchanger	-	90

4 Conclusion

Renewable energy sources can be of great assistance throughout the course of fossil fuel phase-out. Today, mainly owing to major incentives, solar energy has the largest share among various renewable energy sources in Hungary.

The heat pumps presented in this paper have already established themselves in Sweden and Germany. However, even though geothermal energy would be much more affordable in most areas of Hungary, there are still very few examples of established geothermal plants, and heat pumps are far from being widespread. The main reason behind this insensitivity towards heat pumps is the lack of incentives. Were a public domestic tender system available, the various heat pump systems would begin proliferating.

Heat pumps can be powered by renewable energy sources (such as solar or geothermal energy), making their operation much more economical and reducing inherent CO_2 emissions at the same time. Recently there has been a growing tendency to research hybrid heat pump systems, and they are expected to become more common, since they are only slightly more expensive than a simple solar system, but much more versatile.

This research was supported by the European Union and the Hungarian State, co-financed by the European Regional Development Fund in the framework of the GINOP-2.3.4-15-2016-00004 project, aimed to promote the cooperation between higher education and the industry.

References

1. G. Hajdú, A Hőszivattyú a Jövő Energiaforrása a Nap és Föld Hőjének Hasznosítása. | Magyar Kert és Energia Klaszter (2010).
2. B. Darabos, Mi az a Hőszivattyú? | Bio-Solar-Haz (2008).
3. F. Komlós, Z. Fodor, Z. Kapros, and L. Vaszil, Hőszivattyúzás | Csináljuk Jól! Energiahatékonysági Sorozat (2007).
4. J. Mádlné Dr. Szőnyi, A Geotermikus Energiahasznosítás Nemzetközi és Hazai Helyzete, Jövőbeni Lehetőségei Magyarországon (2008).
5. J. Cranston, A. Askalany, and G. Santori, Efficient Drying in Washer Dryers by Combining Sorption and Heat Pumping, Energy 183, 683 (2019).
6. G. Géczi, P. Korzenszky, T. Szabó, L. Bense, and B. Urbányi, Levegő-Víz Hőszivattyú Alkalmazásának Lehetősége az Intenzív Recirkulációs Haltenyésztésben, AWETH 9, 139 (2013).
7. Quadronet, Hőszivattyú az Igazi Rezsistop Gázkazán Helyett. Fűt, Hűt Gazdaságosan (2018).
8. E. Eurostat, Share of Energy from Renewable Sources for Heating and Cooling, 2004-2018 (2020).
9. Z. Kalocsai, Fellendülés Előtt a Hőszivattyú Piac - Magyar Hőszivattyú Szövetség (2010).
10. E. Energiainfo, Drága, De Megtérül a Geotermikus Energia, Energiainfo (2009).
11. NKFIH, Széchenyi Terv Plusz, Magyarország a Kormánya (2021).
12. L. O'CONNOR, Geothermal Energy – New Report Analyses the Sector - European Commission (2016).
13. S. Dr. Björn, Bundesverband Wärmepumpe, BEG Förderprogramm (2020).
14. F. Komlós and Z. Fodor, Nem Szívás! Érvek, Számítások a Hőszivattyús Rendszer Mellett., Mérnök Újság 18, 24 (2011).
15. Mannvit, Tura Geotermikus Erőmű (2018).
16. W. Steel, Novel Hybrid Solar PV-Geothermal Energy System Piloted in Sweden, Renewable Energy World (2019).
17. N.A.S. Elminshawy, A.M.I. Mohamed, K. Morad, Y. Elhenawy, and A.A. Alrobaian: Performance of PV panel coupled with geothermal air-cooling system subjected to hot climatic, Applied Thermal Engineering 148, 1 (2019).

The efficiency of transcritical CO2 cycle near critical point and with high temperature

Sindu Daniarta[1, 2*], *Attila R.* Imre[2, 3], and *Piotr* Kolasiński[1]

[1]Department of Thermodynamics and Renewable Energy Sources, Wrocław University of Science and Technology, Wybrzeże Wyspiańskiego 27, 50-370 Wrocław, Poland.
[2]Department of Energy Engineering, Faculty of Mechanical Engineering, Budapest University of Technology and Economics, Műegyetem rkp. 3, H-1111 Budapest, Hungary
[3]Centre for Energy Research, Department of Thermohydraulics, POB. 49, H-1525 Budapest, Hungary

Abstract. Efficiency is a key parameter used to assess the quality of operation of power generation systems and devices applied for converting one type of energy to the other. Although, in the end, an investment project is mainly evaluated by economic aspect. Furthermore, many researchers have been investigating the possible types of energy conversion systems and devices applied for power generation and utilizing different types of working fluids. This paper presents the inside into transcritical carbon dioxide (CO2) cycle and the gradients of its efficiency. Transcritical CO2 cycle (TCO2C) here refers to a CO2-based thermal power generation cycle absorbing heat from a heat source (ideally with constant pressure) till the supercritical state is reached. It is followed by an expansion to a sub-critical superheated or even two-phase (wet) state. As alternatives, trilateral flash cycle (TFC) and organic Rankine cycle (ORC) utilizing CO2 are also introduced in this paper. The calculation in this study is computed based on MATLAB integrated with thermophysical properties like CoolProp and REFPROP, the mathematical models of the system are built and calculated with the same heat sink temperature of 224.41 K, and the heat source temperature is varied between 274.41 K and 500 K. At a certain temperature, the obtained result shows that the efficiency of the TCO2C is lower than the efficiency of ORC. Another result proves that the quality of working fluid at the end of the expansion process significantly influences the efficiency of the cycle.

1 Introduction

Negative environmental impacts (assessed, for example, by ozone depletion potential (ODP) and global warming potential (GWP)), are crucial issues met during the selection of working fluid in modern power plants and other energy conversion systems which operating principle is based on the implementation of thermodynamic cycles (such as e.g., refrigeration or air-conditioning system). ODP is a measure of the degradation of the ozone layer by given fluid compared to reference working fluid, i.e., R-11 (trichlorofluoromethane). GWP is a parameter to measure contribution to global warming at a period of time (compared to carbon

* Corresponding author: sindu.daniarta@pwr.edu.pl

dioxide). Regarding these issues, many working fluids such as chlorofluorocarbons (CFCs) and hydrochlorofluorocarbons (HCFCs) are not considered as working fluid candidates in any process for further works. Some of them have been already or are currently phased out and forbidden to use by the Montreal and Kyoto Protocols, or the Kigali Amendment [1]. Accordingly, there is an increasing interest observed in research related to natural working fluids like alkanes, alkenes, or carbon dioxide (CO2) which could be applied as working fluids in many processes and technologies [1 – 3].

Among natural working fluids, CO2 is particularly attractive as it is a non-toxic, non-corrosive, inflammable, non-explosive substance, additionally featuring good thermophysical properties. This working fluid is abundant and affordable. Moreover, it might also be environmental-friendly (i.e., its ODP is equal to zero while GWP is equal to 1) compared to other refrigerant working fluids. All these reasons have been driving many researchers to proceed with scientific works on the physical properties of CO2 and its potential application as a working fluid in thermodynamic cycles, like refrigerating systems [3, 4] and power plants (i.e., organic Rankine cycle (ORC), trilateral flash cycle (TFC), transcritical CO2 cycle (TCO2C), and supercritical CO2 Brayton cycle) [5 – 10].

One of the frequently used energy conversion cycles, the ORC, seems to be a promising way for power generation compared to the conventional steam Rankine cycle because it utilizes organic working fluid instead of water. Therefore, it could be installed to utilize heat sources with various operating temperature ranges, even below 100 °C. As an alternative method, TCO2C could be used. It refers to a CO2-based thermal power generation cycle absorbing heat from a heat source (ideally with constant pressure) till the supercritical state is reached, followed by an expansion to a sub-critical superheated or even two-phase (wet) state. There are two important differences between ORC and TCO2C; i.e., the working fluid is different (ORC uses an organic working fluid, while TCO2C uses CO2) and the isobaric heating step is proceeded in a different way (with evaporation for ORC and without evaporation for TCO2C).

Several studies have been conducted to show how working fluid should be selected to obtain a favourable result. Comparison between ORC and TCO2C [11] shows that TCO2C has better economic performance. A comprehensive study about a comparison between Rankine cycle, transcritical cycle and combined cycle with the source temperature up to 500 °C was presented in a study [12] and reported that TCO2C has a good performance when the temperature of heat sources ranges between 500 – 600 °C. Another comparative study between TCO2C with Kalina cycle and ORC were presented in a study [13] and reported that TCO2C outperforms Kalina and ORC for low-temperature heat sources.

Since many researchers believe that TCO2C outperforms these other cycles, in this study, we will present the trendline of efficiency focusing on TCO2C. As a comparison, ORC and TFC are also described in this article to understand the trendline of efficiency and the transition among the cycles. In this study, the temperature of heat sources is ranging between 274.41 K and 500 K. Obtained modelling results proved that under certain conditions, ORC – against all expectations - could outperform TCO2C.

2 Systems, models and mathematical description

TCO2C has a basic layout that is very similar to TFC and ORC, but the course of thermodynamic processes in the cycle is different. The working fluid in the transcritical cycle absorbs heat from heat sources under constant pressure (above the critical pressure) to reach a supercritical state (above the critical temperature) and then is expanded in the expander to superheated or even wet conditions (i.e., two-phase). For TCO2C, TFC and ORC, the thermodynamic parameters of the working fluid in the endpoint of the heating (which is at the same time the initial point of expansion) differ. For TCO2C, it is a supercritical fluid state (neither liquid nor vapour), for TFC, it is a saturated liquid state, while for ORC, it is a saturated vapour state. The comparison between TFC, ORC, partially evaporated ORC (PE-ORC), superheated ORC, and TCO2C in terms of system layout and the course of thermodynamic processes (depicted in T-s diagrams) could be seen in Figure 1.

The difference between TFC and ORC showed in Figure 1 could also be indicated with the striped box area (visualizing the heat input) in the *T-s* diagram. This striped box means that ORC needs more heat to evaporate the working fluid, which is then expanded, while in TFC, that area (indicated in this case by dotted area), visualizes heat input needed only to flash the working fluid. It seems that the ORC could be interpreted as the sum of a TFC (dotted area) and a virtual Carnot cycle (stripped area). Being the Carnot cycle the thermodynamic cycle featuring the highest possible to obtain efficiency within the given temperature range, one could assume that ORC is more efficient than TFC. According to some recent studies, this is true only for cycles using wet or moderately dry working fluids, including CO2 [14, 15].

Sometimes, the intermittent heat sources (e.g., solar thermal power, ocean thermal energy, waste heat, etc.) are found to be utilized for ORC power plant feeding, and this intermittent condition could significantly affect the thermodynamic process and operating conditions of the system. It is often found that the thermal power of these intermittent heat sources is not sufficient to evaporate the working fluid (i.e., the working fluid is not fully evaporated). Therefore, the expansion process starts from this partially evaporated or wet state (i.e., two-phase state or the condition where the working fluid quality is in the range of $0 < x_2 < 1$). It seems that this might lead to the serious issue for the performance of the expansion process and working conditions of the expander (e.g., the erosion of the turbine blade caused by droplet or corrosion of the machine's subassemblies). This process so-called PE-ORC (partially evaporated ORC) refers to Figures 1(c) and 1(e).

Furthermore, to avoid this serious issue on the expander caused by the liquid droplets, some engineers and scientists tend to configure the ORC system by installing an additional heat exchanger (superheater) to heat up the fluid. Therefore, the working fluid at the outlet of this heat exchanger reaches the superheated condition. Then, this superheated working fluid (in the dry state) could be directly expanded. This process is described as a superheated ORC and its layout and T-s diagram could be seen in Figures 1(f) and 1(g), respectively.

As above mentioned, this study focuses on the TCO2C that the layout and *T-s* diagram could be seen in Figures 1(h) and 1(i) respectively. Typically, the schematic layout of this process is similar to TFC (see Figures 1(a) and 1(h)) but the difference is the course of the thermodynamic processes (see Figures 1(b) and 1(i)). The working fluid in TCO2C starts to be expanded in the transcritical, superheated or wet state. Since the expansion process proceeds from supercritical to superheated or subcritical state. Therefore, this process is a so-called transcritical cycle. In TCO2C, only a heat exchanger is installed to change the liquid to supercritical fluid (see Figure 1(i)).

In this study, the cycle utilizes CO2 as a working fluid, and the model is built and calculated using MATLAB. Thermophysical properties of the working fluid are obtained from CoolProp [16]. The temperature of the condenser is set constant at 224.41 K, and for the evaporator, the temperature is varied in the range of 274.41–500 K. In this study, the condenser is set below 0 °C to see the possibility of wider performance of CO2 based power cycle. It is possible to have the heat sink at this condition by utilization of cold energy, such as the cooling potential from a regasification system of liquefied natural gas (LNG, which is needed to regasify from around -160 °C to 30 °C). Some articles have discussed the utilization of this cold energy as power generation [10, 17 – 20].

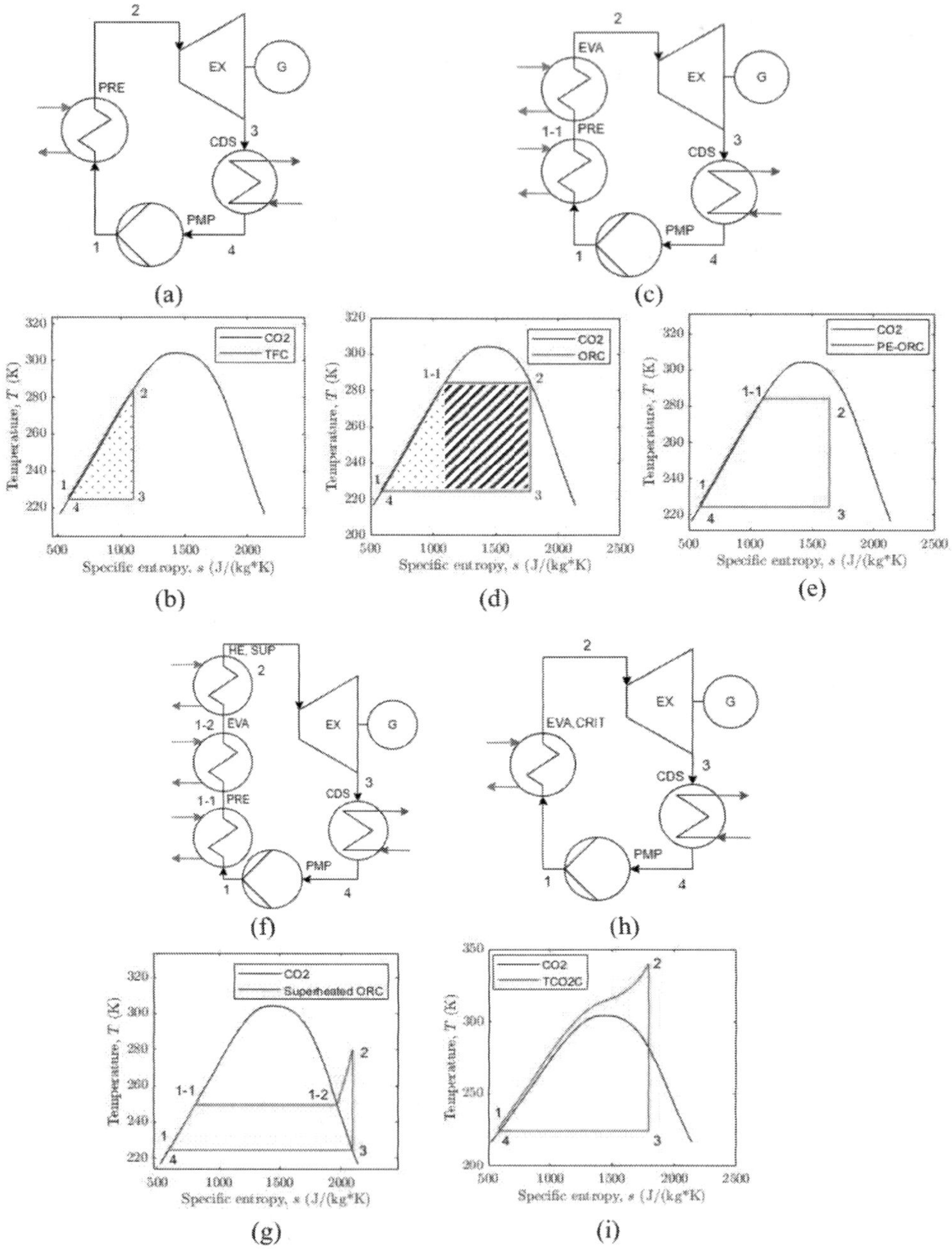

Fig. 1. The schematic layout and the course of the thermodynamic process in *T-s* diagram of (a – b) TFC, (c – d) ORC, (c – e) PE-ORC, and (f – g) superheated ORC, and (h – i) TCO2C.

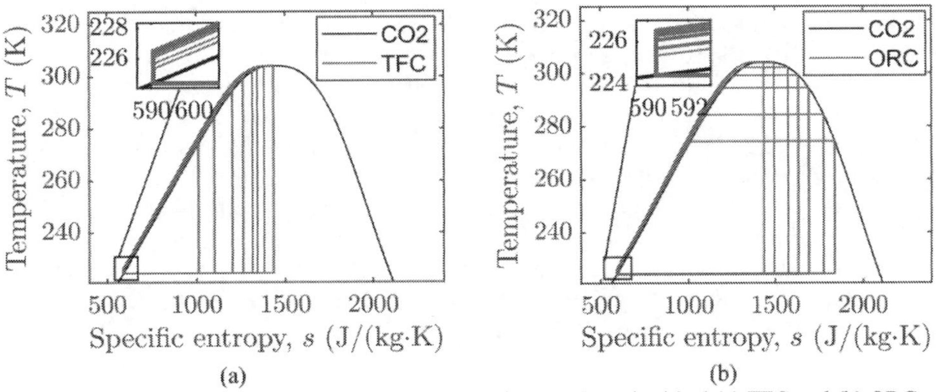

Fig. 2. The steps of modelling calculation in the *T-s* diagram for subcritical (a) TFC and (b) ORC

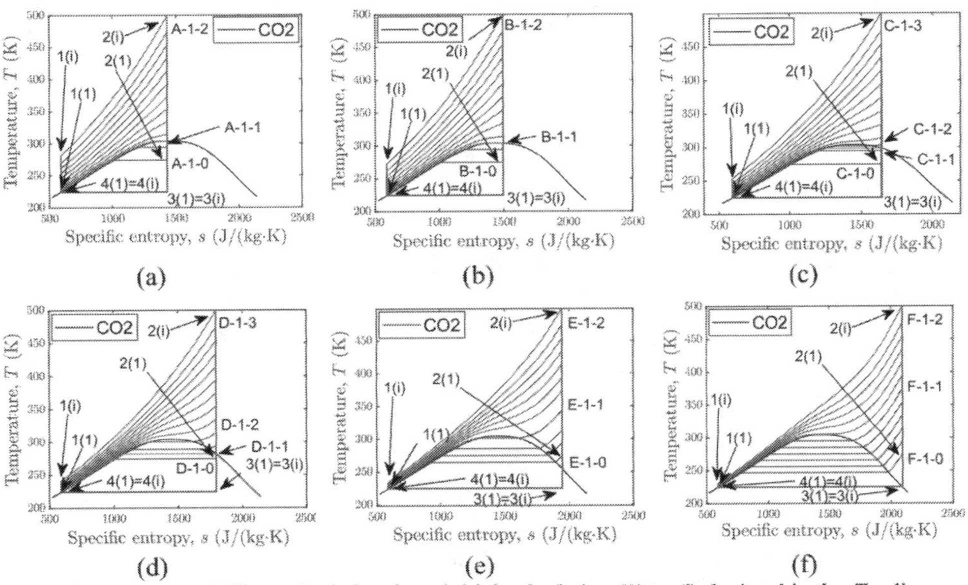

Fig. 3. The step of modelling calculation from initial calculation (1) to (i) depicted in the *T-s* diagram for ideal PE-ORC (purple lines), ORC (green lines), superheated ORC (blue lines), and TCO2C (red lines) where the condensation starts with the vapour quality of (a) x_3 = 0.56, (b) x_3 = 0.60, (c) x_3 = 0.70, (d) x_3 = 0.80, (e) x_3 = 0.90, and (f) x_3 = 1.00. For each figure, the method introduces the temperature increase in beginning of expansion from 2(1) to 2(i) with steps of 0.001 K and the conditions at the inlet and the outlet of the condenser are kept at the same, 3(1) = 3(i) and 4(1) = 4(i). The outlet pumps are varied from 1(1) to 1(i) based on the increase of desired pressure from 2(1) to 2(i).

In the presented study, the cycle is considered ideal to study the effect caused purely by the CO2. It means that the processes in the pump and expander are considered as isentropic and these proceeding in the evaporator and condenser as ideal isobaric (without pressure losses). For the subcritical ORC and TFC, the model (see Figure 2) is developed only below the saturated curve, the so-called subcritical cycle.

Figure 2 illustrates the steps of modelling proceeded for fixed condenser temperature (224.41 K). The starting state for expansion for TFCs follows the saturated liquid curve, and for ORCs, it follows the saturated vapour curve until reaching a critical point. To obtain the precise results, the temperature step for these TFC and ORC models is configured as 0.0001 K as there are small slopes near the critical point. Figure 2 also shows that the pump and the expander are modelled as isentropic. This method is similar to the methods used in studies [14, 15] for understanding the low and maximal efficiencies of ORC near the critical point.

Figure 3 shows the steps of modelling proceeded for PE-ORC, ORC, superheated ORC, and TCO2C where the vapour quality (i.e., the ratio of the mass of vapour to the total mass of working fluid) at the end of the expansion (x_3) is varied from 0.56 to 1.00. The TCO2C with x_3 = 0.56 (see Figure 3(a)) is developed to see the efficiency obtained in the very close vicinity of the critical point (due to the elevated density fluctuations in this region [21], crossing the critical point is technically almost impossible). The temperature of the condenser is fixed at 224.41 K, and the starting temperature of the expansion varies between 274.41 K and 500 K. Only the cycles formed below critical pressure are considered as TFC, PE-ORC, and ORC.

The boundary condition and calculating steps of this model are summarized in Table 1. The mathematical models calculated in MATLAB are described in Equations 1 – 7 for each step of the cycles. The efficiency of the cycles could be defined using Equations 8 – 12.

$$\Delta h_{\text{pre}} = (h_{\text{out,pre}} - h_{\text{in,pre}}) = \frac{\dot{Q}_{\text{pre}}}{\dot{m}_{\text{total}}} \tag{1}$$

$$\Delta h_{\text{eva}} = (h_{\text{out,eva}} - h_{\text{in,eva}}) = \frac{\dot{Q}_{\text{eva}}}{\dot{m}_{\text{total}}} \tag{2}$$

$$\Delta h_{\text{HE,sup}} = (h_{\text{out,HE,sup}} - h_{\text{in,HE,sup}}) = \frac{\dot{Q}_{\text{HE,sup}}}{\dot{m}_{\text{total}}} \tag{3}$$

$$\Delta h_{\text{HE,crit}} = (h_{\text{out,HE,crit}} - h_{\text{in,HE,crit}}) = \frac{\dot{Q}_{\text{HE,crit}}}{\dot{m}_{\text{total}}} \tag{4}$$

$$\Delta h_{\text{ex}} = (h_{\text{in,ex}} - h_{\text{out,is,ex}})\eta_{\text{is,ex}} = \frac{\dot{W}_{\text{ex}}}{\dot{m}_{\text{total}}} \tag{5}$$

$$\Delta h_{\text{cds}} = (h_{\text{in,cds}} - h_{\text{out,cds}}) = \frac{\dot{Q}_{\text{cds}}}{\dot{m}_{\text{total}}} \tag{6}$$

$$\Delta h_{\text{pmp}} = (h_{\text{out,is,pmp}} - h_{\text{in,pmp}}) / \eta_{\text{is,pmp}} = \frac{\dot{W}_{\text{pmp}}}{\dot{m}_{\text{total}}} \tag{7}$$

Where h, $\dot{m}$, $\dot{Q}$, and $\dot{W}$, are specific enthalpy, mass flow rate, heat transfer rate, and power, respectively. The subscript of HE, pre, eva, sup, crit, ex, cds, pmp, in, and out represent heat exchanger, pre-heater, evaporator, superheater, critical, expander, condenser, pump, inlet, and outlet, respectively.

$$\eta_{\text{TFC}} = \frac{\dot{W}_{\text{ex}} - \dot{W}_{\text{pmp}}}{\dot{Q}_{\text{pre}}} \tag{8}$$

$$\eta_{\text{ORC}} = \frac{\dot{W}_{\text{ex}} - \dot{W}_{\text{pmp}}}{\dot{Q}_{\text{pre}} + \dot{Q}_{\text{eva}}} \tag{9}$$

$$\eta_{\mathrm{ORC,sup}} = \frac{\dot{W}_{\mathrm{ex}} - \dot{W}_{\mathrm{pmp}}}{\dot{Q}_{\mathrm{pre}} + \dot{Q}_{\mathrm{eva}} + \dot{Q}_{\mathrm{HE,sup}}} \tag{10}$$

$$\eta_{\mathrm{TCO2C}} = \frac{\dot{W}_{\mathrm{ex}} - \dot{W}_{\mathrm{pmp}}}{\dot{Q}_{\mathrm{HE,crit}}} \tag{11}$$

Table 1. The boundary condition and calculating steps of the model using CO2 as a working fluid with the temperature at the inlet to the expander varied between 274.41 K and 500 K.

The cycle	The process in *T-s* diagram	T_2 (K)	T_3 (K)	x_3 (-)	Steps for $T_{in,ex}$ (K)
TFC	Figure 2(a)	$274.41 \leq T_2 \leq T_{crit}$	224.41	0.28 – 0.56	0.0001
ORC	Figure 2(b)	$274.41 \leq T_2 \leq T_{crit}$	224.41	0.56 – 0.83	0.0001
PE-ORC	Figure 3(a)	$274.41 \leq T_2 < T_{crit}$	224.41	0.56	0.01
ORC	Figure 3(a)	T_{crit}	224.41	0.56	-
TCO2C	Figure 3(a)	$T_{crit} < T_2 \leq 500$	224.41	0.56	0.01
PE-ORC	Figure 3(b)	$274.41 \leq T_2 < 303.94$	224.41	0.60	0.01
ORC	Figure 3(b)	303.94	224.41	0.60	-
Superheated ORC	Figure 3(b)	$303.94 < T_2 \leq 304.15$	224.41	0.60	0.01
TCO2C	Figure 3(b)	$304.15 < T_2 \leq 500$	224.41	0.60	0.01
PE-ORC	Figure 3(c)	$274.41 \leq T_2 < 298.59$	224.41	0.70	0.01
ORC	Figure 3(c)	298.59	224.41	0.70	-
Superheated ORC	Figure 3(c)	$298.59 < T_2 \leq 306.90$	224.41	0.70	0.01
TCO2C	Figure 3(c)	$306.90 < T_2 \leq 500$	224.41	0.70	0.01
PE-ORC	Figure 3(d)	$274.41 \leq T_2 < 281.73$	224.41	0.08	0.01
ORC	Figure 3(d)	281.73	224.41	0.08	-
Superheated ORC	Figure 3(d)	$281.73 < T_2 \leq 321.06$	224.41	0.08	0.01
TCO2C	Figure 3(d)	$321.06 < T_2 \leq 500$	224.41	0.08	0.01
Superheated ORC	Figure 3(e)	$274.41 \leq T_2 \leq 350.31$	224.41	0.09	0.01
TCO2C	Figure 3(e)	$350.31 < T_2 \leq 500$	224.41	0.09	0.01
Superheated ORC	Figure 3(f)	$274.41 \leq T_2 \leq 393.00$	224.41	1.00	0.01
TCO2C	Figure 3(f)	$393.00 < T_2 \leq 500$	224.41	1.00	0.01

Where η is the efficiency of the cycle. The formulas (see Equations 7 – 10) to calculate the efficiency are typically similar for all cycles, representing the ratio of power generation divided by the heat input. For the TFC, the heat input source is only the liquid pre-heater; for a transcritical cycle, the situation is almost the same, although a pseudo-phase transition (pseudo-boiling) could happen, but it differs from traditional evaporation to a great extent [22, 23]. In the ORC there are two possibilities of the expansion depending on the expansion starting parameters. If the expanded working fluid is in a saturated vapour state, the heating is provided by a liquid pre-heater and an evaporator. If the expanded working fluid is dry (superheated) vapour, a third heat exchanger (superheater) is also needed (see Equation 10).

3 Results and discussion

The obtained modelling results are illustrated in the form of a cycle efficiency-expander inlet temperature diagram (i.e., the working fluid temperature at the expander inlet) which could be seen in Figure 4. Different x_3 values represent the vapour quality at the end of the expansion.

Figure 4 shows that the efficiency of the ORC and TFC are similar to the ones presented in Refs. [14, 15] for the subcritical cycles. However, the data is more precise near the critical point as the temperature steps are small, $\Delta T_{\text{in,ex}} = 0.0001$ K. This diagram also illustrates that there is maximal efficiency for ORC at a certain temperature. Moreover, for TFC, the maximal efficiency is found at the critical temperature.

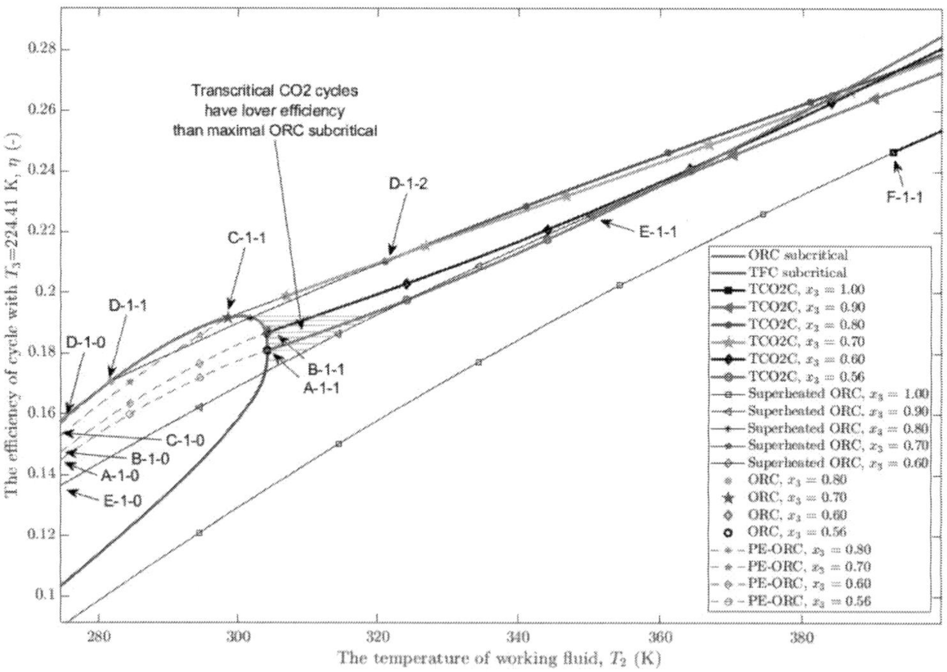

Fig. 4. The area where TCO2C have lower efficiency than maximal ORC under subcritical. The marks in this figure refer to Figure 3.

Some marks (A, B, C, D, E, and F) in Figure 4 refer to the marks in Figure 3 that indicates the transition of one process to the other. For example, the efficiency curve of the simulated model from D-1-0 to D-1-1 ($274.41 \leq T_2 < 281.73$) in Figure 4 is obtained based on PE-ORC with $x_3 = 0.80$ referring to Figure 1(d) that the condenser is at a constant temperature of 224.41 K. When the expansion starts from the temperature of 281.73 K (see Figure 1(d)), the process is ORC as expansion starts with initial vapour quality of x_2=1. Furthermore, at the temperature range of $281.73 < T_2 \leq 321.06$, the process will be superheated ORC as an additional heat exchanger is needed to heat up the working fluid until it reaches a superheated state. The differences between PE-ORC, ORC, and superheated ORC are described by the evaporation process and superheating. Using the same method in Figure 1(d), the process will be TCO2C ($T_2 > 321.06$) that is seen in Figure 4 (indicated with blue dot lines).

The modelled result also found that at a certain temperature, the efficiency of TCO2C is lower than ORC (maximal efficiency) as is indicated with the cyan area highlighted in Figure 4. It is plausible to expect that the maximal efficiency of ORC should play a key role in thermal power plant design. Moreover, the transcritical cycles going close to the critical point should be avoided because of the low efficiency (compared to ORC maximal efficiency) and uncertainties in supercritical conditions.

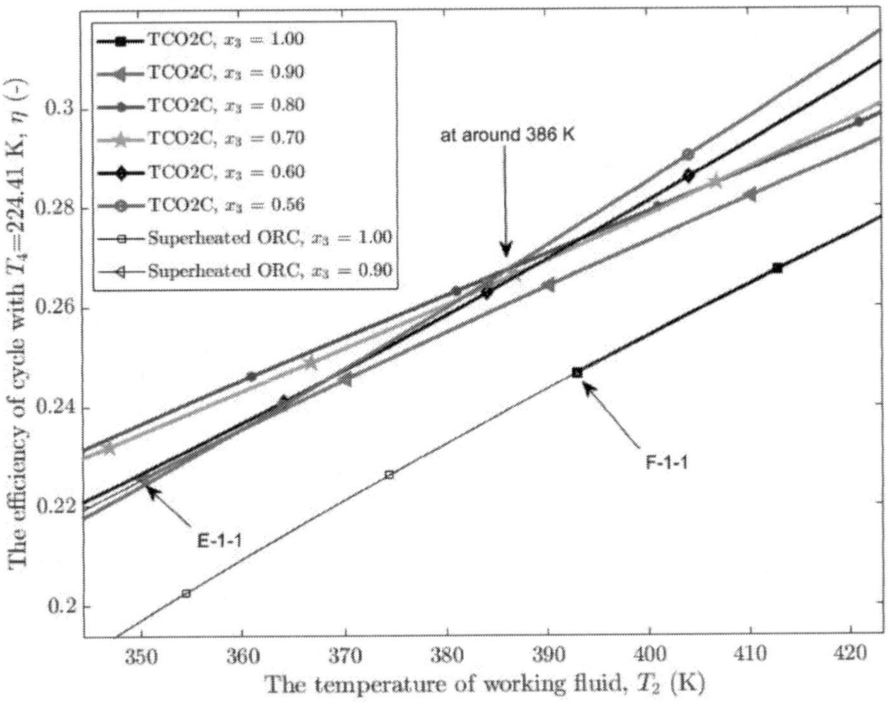

Fig. 5. The cycle efficiency-expander inlet temperature diagram valid for the temperature range of 344 – 422 K. The marks E-1-1 and F-1-1 refer to Figures 3(e) and 3(f).

The position of the curves in Figure 4 illustrates that the superheated ORC cycle might not be a good choice to utilize low-temperature heat sources because of the efficiency, economic aspect, etc. However, many researchers tend to design the ORC with a superheater to avoid the serious issue caused by the liquid droplets that could damage the turbine blade, and cause erosion. The selection of proper expanders has to be taken into full consideration because it plays a significant role in the choice of layout of the cycle and the design of thermodynamic processes. Nowadays, volumetric expanders tend to be a promising technology to be employed in low-temperature power systems [24]. The experimental test proved that this type of expander could handle the wet condition. These expanders offer a simple design, low level of noise and vibrations, low rotational speed and compactness.

Considering these devices as prime movers might be favourable for ORC and other energy conversion systems in the future.

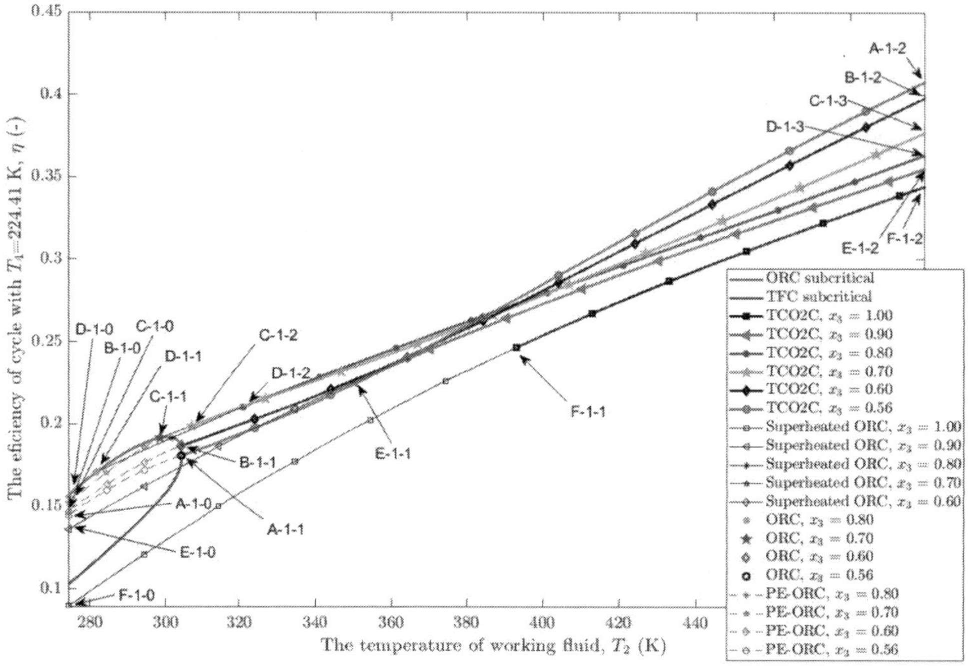

Fig. 6. The cycle efficiency-expander inlet temperature diagram, the mark refers to Figure 3.

The cycle efficiency-expander inlet temperature diagram plotted for the working fluid temperature range of 344 K – 420 K is illustrated in Figure 5. It seems that all efficiencies are changing smoothly. However, many crossings are visible in the diagram. It means that depending on the maximal cycle temperature, one could always find a cycle, which could outperform the others and the best cycle could differ for different temperatures. For example, below the temperature of ca. 386 K, the efficiency of the transcritical cycle with x_3=0.80 outperforms the others. However, above the temperature of 386 K, the transcritical cycle with the wet expansion x_3=0.56 has the best efficiency. Due to the special shape of CO2 isobars in the supercritical region, crossing of efficiency-lines (representing different x-values) exist.

The full results of this study could be seen in Figure 6 where the efficiency of superheated ORCs with x_3=1.00 at the end of the expansion process is illustrated. It is reported that the efficiency of this cycle is the lowest among the reported ones.

4 Conclusion

In this paper, the trendline of the efficiency for TFC, ORC, PE-ORC, superheated ORC, and TCO2C has been introduced and presented by using a diagram of cycle efficiency-expander inlet temperature. This study could be used as consideration in the preliminary design stage of CO2 based thermal power plant. It is reported that transcritical cycles with maximal cycle temperature close to the critical one should be avoided because they have lower efficiency than the maximal efficiency of subcritical ORC with smaller maximal cycle temperature (see Figure 4). At a certain temperature range of inlet expansion, TCO2C with the expansion process ending at x_3=0.56 (see Figure 6), x_3=0.70 (see Figures 4 and 6), and x_3=0.80 (see Figures 5 and 6) outperform the other types of the cycles reported in this study. Since there are many types of expanders (turbines and volumetric expanders), the results of the present study could be also used during the selection of proper expanders that have to be taken into consideration to obtain the desired operating conditions of the power plant.

Since the condenser in this study was set at 224.41 K, it is possible to have the heat sink below 0 °C by utilizing the cold energy from LNG or other refrigerant fluids. The utilization of this cold energy as power generation is good in the sustainable point of view.

Part of the research reported in this paper and carried out at BME has been supported by the NRDI Fund (TKP2020 NC, Grant No. BME-NC) based on the charter of bolster issued by the NRDI Office under the auspices of the Ministry for Innovation and Technology.

References

1. B.K. Sovacool, S. Griffiths, J. Kim, and M. Bazilian, Renew. Sustain. Energy Rev. **141**, 110759 (2021).
2. A. Chesi, F. Esposito, G. Ferrara, and L. Ferrari, Appl. Energy **135**, 274 (2014).
3. B. Dai, S. Liu, K. Zhu, Z. Sun, and Y. Ma, Energy **122**, 787 (2017).
4. A. Yu, W. Su, X. Lin, N. Zhou, and L. Zhao, Energy Convers. Manag. **221**, 113214 (2020).
5. S. Khatoon, S. Ishaque, and M.-H. Kim, Energy Convers. Manag. **232**, 113895 (2021).
6. R. Chen, M. Romero, J. González-Aguilar, F. Rovense, Z. Rao, and S. Liao, Energy Convers. Manag. **232**, 113870 (2021).
7. W. Su, Y. Hwang, S. Deng, L. Zhao, and D. Zhao, Energy Convers. Manag. **165**, 720 (2018).
8. L. Pan, B. Li, W. Shi, and X. Wei, Appl. Energy **253**, 113608 (2019).
9. D. Zhao, R. Zhao, S. Deng, L. Zhao, and M. Chen, Energy Convers. Manag. **226**, 113578 (2020).
10. S. Daniarta and A.R. Imre, Period. Polytech. Mech. Eng. (2020).
11. M. Li, J. Wang, S. Li, X. Wang, W. He, and Y. Dai, Geothermics **50**, 101 (2014).
12. C. Li and H. Wang, Appl. Energy **180**, 707 (2016).
13. F. Meng, E. Wang, B. Zhang, F. Zhang, and C. Zhao, Energy Convers. Manag. **195**, 1295 (2019).
14. A. Mohammed Ahmed and A.R. Imre, Pollack Period. Pollack (2021).
15. A.M. Ahmed, L. Kondor, and A.R. Imre, Energies **14**, (2021).
16. I.H. Bell, J. Wronski, S. Quoilin, and V. Lemort, Ind. Eng. Chem. Res. **53**, 2498 (2014).
17. I.-H. Choi, S. Lee, Y. Seo, and D. Chang, Energy **61**, 179 (2013).
18. O. Koku, S. Perry, and J.-K. Kim, Appl. Energy **114**, 250 (2014).
19. R.F. García, J.C. Carril, J.R. Gomez, and M.R. Gomez, Energy **105**, 16 (2016).
20. B.B. Kanbur, L. Xiang, S. Dubey, F.H. Choo, and F. Duan, Renew. Sustain. Energy Rev. **79**, 1171 (2017).
21. A.R. Imre, A. Groniewsky, G. Györke, A. Katona, and D. Velmovszki, Period. Polytech. Chem. Eng. **63**, 276 (2019).

22. D.T. Banuti, J. Supercrit. Fluids **98**, 12 (2015).
23. A.R. Imre, C. Ramboz, U.K. Deiters, and T. Kraska, Environ. Earth Sci. **73**, 4373 (2015).
24. P. Kolasiński, *Układy Organic Rankine Cycle (ORC) Małej i Średniej Mocy Wykorzystujące Rozprężarki Wyporowe* (Oficyna Wydawnicza Politechniki Wroclawskiej, Wrocław, 2021).

Turbulent jet stability increased by ribs inside the nozzle – Stereo PIV measurement one diameter past the nozzle

Daniel Duda[1*], *Vladimír* Abrhám[1], *Václav* Uruba[1,2], and *Vitalii* Yanovych[1]

[1]University of West Bohemia in Pilsen, Univerzitní 22, Pilsen, Czech Republic
[2]Intstitute of Thermomechanics of the Czech Academy of Sciences, Dolejškova 5, Prague, Czech Republic

Abstract. We observe that decreasing the inner nozzle surface by adding longitudinal ribs increases the jet stability in terms of the amount of turbulent kinetic energy in the near shear layer. We try to explain our observation as a stabilization effect of secondary flow vortices emerging in the corners of the ribs. These stream-wise vortices damage the development of larger-scale structures in the near shear layer. This explanation is supported by autocorrelation function of the stream-wise velocity component, which displays slightly smaller integral length-scale in the case with ribs than in the case of smooth nozzle. The experiment is performed at Reynolds number 2.2×10^5 (based on the nozzle diameter 50 mm); the Stereo-PIV (Particle Image Velocimetry) measurement takes place at the plane perpendicular to the jet axis one diameter past the nozzle exit. Optical 3D scanner controls the real nozzle geometry. This article presents preliminary measurement at single position and single velocity only; further exploration of this problem is needed.

1 Introduction

Jet is the flow structure created by fluid leaving some nozzle or orifice into surroundings with different velocity. In the near stage, we distinguish the *jet core* surrounded by the growing shear layer, in later stages, the jet core vanishes, the stream-wise velocity profile converges to Gaussian growing in space and weakening in velocity [1,2]. The character of turbulence converges to self-similar isotropic and homogenous (at least locally) turbulence of Kolmogorovy style until the dissipative length-scale does not overgrow the integral length-scale. Back to the beginning: the shear layer near the nozzle is often though to display the clearest natural living environment for *Kelvin-Helmholtz instability*, which is one of the basic flow instabilities, because it takes places in all shears and transverse velocity gradients everywhere within the flowing fluids. It occurs even inside the vortices [3] leading to the *fractal nature of turbulence* [4].

Jet flows have many engineering applications from chemical mixing (we want high turbulence without strong winds) up to propulsion (we want maximum velocity without

* Corresponding author: dudad@kke.zcu.cz

turbulent loses). Therefore, it would be nice to control the jet stability by manipulating the last solid part in the problem – the nozzle.

2 Experimental setup

The air jet is powered via the radial compressor with inlet from the atmosphere. The outlet flow is accelerated by using a smooth convergent nozzle from diameter 200 mm to 50 mm. The airflow is further modified by using the studied nozzle with or without ribs on its inner surface (see Fig. 1 and 2 for easier imagination). Diameter of this second nozzle is constant $D = 50$ mm and its length is 100 mm, i.e. $2D$. This nozzle has been created in three variants: (a) with a smooth inner cylindrical surface, (b) with 6 longitudinal ribs and (c) with 12 longitudinal ribs with half thickness than in the (b) case in order to keep the *blockage ratio* constant. The height of the ribs linearly increases from 0 at the nozzle inlet to 2.5 mm (i.e. 5 % of D) at the nozzle outlet. The blockage ratio is 1.5 % in both variants with ribs; this has been compensated during the measurement by increasing the compressor power by this percentage, when measuring the flow past the smooth nozzle in order to keep the core velocity comparable. All the nozzles have been created by using 3D printer [6] Raise 3D from polymerized lactic acid PLA [5].

The accuracy of 3D printing procedure has been checked by using the optical 3D scanning system GOM Atos Core 300. This measuring is based on the projecting fringes of varying width and observing their displacement [7, 8, 9]. Results in Fig. 1 show reasonable accuracy with deviations up to 0.3 mm from the planed geometry. One of the limitation of the 3D scanning technology is the problematic scanning of details inside voids [8]; therefore, we are not able to check the quality of the ribs in the middle of the nozzle length. However, at the outlet, they are scanned well. The systematic error of printing lies in the sharpness of the edges, which display radius around half millimeter (see Fig. 1 right), although they was planned to be sharp. The nozzle diameter was smaller by 0.08 mm (i.e. 0.16 %) than the planned one. The rib thickness is larger by 7 %. On the other hand, the circularity of cylindrical surfaces is very good (see Fig. 1 left panel).

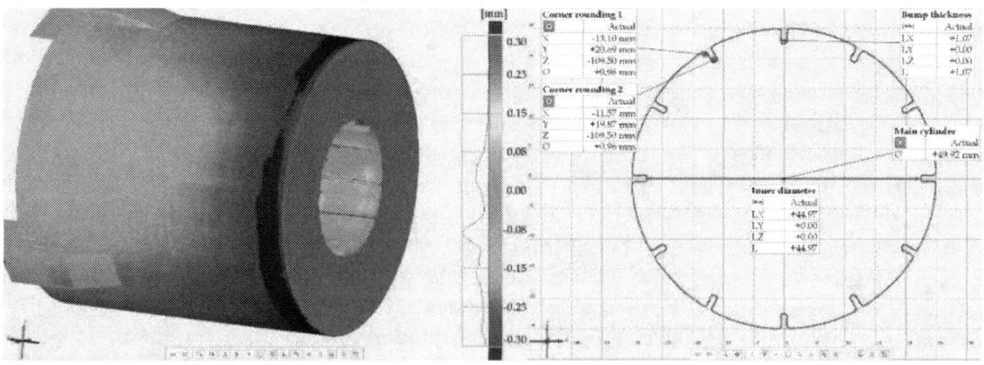

Fig. 1. The result of optical 3D scanning of the nozzle modification (c) with 12 ribs. Left panel shows the deviation from ideal geometry represented by the CAD (computer-aided design) model used for the 3D printing. Right panel shows the section at the exit (0.5 mm from it in order to avoid noise) with denoted actual dimensions. In a case of ideal 3D printer, the corner roundings should have diameter 0, the rib (bump) thickness might be exactly 1 mm, the main cylinder diameter D should be 50.0 mm and inner diameter might be 45.0 mm (D – 2× rib height).

The flow in the jet produced by the described nozzles is measured by using Stereo PIV method [10]. In the fluid, there are small droplets of diameter around 1 μm, thus the inertial (volume) forces are much smaller than the viscous (surface) forces affecting trajectories of such a small particles and therefore their movement can be interpreted as the movement of the fluid around them [10]. Double-pulse solid-state laser defocused via cylindrical lenses into a sheet illuminates these particles. The illuminated area is normal to the jet direction; it lies 50 mm, i.e. 1*D* past the nozzle outlet. Particles in this area are observed by pair of cameras from two directions rotated by 45° against the jet direction. The position of observed plane is displayed in Fig.1 top. We use commercial the Stereo PIV system from the company Dantec, whose software Dantec Dynamic Studio is used for operating the equipment and for processing the data in terms of converting the pairs of double-snapshots gray-scale photographs into single 3D velocity vector fields; the details can be found elsewhere [10, 11]. The air velocity is 62.7 m/s corresponding to diameter-based Reynolds number $2.22 \cdot 10^5$.

3 Results

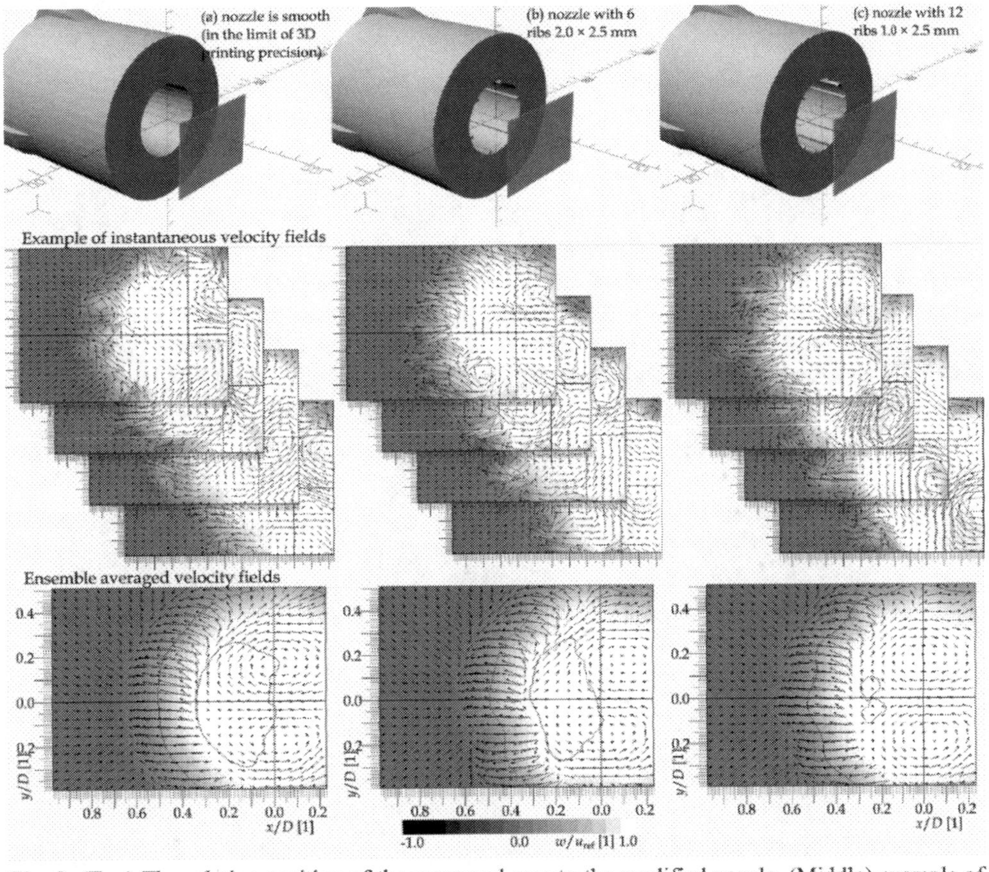

Fig. 2. (Top) The relative position of the measured area to the modified nozzle. (Middle) example of few instantaneous velocity fields. (Bottom) Ensemble average of 628 snapshots. The in-plane velocity components are displayed as arrows (only every second vector is displayed). Stream-wise component (normal to the measured plane) is represented by the gray-scale.

The studied area covers partly the jet core, the shear layer and the surroundings as well. The ensemble average velocity fields for the three nozzles are displayed in Fig. 2 bottom. We see the jet core area with "table-top" in the right-hand side of the figures, in the middle, there is the bow of the shear layer, where the stream-wise velocity component perpendicular to the measured plane (denoted *w* in this paper) decreases. Due to the mass conservation, the direction of the flow deflects away from the jet axis, which is observable as increase in the magnitude of in-plane velocity components (denoted *u* and *v*). The ribs in the nozzle slightly affect the shape of the average velocity creating waves on the isotach at ½ u_{ref}. We see an imperfectness in the average velocity – a pair of stream-wise vortices in the jet core. The used wind tunnel is not fully completed yet, it has not the grids in the settlings chamber. We hope that for the future experiments, the chamber will be completed and these large-scale vortices will no longer perturb our measurements.

The shear layer is home of the turbulence created firstly via the Kelvin-Helmholtz instability and later via reconnections of larger-scale vortices, vortex stretching and other mechanisms typical for turbulent flows. The energy source and entropy bath for this show is the velocity gradient between the jet core and the ambient. The turbulent motion appears *chaotic* for a first look [4], and its amount can be expressed in terms of the *turbulent kinetic energy* TKE

$$\mathrm{TKE} = \frac{1}{2}\langle u'^2 + v'^2 + w'^2 \rangle \tag{1}$$

where $\langle\cdots\rangle$ means ensemble averaging, *u, v* and *w* are the fluctuating velocity components

$$u' = u - \langle u \rangle \tag{2}$$

and other components analogically. This operation is known as Reynolds decomposition [12].

The spatial distribution of TKE is in Fig. 3 and Fig. 4 displays its development along the horizontal axis depicted in the spatial maps. Here we finally come to the effect mentioned in the title – roughening the nozzle surface leads to decrease in turbulent kinetic energy in the near shear layer. Therefore one can estimate the decrease in the turbulent transport of momentum and thus to better stability of the jet in later stages. (Although this is only assumption, we did not measure it yet.)

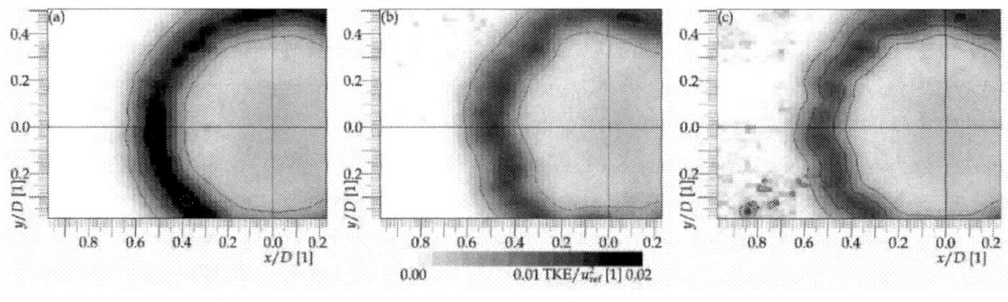

Fig. 3. Spatial distribution of turbulent kinetic energy TKE past three nozzle variants.

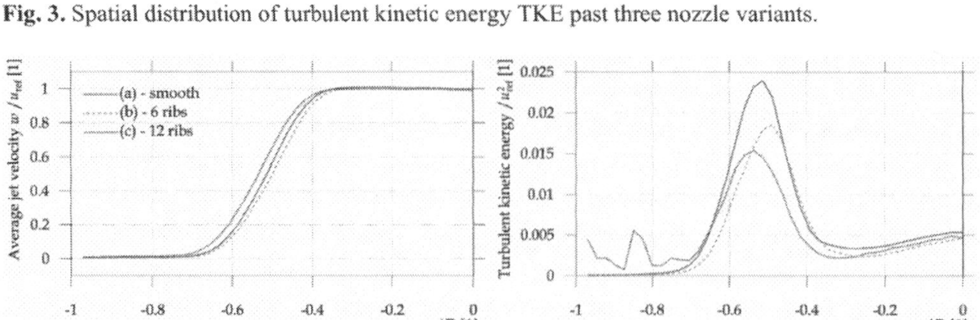

Fig. 4. Radial profile of the ensemble-average stream-wise velocity w (left) and Turbulent kinetic energy (right) along the x-axis depicted in other figures.

4 Discussion

We venture to explain this phenomenon by the presence of *secondary flows* [13] along the ribs [14, 15]. This secondary flow takes the form of longitudinal vortices [16, 17, 18], which, although are weak and disturbed by the surrounding turbulence, can prevent or, at least, delay the development of large-scale toroidal vortices of Kelvin-Helmholtz instability. The presence of stream-wise vortices past the ribs may appear in the averaged vorticity field even in higher distances from the nozzle, see Fig. 5. However, the in-plane fluctuations of the entire jet blur them in the ensemble-averaged data disturbed by experimental noise.

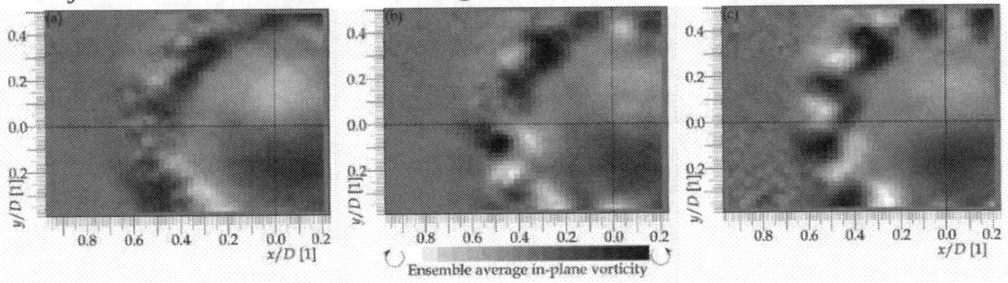

Fig. 5. The ensemble average of in-plane vorticity component (other vorticity components are not calculable as we do not know the gradients in stream-wise direction).

The vorticity ω displayed in gray-scale in Fig. 5 is calculated as the rotation of two-dimensional velocity field

$$\omega = \nabla \times \begin{pmatrix} u \\ v \end{pmatrix} \tag{3}$$

which is calculated on grid with spatial resolution Δx and Δy by using simple symmetrical differential scheme [18]

$$\omega(x, y) = \frac{u(x, y + \Delta y) - u(x, y - \Delta y)}{2\Delta y} - \frac{v(x + \Delta x, y) - v(x - \Delta x, y)}{2\Delta x} \tag{3b}$$

Thus, it is only the z-component of vorticity as we do not know the velocity gradients in z-direction.

Although the instantaneous velocity fields display turbulent behaviour at the shear layer, the presence of regular weak stream-wise vortices leads to smaller length-scale of turbulence there [2]. The autocorrelation function R_{ww} is widely used in flow research [19]. It represents somehow the "memory" of the system, i.e. a statistical connection of events separated by distance Δr

$$R_{ww}(\vec{x}, \Delta\vec{r}) = \frac{1}{\sigma_w(\vec{x})} \cdot \frac{1}{\sigma_w(\vec{x} + \Delta\vec{r})} \langle w'(\vec{x}) \cdot w'(\vec{x} + \Delta\vec{r}) \rangle \tag{4}$$

Fig. 6 shows the correlation function of stream-wise velocity component w. The localization of reference point x has been chosen in four different locations: (1) in the surroundings near the shear layer; (2) in the center of shear layer; (3) at the beginning of shear layer (from the core) and (4) in the jet core axis. The case (2) is shown in Fig. 6 for all possible Δr displaying the *shape* of correlation spot as well. Fig. 7 displays the comparison of the correlation function R_{ww} along the vertical lines depicted in both figures.

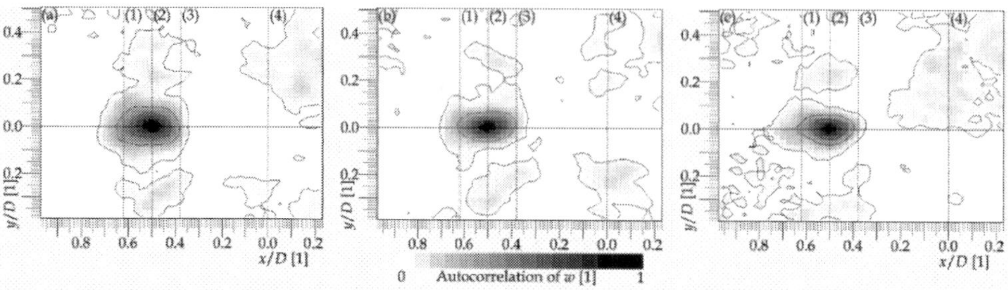

Fig. 6. The spatial correlation function of stream-wise fluctuation velocity R_{ww}. The cross-section points of lines (1) – (4) with the horizontal x-axis are the reference points of R_{ww}, the case (2) applies for this figure. The other cases are displayed in Fig. 7 only along the lines denoted (1) – (4).

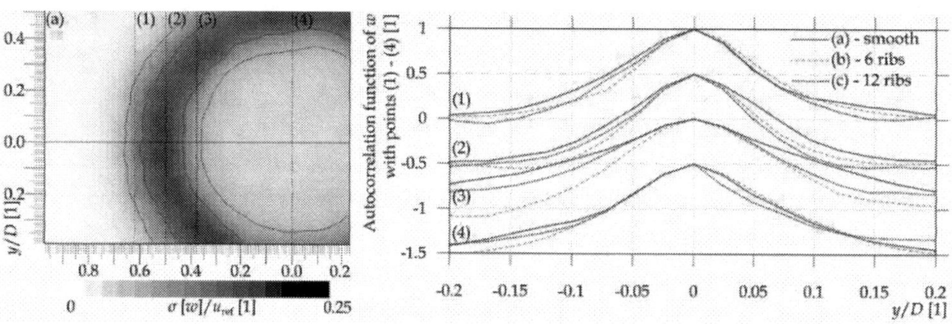

Fig. 7. (Right) the correlation function R_{ww} along the lines (1) – (4) depicted in the left-hand side in the spatial distribution of standard deviation of w.

The autocorrelation function R_{ww} clearly displays that the fluctuations in the smooth case (a) are more spacious. This effect is stronger inside the shear layer, while in the jet core (4) and ambient (1) their spatial size is comparable.

5 Conclusion

We have observed that adding the ribs into the nozzle *decreases* the turbulent kinetic energy in the shear layer in the distance of 1 diameter past the nozzle exit. We suggest explanation, that the secondary corner vortices formed along the ribs may stabilize the shear layer against the large-scale vortices formed via the Kelvin-Helmholtz instability. Therefore, the fluctuations are slightly smaller in spatial terms.

There remain many questions to answer in the future: e.g., what happens later? Will the stabilization effect continue or not? In addition, what happens, when we twist the ribs? The swirling jets generally opens faster due to centrifugal forces [20, 21].

The scanning was supported from ERDF under project Research Cooperation for Higher Efficiency and Reliability of Blade Machines (LoStr) No. CZ.02.1.01/0.0/0.0/16_026/0008389.

We thank in advance to the Grant Agency of the Czech Republic for supporting our project proposal, which will allow us to explore these fascinating phenomena deeper.

References

1. E. J. List, Turbulent jets and plumes, Annual Review of Fluid Mechanics 14, 189-212 (1982), doi: 10.1146/annurev.fl.14.010182.001201

2. D. Duda, V. Uruba, Spatial spectrum from Particle Image Velocimetry data, Journal of Nuclear Engineering and Radiation Science, 5, 031901 (2019), doi: 10.1115/1.4043319
3. E. Amromin, Analysis of vortex core in steady turbulent flow, Physics of Fluids 19, 118108 (2007); doi: 10.1063/1.2813045
4. J. M. Redondo, A. K. Platonov, Self-similar distribution of oil spills in European coastal waters, Environmental Research Letters, 4, 014008 (2009), doi: 10.1088/1748-9326/4/1/014008
5. Inkinen, S., Hakkarainen, M., Albertsson, A.., Södergøard, A. From lactic acid to poly(lactic acid) (pla): Characterization and analysis of pla and its precursors. Biomacromolecules 12, 523–532 (2011) doi:10.1021/bm101302t.
6. Alsoufi, M.S., Elsayed, A.E., 2018. Surface roughness quality and dimensional accuracy – a comprehensive analysis of 100per-sonal/desktop cost-effective FDM 3D printer. Materials Sciences and Applications 09, 11–40. doi:10.4236/msa.2018.91002.
7. Zhang, C., Zhao, H., Gu, F., Ma, Y., 2015. Phase unwrapping algorithm based on multi-frequency fringe projection and fringe background for fringe projection profilometry. Measurement Science and Technology 26. doi:10.1088/0957-0233/26/4/045203.
8. Mendricky, R., 2016. Determination of measurement accuracy of optical 3d scanners. MM Science Journal 2016, 1565–1572. doi:10.17973/MMSJ.2016_12_2016183.
9. Vagovský, J., Buranský, I., Görög, A., 2015. Evaluation of measuring capability of the optical 3d scanner, in: Procedia Engineering, pp. 1198–1206. doi:10.1016/j.proeng.2015.01.484.
10. Tropea, C., Yarin, A., Foss, J.F., 2007. Springer Handbook of Experimental Fluid Mechanics. Springer, Heidelberg, DE.
11. V. Kopecký, Laserová anemometrie v mechanice tekutin (2008)
12. D. Duda, V. Uruba, PIV of air flow over a step and discussion of fluctuation decompositions, AIP Conference Proceedings, 2000, 1, 020005 (2018) doi: 10.1063/1.5049912
13. J. Nikuradse, Untersuchungen über die Geschwindigkeitsverteilung in Turbulenten Strijmungen (Ph.D. thesis), Univesty of Götingen, Götingen, 1926.
14. Z. Wang, N. Cheng, Time-mean structure of secondary flows in open channel with longitudinal bedforms, Adv. Water Resour. 29 (11) (2006) 1634–1649, doi: 10.1016/j.advwatres.2005.12.002.
15. M. Uhlmann, A. Pinelli, G. Kawara and A. Sekimoto, Marginally turbulent flow in a square duct, J. Fluid Mech. 588: 153-162 (2007) doi: 10.1017/S0022112007007604
16. D. Duda, J. Béma, V. Yanovych, P. Pavlíček, V. Uruba, Secondary flow of second kind in a short channel observed by PIV, European Journal of Mechanics / B Fluids 79, 444-453 (2020), doi: 10.1016/j.euromechflu.2019.10.005
17. D. Duda, M. La Mantia and L. Skrbek, Streaming flow due to a quartz tuning fork oscillating in normal and superfluid He 4, Physical Review B 96 (2) 024519 (2017) doi: 10.1103/PhysRevB.96.024519
18. J. Bém, D. Duda, J. Kovařík, V. Yanovych, V. Uruba, Visualization of Secondary Flow in a Corner of a Channel, AIP Conference Proceedings 2189, 020003 (2019), doi: 10.1063/1.5138615
19. E. O. Schulz-DuBois and I. Rehberg, Structure function in lieu of correlation function, Applied Physics **24**: 323-329 (1981)
20. C. E. Cala, E. C. Fernandes, M. V. Heitor, S. I. Shtork, Coherent structures in unsteady swirling jet flow, Experiments in Fluids volume 40, 267–276 (2006), doi: 10.1007/s00348-005-0066-9
21. D. Duda, The PIV study of air flow past the counter-swirler 53983, MATEC Web of Conferences 168, 05004 (2018), doi: 10.1051/matecconf/201816805004

CFD evaluation of the influence of the parts of a shell and tube heat exchanger on heat transfer

Jiří Frank[1*], *Michal* Volf[1], and *Stefan* Bajić[1]

[1]University of West Bohemia, Faculty of Mechanical Engineering, Department of Power System Engineering, Univerzitní 22, Pilsen, Czech Republic

Abstract. This article deals with numerical fluid flow and heat transfer simulations of a shell and tube heat exchanger in which cooling water is heated by hot exhaust fumes. This heat exchanger plays a major role in a cogeneration unit, since it is responsible for the effective use of residual heat. The objective of the simulations is to evaluate the effects of various design changes made to the heat exchanger and their influence on the temperature fields and thus on the overall performance and efficiency of the system. In our analysis we looked at the baffles which cause the cross-current flow of water outside the tubes and at the placement of the gas inlet, i.e., on the distribution of the mass flow rate of the exhaust fumes inside these tubes.

1 Introduction

Cogeneration of electricity and heat is a very attractive way of using primary sources of energy as it allows for highly efficient conversion. The bottleneck of this approach is obviously that at the place of its production, there must be demand for the heat energy, either for heating purposes or for industrial use. The principle of cogeneration is not new or ground-breaking as there are many power plants that produce both electricity and heat. However, over the last decade this principle is being used in much smaller units, cogeneration units. Their advantage is that they can be placed directly in industrial facilities, residential buildings, and many other places where both electricity and heat are needed. This greatly increases the opportunities for implementing cogeneration.

A cogeneration unit consists of many components: the engine (or generally a device for converting primary energy into mechanical energy, i.e., the rotary motion of a shaft), a generator for converting the mechanical energy into electrical energy and finally a heat exchanger in which the waste heat is transferred into a secondary circuit and used further.

Such a heat exchanger, which plays a major role in the effective use of waste heat, is the focus of our research. We investigate a shell and tube heat exchanger, in which hot exhaust fumes (coming out of an engine) are heating up cooling water flowing in a secondary circuit. The hot water is then further used by a customer. The goal of our research is to evaluate the effects of various design changes made to the heat exchanger and their influence on the temperature fields and thus on the overall performance and efficiency of the system. In our analysis we focus on the baffles which cause the cross-current flow of water outside the tubes

* Corresponding author: frankji@students.zcu.cz

and on the placement of the gas inlet, i.e. we investigate the distribution of the mass flow rate of the exhaust fumes inside the tubes.

In the first part of the article, the design of the heat exchanger is briefly described and the main parts which are of interest are discussed in detail. Secondly, the proposed modifications of the geometry, which include changes in the position of the gas inlet and the design of the baffles, are described. Then the simulation setup is discussed. Finally, the simulations are evaluated and we present our results. It should also be noted that all presented simulations were carried out using the Ansys CFX 2019 R1 [1].

Finally, it should be noted that similar studies have also been conducted by various scientists, especially regarding the shape and dimensions of the baffles. One can take a look for example at the following comprehensive studies [2] or [3]. The presented research, however, deals also with the inlet and outlet chambers.

2 Shell and tube heat exchanger

A shell and tube heat exchanger (see Fig. 1) is a device used to transfer heat between two separated fluids. It consists of several tubes mounted inside a cylindrical shell. In a flue gas heat exchanger water flows on the outside of the tubes while the exhaust gas flows inside them. Other main parts of the heat exchanger are the inlet and outlet chambers, tubesheets and baffles. Tubesheets are rigidly connected with the tubes and thus determine their exact placement. The inlet chamber is responsible for distributing the exhaust gases to individual tubes.

According to the source [2], the most important geometric parameters are the tube arrangement, the type, the orientation and the opening of the baffles, the space between the baffles. However, we focus mainly on the baffles that are present in the heat exchanger for adequate circulation of working fluid inside the shell [3], It follows that they make a barrier to the flowing water and therefore constrain the direction in which the water flows.

Bear in mind that such a shell and tube heat exchanger consists of many other parts that are not mentioned in the article since they are not relevant for the purposes of our research. [4] [5]

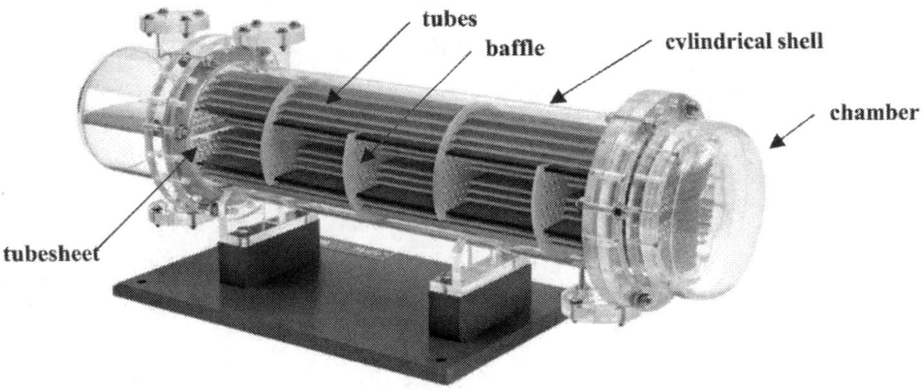

Fig. 1. Shell and tube heat exchanger (generic illustration) [7]

3 Geometric modifications of selected parts

3.1 Inlet and outlet chambers

In order to evaluate the influence of the gas chambers, their inlets and outlets were modified. The basic configuration is depicted in Fig. 2 where both gas inlet and outlet are radial.

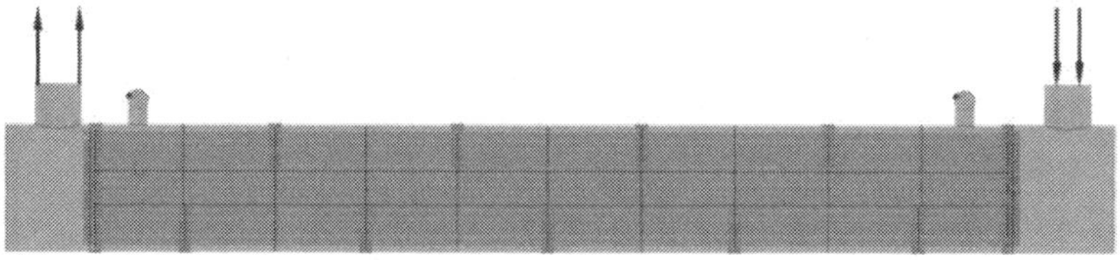

Fig. 2. Basic configuration of gas inlet and outlet

Radial air intake causes the gas in the chamber to whirl and thus the mass flow rate distribution in the tubes is not homogenous as some of the tubes are left outside the main gas flow.

To evaluate the influence of this air whirling in the chambers, several modifications were designed (Fig. 3). The basis of these modifications is changing the radial inlet or outlet to an axial one and their different combinations.

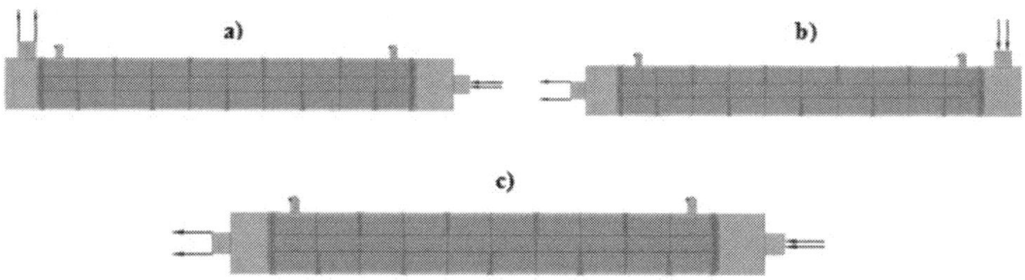

Fig. 3. Proposed modifications to chambers: a: axial inlet & radial outlet; b: radial inlet & axial outlet; c: axial inlet & axial outlet

3.2 Baffles

Two parameters were used for optimizing the baffles. First, the **length of the baffles** was modified to cause changes in the flow area. The idea behind this change is that the lower the flow area, i.e., the greater the length of the baffles, the higher the water velocity. Higher water velocity means that there will be a greater inertia effect causing the water to not so easily change its flow direction. On the other hand, if the flow area is too large, meaning it would not be significantly restricted, there would be no substantial change in the flow direction (i.e., significant reduction in pipe cross-circulation) and thus the effect of the baffles would likely be negligible. It follows that a significant reduction in pipe cross-circulation could even lead to a stoppage of the flow and vortex formation in some areas which would greatly reduce the thermal efficiency of the exchanger. Finally, it should also be noted that a smaller cross-sectional area can result in a large increase in pressure drop. Keeping this in mind, two lengths for the baffles were selected.

In the first variant (Fig. 4a) the baffles go slightly above the plane of symmetry which will prevent the water from flowing purely in the axial direction; however, it will not significantly decrease the flow area and thus will not greatly increase the inertia effect. For the second modification (Fig. 4b), the length of the baffles was increased considerably in order to decrease the flow area and thus to evaluate the effect of inertia. The length of the baffles was selected based on the literature, where it is reported that the most used cut-out height is between 20% - 25% of the shell diameter (Fig. 4). Bear in mind that for our analysis we have chosen a value of 20%. Since the base cut-out height was 126 mm, it has been reduced to 69.6 mm.

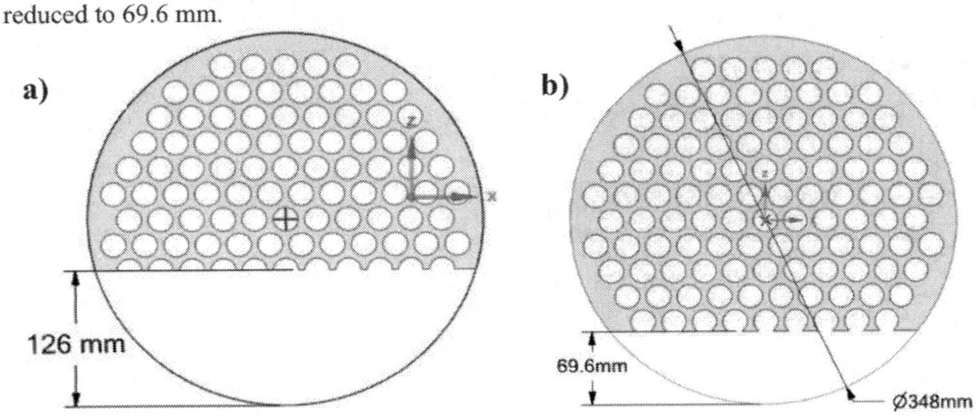

Fig. 4. Modification of the length of baffles

The distance between the baffles was used as the second parameter. Bear in mind that with the changing distance, the number of baffles changes as well. For our analysis we have reduced the base distance and increased the number of baffles from 9 to 13. This modification can be seen in Fig. 5.

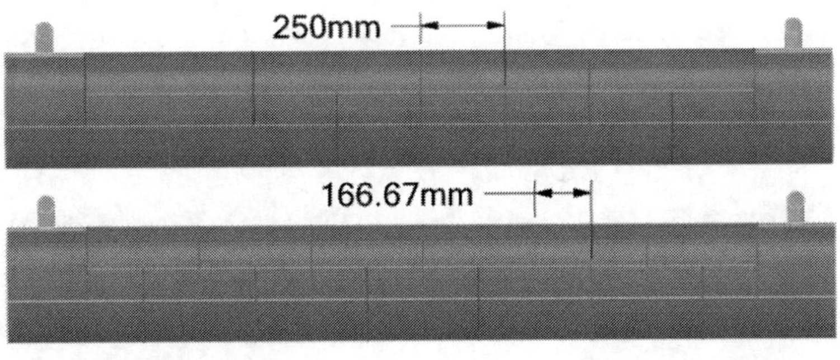

Fig. 5. Modification of distance between baffles

4 CFD simulations

The analyses presented in this paper were conducted using the same simulation setup and element sizes of computational meshes, both are discussed in more detail in the following subchapters. Bear in mind that the mathematical model consists of two fluid regions (water and gas) and one solid region (tubes with tubesheets).

4.1 Computational meshes

Computational meshes consist of both tetrahedral and hexahedral elements. Chambers, tubesheets and both input and output pipes for cooling water were discretized using tetrahedral elements in order to locally decrease element sizes and thus to get denser meshes. Other parts were meshed using hexahedral elements. One can see the examples of computational meshes in Fig. 6 and Fig. 7.

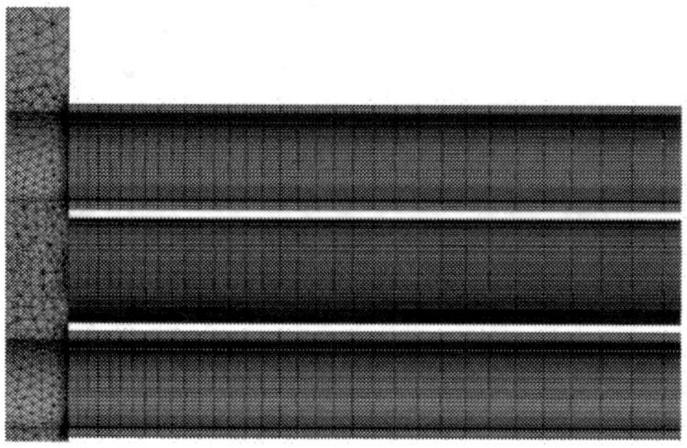

Fig. 6. Mesh for the solid region

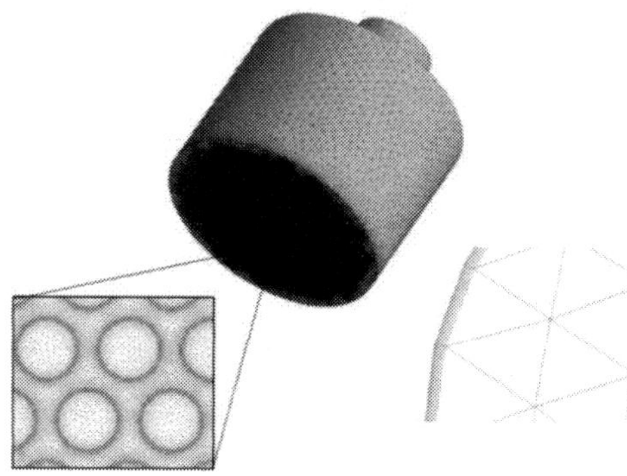

Fig. 7. Mesh of the inlet chamber of the exhaust fumes with a detail of inflation layer

The length of hexahedral elements was chosen to be globally 8 mm at maximum. Regarding the tubes, the thickness of these elements was 0.5 mm to get at least three cells across the wall. Element sizes of tetrahedral elements for tubesheets and chambers were 4 mm and 10 mm respectively. Concerning the inflation layer, the thickness of the first element was chosen to be 0.05 mm so that the Y^+ does not exceed 2.5. In total, the mesh contains approx. 25 million of elements.

For the purposes of the simulation, the baffles were simplified and they were assumed to be of zero thickness, see Fig. 8. Thus, there is no computational mesh prepared for them. This simplifies the model and decreases the number of elements; however, it does not neglect the heat transfer through the baffles because their thickness is defined. Thus, the simulation software (ANSYS CFX) can calculate the heat flux properly.

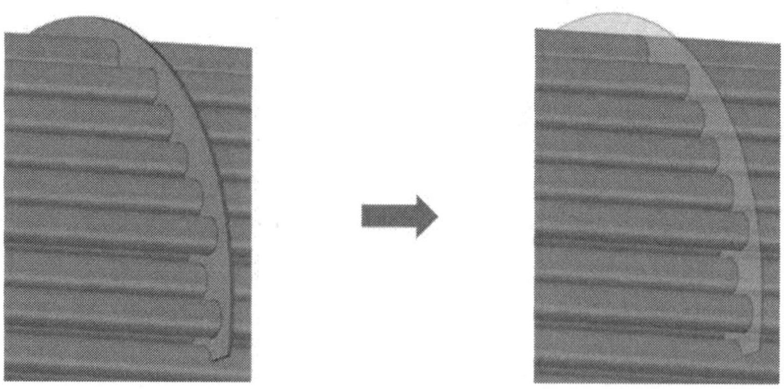

Fig. 8. Simplification of baffles

5 Simulation setup

The presented simulations were done as steady-state ones and thus only boundary conditions were needed, see Table 1.

Table 1. Boundary conditions

Name	Type of condition	Value
Exhaust gas inlet	Average pressure	10 330 [Pa]
	Temperature	520 [°C]
Exhaust gas outlet	Mass flow	0.31 [$kg \cdot s^{-1}$]
Water inlet	Mass flow	5 [$kg \cdot s^{-1}$]
	Temperature	88 [°C]
Water outlet	Average pressure	250 000 [Pa]

Bear in mind that the pressure boundary conditions are relative to the reference pressure which is set to the atmospheric pressure of 101 325 Pa. The exhaust fumes flowing inside the tubes are modelled as an ideal gas with the properties of carbon dioxide, see Table 2. The water properties are taken from IAPWS and the material properties of X6CrNiTi1810 used for the tubes with tubesheets are given in Table 3.

Table 2. Properties of exhaust fumes (modelled as the ideal gas)

Property	Value
Molar mass	44.01 [$kg\ kmol^{-1}$]
Specific heat capacity	851 [$J\ kg^{-1}\ K^{-1}$]
Thermal conductivity	0.0145 [$W\ m^{-1}\ K^{-1}$]
Dynamic viscosity	1.49 E-5 [$kg\ m^{-1}\ s^{-1}$]
Density	*ideal gas equation*

Table 3. Material properties of X6CrNiTi1810

Property	Value
Density	8030 [$kg\ m^{-3}$]
Specific heat capacity	502.48 [$J\ kg^{-1}\ K^{-1}$]
Thermal conductivity	16.27 [$W\ m^{-1}\ K^{-1}$]

Last but not least, a two-equation SST k-ω turbulence model is used with a generic turbulence intensity of 5% at the inlet. Moreover, gravity is considered and it is assumed that the heat exchanger is placed horizontally. All walls are assumed to be hydraulically smooth. All outer walls of the shell and both inlet and outlet gas chambers are adiabatic and thus no heat transfer between the heat exchanger and ambient environment is modelled.

6 Evaluation of CFD analysis

In the following paragraphs, we look at the analysis of the CFD results. First we discuss the influence of the design of the inlet and outlet gas chambers on the performance of the heat exchanger. Then we focus on the evaluation of the baffle design. The contours of the temperatures, mass flow rates and velocities are described.

6.1 Evaluation of the chamber variants

Table 4 shows the differences in the input and output values of the monitored parameters for the exhaust gas. It is evident that in the variants with an axial inlet there is less pressure loss than in the variants with a radial inlet. The smallest pressure drop is shown by the variant with both axial inlet and outlet volumes. These changes are not very important for the final temperature of the exhaust gas.

Table 4. Results of modifications to the design of the chambers for the exhaust gas

EXHAUST GAS					
	Base	**Axial inlet & radial outlet**	**Radial inlet & axial outlet**	**Axial inlet & axial outlet**	**Units**
Pressure difference	484.0	385.0	480.6	376.8	[Pa]
Temperature difference	409.5	409.9	410.7	410.1	[°C]

The resulting values for cooling water for all the variants are listed in Table 5. As can be seen, changes in the geometry of the inlet and outlet cause differences in pressure losses in the range of ± 0.4%. For the cooling water velocity, these differences are approx. 5%. Only in the first new variant will the cooling water heat up significantly, by up to 19% compared to the original design.

Table 5. Results for modifications to the design of the chambers for cooling water

COOLING WATER					
	Base	**Axial inlet & radial outlet**	**Radial inlet & axial outlet**	**Axial inlet & axial outlet**	**Units**
Pressure difference	8 540.0	8 570.3	8 568.8	8 575.9	[Pa]
Temperature difference	3.9	4.6	4.3	4.2	[°C]

It can be concluded that the modifications to the chambers do not have too much influence on the resulting flue gas temperature. For the configuration with the axial inlet and radial outlet and the configuration with the radial inlet and axial outlet, the pressure losses in the exhaust gas were reduced due to the axial inlet volume, which causes a more homogeneous distribution of exhaust gases in the tubes. In the axial inlet & radial outlet configuration, the change to the inlet chamber results in a more uniform heating of the cooling medium. This value is up to 19% higher than in the basic configuration. This temperature rise is due to a more homogeneous distribution of the exhaust gas in the tubes and a more uniform heating of the cooling medium throughout the heat exchanger (see Fig. 9). Thus, the configuration with axial inlet and radial outlet seems to be the most advantageous in terms of thermal energy transfer.

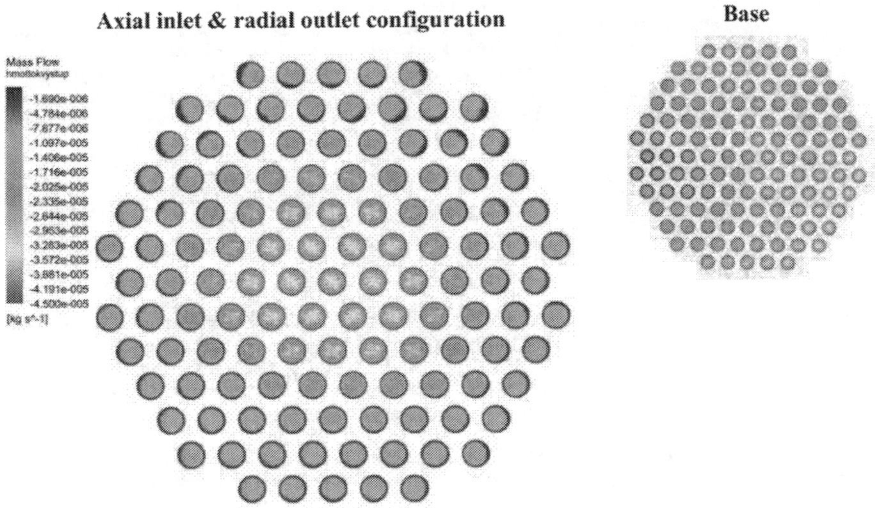

Fig. 9. Mass flow distribution at the outlet from tubes [6]

6.2 Evaluation of the baffle design

Fig. 10 shows the velocity contours (left) and the temperature contours (right). All the contours represent the water side of the heat exchanger since this is the main area of interest for these analyses.

It can be seen that in the basic configuration, there is a rather significant region behind the baffles where the flow of water is slowed down and a vortex region is established. This effect is undesirable for heat transfer since water accumulates here which heats up and thus the temperature difference decreases in these regions. The lower the temperature difference, the lower the heat flux and thus the cooling effect is reduced. Thus, we need to modify the geometry of the heat exchanger to get rid of these regions.

The configuration with a smaller distance between the baffles is shown in the middle diagram in Fig. 10. One can see that this configuration has a smaller region of vortices behind the baffles, which is good for the heat transfer. The shorter baffle length in the third variant creates a much greater vortex region, as shown in the last diagram in Fig. 10.

Table 6 and Table 7 sum up the results of these CFD simulations. It is evident that the configuration with the modified distance between the baffles has the highest temperature difference between the inlet and outlet on the water side; however, this improvement is accompanied by increased pressure loss.

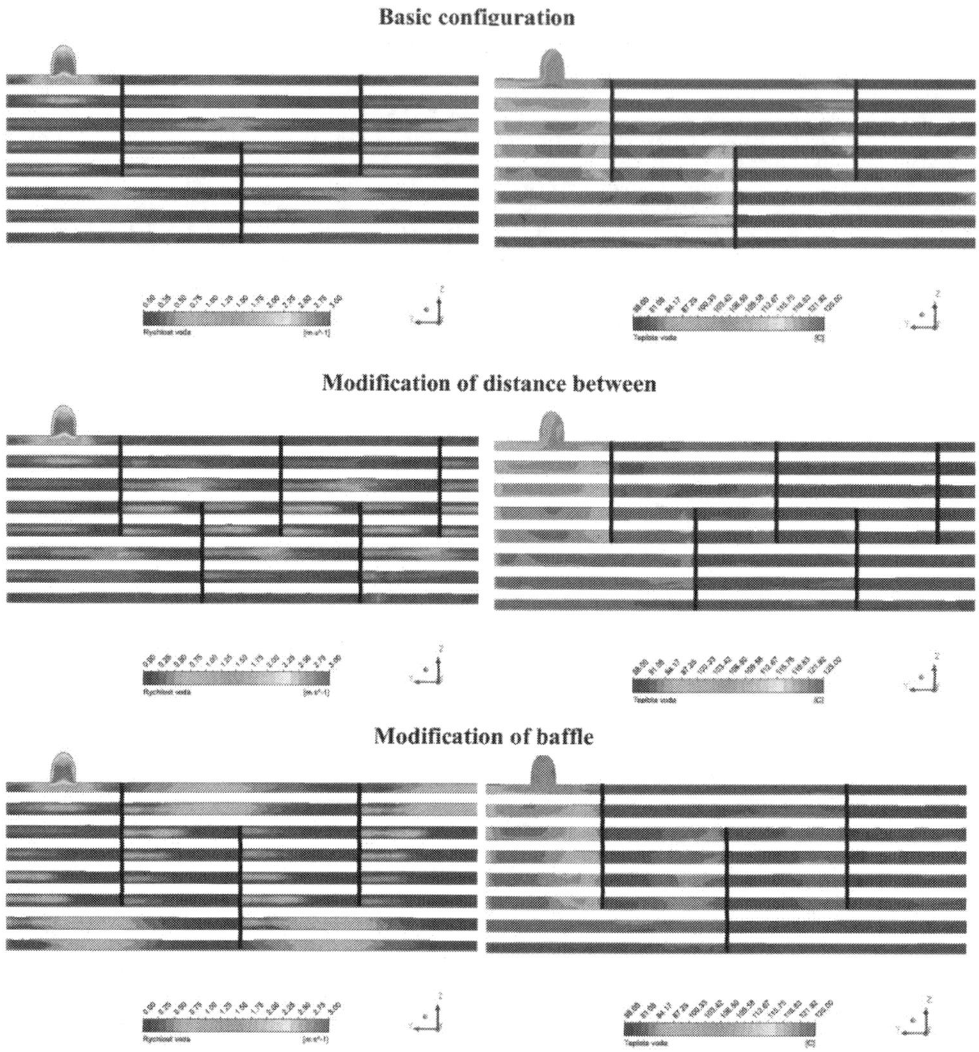

Fig. 10. Current and temperature field of water

Table 6. Results for modifications to the baffle design for the exhaust gas

EXHAUST GAS				
	Base	**Modification of distance between baffles**	**Modification of baffle length**	**Units**
Pressure difference	596.0	599.0	549.0	[Pa]
Temperature difference	391.1	394.9	389.5	[°C]

Table 7. Results for modifications to the baffle design for the cooling water

COOLING WATER				
	Base	**Modification of distance between baffles**	**Modification of baffle length**	**Units**
Pressure difference	9 179.0	11 095.0	12 004.0	[Pa]
Temperature difference	5.1	5.6	5.0	[°C]

In both modified configurations there is a greater pressure loss on the water side, which is due to the more complicated flow path than in the basic configuration. Regarding the temperature difference, i.e., the heat transfer efficiency, the modification of the distance between the baffles seems to be suitable, as the exhaust gas cools down to a lower temperature and thus transfers more thermal energy to the cooling water (in comparison with the basic modification). From this point of view, it could be concluded that this configuration is better; however, from the point of pressure losses, it is fair to say that decreasing the distance between the baffles results in an increased pressure loss, which is undesirable.

6.3 Summary of results

In the following tables, there are summaries of the heat transferred between exhaust fumes and the water. It goes without saying that the higher the value, the better overall design of the exchanger.

Table 8. Summary of heat transferred between exhaust fumes and water (chambers variants)

Configuration of the heat exchanger	Transferred heat [W]
Basic	81 510
Axial inlet & radial outlet	96 140
Radial inlet & axial outlet	89 870
Axial inlet & axial outlet	87 780

Table 9. Summary of heat transferred between exhaust fumes and water (baffle design)

Configuration of the heat exchanger	Transferred heat [W]
Basic	106 590
Modification of distance between baffles	117 040
Modification of baffle length	104 500

To sum up, selected modifications of baffles had slightly positive effect to the value of temperature difference between inlet and outlet of cooling water, but they proved to have quite a large negative effect to the pressure loss in comparison with the base heat exchanger configuration. The pressure loss was higher than 20% in both modifications and thus it can be concluded that these modifications of baffles are not appropriate.

On the other hand, the modifications in the position of inlet and outlet of chambers seems to result in a significant improvement of the exchanger as, for all modifications, the value of temperature difference between inlet and outlet of cooling water has increased (more than 17 %). Moreover, the pressure loss of the flowing gas decreased compared with the base variant (more than 20 %). The configuration with the most significant improvement on the performance of the heat exchanger is the variant with axial inlet and radial outlet, which had the highest temperature difference of cooling water and the lowest pressure loss among all considered configurations.

7 Conclusion

This article describes numerical simulations of the exhaust gas and cooling water flow in a shell and tube heat exchanger. The aim of these simulations was to examine the influence of geometric modifications to the inlet and outlet chambers, baffle length and the distance between the baffles on the overall performance of the heat exchanger. Three variants were designed for the inlet and outlet exhaust gas chambers and two variants for the modification of the baffles. These variants were compared with each other and with the basic configuration.

The simulations with modified exhaust gas chambers show that the configuration with the axial inlet and radial outlet seems to be the most advantageous in terms of thermal energy transfer. Regarding the design of the baffles, the simulations show that shortening the distance between them results in better heat transfer; however, it negatively effects the pressure loss on the side of the water. Finally, comparing the modifications to the chambers and baffles, it can be said that the modifications of the inlet and outlet gas chambers have a more positive effect on the heat transfer and thus on the overall heat exchanger performance.

It follows that improvements gained by modifications in inlet and outlet position of chambers could also be beneficial for practical use; however, experimental validation of numerical results (at least partial) must be done.

The presented work was financially supported by the student project SGS-2019-021 (Improving the efficiency, reliability and service life of power machines and equipment 5).

Computational resources were supplied by the project "e-Infrastruktura CZ" (e-INFRA CZ ID:90140) supported by the Ministry of Education, Youth and Sports of the Czech Republic.

References

1. Analysis of the influence of parts of a heat exchanger on the current and temperature fields in the heat exchanger using CFD. Bajic, Stefan. 020001, Pilsen : AIP Conference Proceedings, 2019, Vol. 2189.
2. ANSYS®. CFX 2019R1. Release date: January 29th, 2019.
3. Shell Side CFD Analysis of a Model Shell-and-Tube Heat. Máté Petrik, Gábor L. Szepesi. s.l. : AIDIC Servizi S.r.l., 2018, CHEMICAL ENGINEERING TRANSACTIONS.
4. CFD Analysis of a Shell and Tube Heat Exchanger with Single Segmental Baffles. Shuvam Mohanty, Rajesh Arora. 2020, International Journal of Automotive and Mechanical Engineering.
5. Brogan, R.J. Thermopedia. Heat exchangers. [Online] 2 February 2011. [Cited: 6 July 2021.] https://www.thermopedia.com/content/832/.
6. Thulukkanam, Kuppan. Heat exchangers design handbook. 6000 Broken Sound Parkway NW, Suite 300 : Taylor & Francis Group,, 2013. 13: 978-1-4398-4213-3.
7. Multi-Pass Floating Head Heat Exchanger Training Model. [Online] Technical Laboratory Systems, Inc. [Cited: 23 6 2021.] https://tech-labs.com/products/multi-pass-floating-head-heat-exchanger-training-model.

MATEC Web of Conferences **345**, 00008 (2021) https://doi.org/10.1051/matecconf/202134500008
Power System Engineering 2021

On-line pump for very low power heating or cooling system

Zbigniew Goryca[1], *Grzegorz* Peczkis[2], *Tomasz* Synowiec[3] and *Piotr* Wiśniewski[2*]

[1] Kielce University of Technology, Department of Building Physics and Renewable Energy, Tysiąclecia Państwa Polskiego 7, 25-314 Kielce, Poland
[2] Silesian University of Technology, Department of Power Engineering and Turbomachinery, Konarskiego 18, 44-100 Gliwice, Poland
[3] WIROMET S.A., Wyzwolenia 27, 43-190 Mikołów, Poland

Abstract. The article presents the concept of a new circulation pump for low power heating or cooling system, dedicated to systems where it is necessary to ensure a slight liquid flow during periods of operation at a significantly low efficiency of the systems. The assumption for the construction of the new pump was the possibility of its installation in the pipeline axis. The proposed prototype design includes the application of a patented solution to relieve the axial force acting in a flow system [1,2]. Additionally, a new permanent magnet DC motor designed and manufactured for this pump was used. The design of the pump takes into account the requirement of maximum silence, which results from its intended use in close proximity to human habitation. The design of the pump has been tested numerically. The results of numerical tests are presented. The pump was made in the rapid prototyping technique in a 3D printer. The pump was built into a dedicated test stand with measuring instruments verified in terms of the quality of measurements. A series of measurements was made on the basis of which the flow characteristics of the pump were determined. During the research, risks for the operation of the new structure were identified, which will be taken care of in a further stage of the research, and the structure, after removing the defects, will be directed to production [3].

1 Introduction

The heating / cooling medium circulation pumps for individual applications available on the market, especially in district heating, constitute a wide range of products. Dominating devices with an electric drive motor operating directly in the pumped medium. The motor is rarely separated from the pump flow system due to the significant cost of the gland packing, maintenance-free, for the entire structure [4]. An important gap in the series of types, indicated by a domestic producer of complete heating systems, is a pump with a low head and low flow, which protects the circulation systems against damage during periods of its shutdown. Two practical failure-generating cases are particularly indicated. The first is the

* Corresponding author: piotr.wisniewski@polsl.pl

strong local heating of the elements of solar systems, shut down during the summer period when the demand for heat is insufficient. Second, the deposition of mineral particles and the effects of corrosion on components immersed in the liquid, which are not moved, and through which the liquid does not flow. The flow parameters were specified by the manufacturer, who also set the task of minimizing the dimensions of the pump and the possibility of including it in the axis of the pipeline, the so-called on-line [5]. The presented construction and research results are a pump that fulfills the assumptions of operation in systems that are periodically shut down.

2 Pump design

The design of the impeller of the flow system was based on the known algorithm for designing centrifugal centrifugal pumps available in the literature [6,7]. It has been proposed to reverse the direction of the rotor in relation to the drive motor, i.e. to place the inflow to the pump rotor from the side of the driving motor and to make a new motor that, when seated on the liquid supply channel to the pump, will rotate directly with the rotor [8]. The impeller therefore does not have a drive shaft, the function of which is taken over by the pump's influence channel [9]. The pump discharge channel has been divided into two streams and, after changing the direction from radial to axial, it is connected in the discharge stub pipe. [10] The authors of the article successfully tested and applied such a solution to the previously implemented mining pump. The three-phase alternating current electric motor with permanent magnets presented in the construction was designed and made by professor Zbigniew Goryca. In addition, a small electronic system powered by DC voltage was made, from which the 3-phase voltage was supplied to the pump motor. The frequency of the alternating voltage on the basis of which the pump rotational speed was calculated using an oscilloscope was later measured on the 3-phase voltage. The constant magnetic force generated by the motor, independent of the applied voltage, keeps the rotating system suspended in the slide bearings in balance with respect to possible axial displacements [11,12]. The displacement of the rotating system along the pump axis at rest meets the resistance of the magnetic force. During pump operation, the rotor system creates an axial force whose direction is opposite to the flow direction of the pumped liquid at the inlet to the rotor. [13] This force is the result of an imbalance in the pressure distribution acting between the stationary fuselage and the rotating walls of the cover plate and the rotor support plate. The figure shows schematically the pressure imbalance acting in the described spaces on the flow system, resulting in an axial force. In order to balance the resulting axial force, a proprietary, patented oblique gap solution was used, which creates a quasi-bearing operating on the pumped liquid and finally provides friction-free. This solution is depicted on Figure 1.

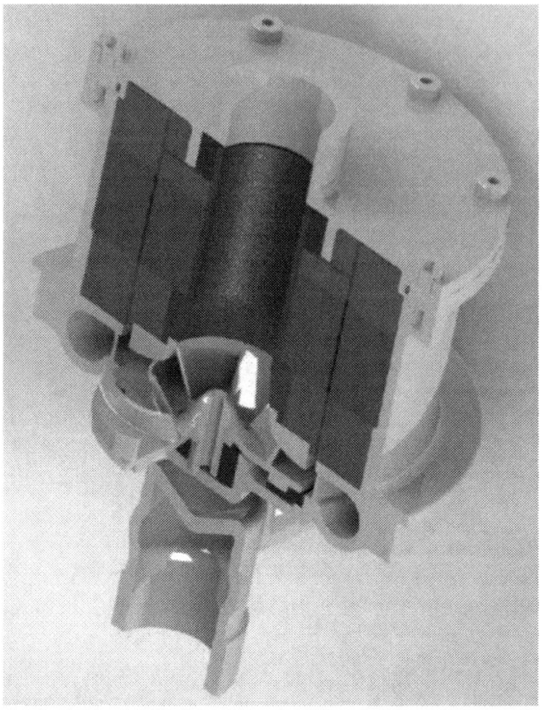

Fig. 1. Pump final design

3 Numerical Analysis

The geometric model of the pump was subjected to numerical analysis in order to verify the design assumptions within the flow system. For this purpose, using ANSYS Fluent software, a numerical polyhedral mesh consisting of about 2 million elements was generated. The mesh was densified in the boundary layer area to provide a y + value of about 1, the successive cell growth rate in the layer is 1.2, which allows high quality results to be obtained using a reasonable number of elements. The turbulent effects of the flow were modeled using the k-ω shear-stress transport (SST) model proposed by Mentner [14,15]. The remaining grid parameters were determined on the basis of data available in the literature for modeling water flows in turbines [16]. It was assumed that the flow is isothermal and the water temperature is 25 ° C. The conditions defining the total pressure at the inlet and the static pressure of the fluid at the outlet from the system were used, the condition defining the rotational speed of the rotor was added in the rotor part, the calculations were carried out in a steady state.

Numerical calculations were performed for various useful pump power. Figure 2 shows the characteristics of the system as a function of the flow rate at constant power on the pump shaft. The efficiency presented on Figure 2 does not include the gearing and electric engine efficiency, it is computed as follows:

$$\eta = \frac{HQ}{P_m} \tag{1}$$

where η is efficiency, H is lifting, Q is volume flow rate and P_m is mechanical power. The results of the numerical analysis are in line with the pump design assumptions and are a good reference to experimental work.

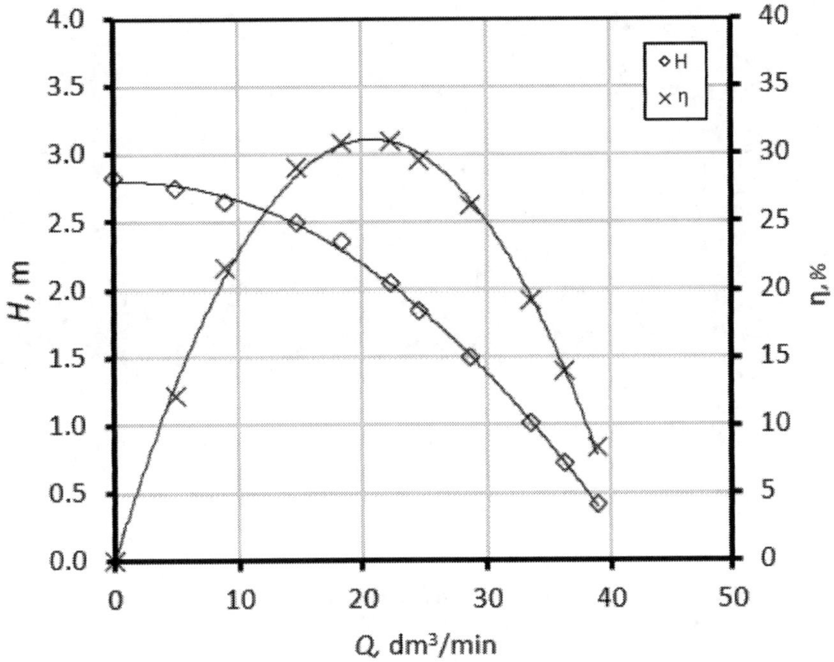

Fig. 2. Pump lifting and efficiency characteristic based on the numerical study

4 24, 26 V experimental study

The constructed pump was subjected to numerical tests at the 3D construction stage. Further work was aimed at printing individual pump elements from plastic and developing a technique for machining the print elements in order to give them matching dimensions. The 3D printout is burdened with surface unevenness, and the external, overall dimensions make it impossible to fit precise elements, e.g. bearing elements. In practice, it is possible to mount the bearings by pressing, but they deform and change other dimensions of the printed element. The metal ring visible in the photo into which the pump housing is screwed enables dimensionally repeatable mounting in the jaws of the machine tool and precise machining in order to adjust, for example, the fitted winding of the motor, both the stator and the rotor. The individual elements of the engine and the power cables led out of the fuselage were filled with epoxy resin, the composition of which does not chemically react with the material used for 3D printing. After the filling was made and it dried, and it was re-machined, it was tested for the separation of the electrical system from the pumped liquid. The pump tests were performed on the stand shown on figure 3. The measuring system was made of instruments subjected to calibration in the laboratory.

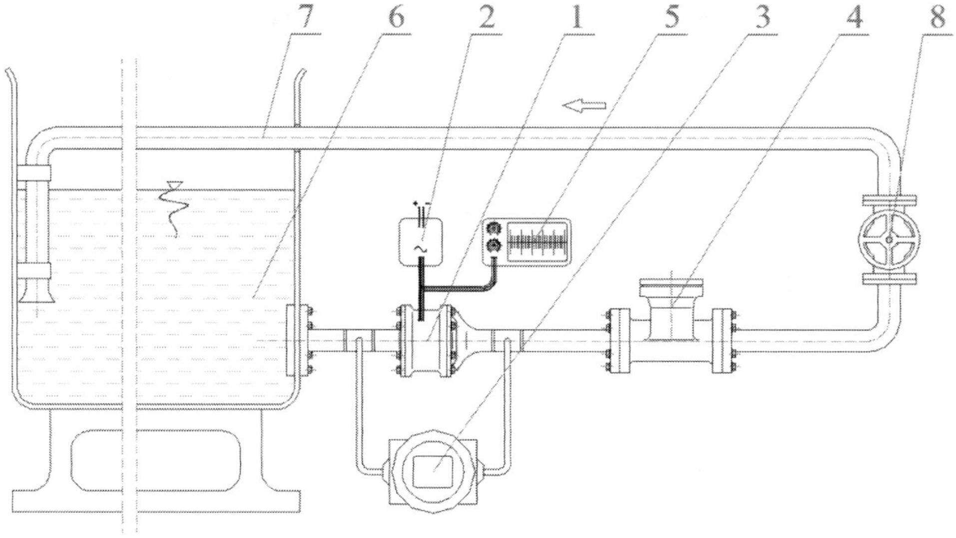

Fig. 3. Scheme of the measuring system. Positions: 1-tested pump 2-power supply system, generation of alternating current from direct current 3- differential manometer 4-electromagnetic flowmeter 5-oscilloscope 6- circulation liquid tank 7-circulation pipeline system 8 - throttle valve

The pressure sensor was checked on a hydraulic press, the electromagnetic flow meter was verified for the correctness of indications by the volumetric method. The oscilloscope attachment used for electrical measurements was not marked. It was used to determine the actual rotational speed of the pump rotating assembly. Further the pump lifting characteristics was designated for different voltage, see figure 4. The pressure behind the impeller and in front of the steering wheel on the discharge of the pump was measured in order to reduce the amount of losses to the discharge port. The characteristic of the pump head and the pressure drop between the pump's suction port and the measurement point behind the impeller, presented on figure 5, are consistent with the expected energy parameters of the pump. The following characteristics are systematized depending on the DC supply voltage of the system generating the 3-phase alternating current supplying the motor.

During the tests, it was found that the pump, together with the electric motor and the electronic system for converting the DC voltage of the system supply to AC voltage, works with unsatisfactory efficiency close to 20%. Observation of the pump operation and additional measurements performed showed insufficient magnetic force of the motor, which did not keep the rotating assembly in the balance of axial forces. Also, the first prepared variant of the oblique slot, tested computationally, did not faithfully reflect the assumptions in practice. This resulted in a slight displacement of the rotating assembly and its elements rubbing against the fuselage elements during operation. Hence the unsatisfactory efficiency obtained in the first experiments. Further works are aimed at eliminating losses by rebuilding the bearing system. Due to the high degree of saturation of the structure with innovative solutions, this should not be seen as a disqualification of the structure, and the currently obtained results of the prototype tests as an introduction to a new, stable design solution in line with the previous assumptions.

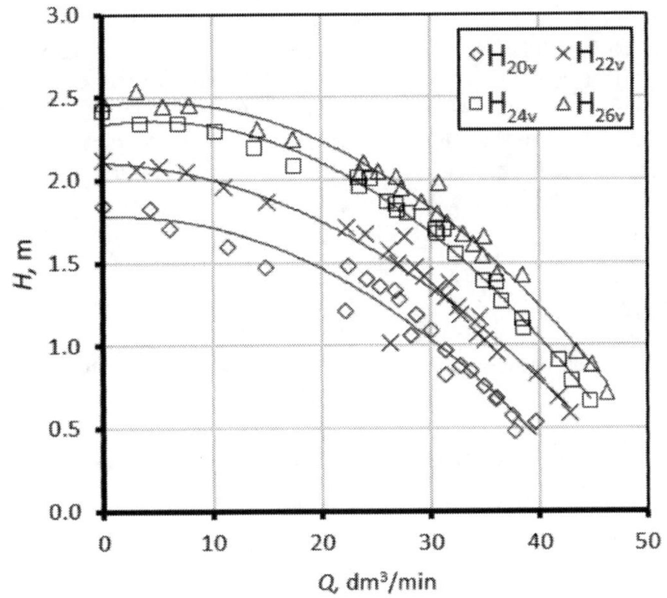

Fig. 4. Pump lifting characteristics intended at the laboratory stand for successive supply voltages of 20, 22, 24, 26 V

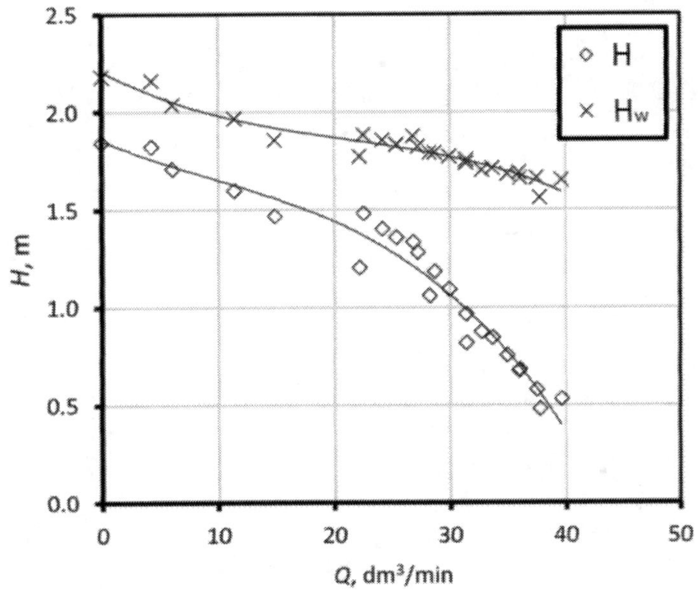

Fig. 5. Comparison of the pressure measurement behind the impeller H_w (Q) to the pressure drop H (Q) realized on the pump for the supply voltage of 20V

5 Summary

The course of construction works, numerical tests, execution and laboratory tests presented in the article confirmed in practice the achievement of the assumed useful lifting height met at the desired volumetric flow rates. The multitude of innovative solutions used in one structure is the reason for the need to improve the presented pump. The conducted works require continuation prior to production improvement of the efficiency of the centrifugal pump unit, the motor and the three-phase DC current generation system up to the values obtained in numerical calculations. In particular, the shape of the double-stream channel requires optimization, which is confirmed by the comparison of the pressure drop across this element presented in the article to the characteristic of the effective head of the pump. These works are ongoing and their effect will be reflected in the near future. The success of the task performed by the presented pump at the present stage does not allow doubting the success of the continued work.

References

1. Korczak A. Badania układów równoważących napór osiowy w w wielostopniowych pompach odśrodkowych. Pierwsze ed. Gliwice: Wydawnictwo Politechniki Śląskiej; (2005).
2. Альбінович МВ. Щелевые Уплотнения Сумы: Изд-во Сумского; (2005).
3. Ostojic P. Pumps and circular economy. World Pumps. 2016 Volume (2016), Issue 1: p. 30-33.
4. Waldemar J. Pompy Wirowe Warszawa: PWN; (2001).
5. Jęrdal W. Efektywne energetycznie układy pompowe. I ed. Warszawa: Oficyna Politechniki Warszawskiej; (2018).
6. Palgrave R. Centrifugal pump basics. Troubleshooting Centrifugal Pumps and their Systems. (2020): p. 13-60.
7. Wenjie Zhou YWCLWZGW. Analysis of fluid-induced force of centrifugal pump impeller with compound whirl. Alexandria Engineering Journal. (2020): p. 4247-4255.
8. Nader Moazami KFMKNGSKJHAMSLDJHRCS. Axial and centrifugal continuous-flow rotary pumps: A translation from pump mechanics to clinical practice. The Journal of Heart and Lung Transplantation. (2013) Volume 32, Issue 1,: p. 1053-2498.
9. Younes YK. A revised design of circular hydrostatic bearings for optimal pumping power. Tribology International. (1993) Volume 26, Issue 3: p. 195-200.
10. Bin Hu SXRZWHLLHZZHL. Investigation on advanced heat pump systems with improved energy efficiency. Energy Conversion and Management. (2019): p. 161-170.
11. Zhichao Zou FWZYRTRXHL. Impeller radial force evolution in a large double-suction centrifugal pump during startup at the shut-off condition. Nuclear Engineering and Design. (2016) Volume 310: p. 410-417.
12. Kalinichenko P, Suprun A. Effective Modes of Axial Balancing of Centrifugal Pump Rotor. In XIIIth International Scientific and Engineering Conference "HERVICON-2011"; (2012); Sumy: Procedia Engineering.
13. Bin Hu HLRZWHLZZSW. A high-efficient centrifugal heat pump with industrial waste heat recovery for district heating. Applied Thermal Engineering. (2017) Volume 125: p. 359-365.
14. Menter FR. Two-equation Eddy-Viscosity Turbulence Models for Engineering Applications. AIAA Journal. (1994): p. 1598-1605.
15. ANSYS Fluent Theory Guide, (2021)R1. [Online]. [cited 2021 July 1.
16. Peczkis G, Wiśniewski P, Zahorulko A. Experimental and Numerical Studies on the Influence of Blade Number in a Small Water Turbine. Energies. (2021).

Analysis of unstable mode of a free gas turbine of turbo-compressors engines

Marián Hocko[1]*, and *Samer* Al-Rabeei [2]

[1]Technical University of Kosice, Faculty of Aeronautics, Department of Aerospace Engineering, Rampová 7041 21 Kosice, Slovakia.

Abstract. This paper analyses unstable mode of a free gas turbine of turboshaft helicopter engine TV3-117. The analysis is focused on the conditions of this phenomenon and the possibilities of its solution in a turboshaft helicopter engine and an industrial turbocharger engine with a free gas turbine. Knowing the causes of the unstable mode of operation of a free gas turbine will allow helicopter pilots to prevent accidents and increase the level of flight safety.

1 Introduction

On May 10, 2017, a Bell 429 helicopter crash occurred at the airport in Prešov Fig. 1, killing two firefighters. The helicopter fell from a height of 100 meters during the training of planting firefighters using ropes. The inquiry commission proceeded that the non-compliance of the helicopter was an incorrect reaction of the pilot to the behaviour of the helicopter in the future when the free gas turbine of the engine entered the mode of unstable work [1].

Fig.1. Helicopter Bell 429 crash near Prešov

* Corresponding author: marian.hocko@tuke.sk

The issue of unstable operation of a free gas turbine of turboshaft helicopter engines is not given sufficient attention in the professional literature or in the training of pilots, despite the fact that it directly affects the safety of flights. For this reason, this article analyses the physical nature of this mode of a turboshaft engine with a free gas turbine.

Since the Bell 429 helicopter engine for Pratt & Whitney Canada PW207D helicopter engines did not release all the technical data, the TV3-117 Mi-24 helicopter engine was used for analysis, which in terms of concept and principle of operation corresponds to the Pratt & Whitney Canada PW207D Bell 429 helicopter engine. the turbine of the TV3-117 turboshaft engine could be realized thanks to the detailed knowledge of the construction and control system of the TV3-117 helicopter as well as the availability of its detailed technical documentation [2, 3].

2 Regulatory system turboshaft engine helicopters

The main task of the helicopter power unit, which currently usually consists of two turboshaft engines (Mi-2, Mi-17, Mi-24, Aerospatiale SA 330 Puma, NH-90, etc.), is to ensure the drive of the helicopter's supporting rotor during its flight. In any mode and maneuvers, as well as when starting the engines, turning and stopping the helicopter's supporting rotor. In steady-state modes of operation of the turboshaft engine functions, the automatic drive unit maintains the set speed of the supporting rotor. When switching to reduced engine power (reducing the setting angle of the carrier rotor blades, or when adjusting the correction handle to the left), insufficient torque (insufficient power) is generated on the free turbine of the turboshaft engine and the carrier rotor speed is reduced. When moving the collective control lever of the helicopter while the turboshaft engines are operating in steady state modes (travel, nominal or take-off), the pilot controls the change of the setting angle of the supporting rotor blades by means of the collective control lever. The control unit adjusts the fuel supply (Q_{fuel}) to the engine's main combustion chamber and maintains the selected speed.

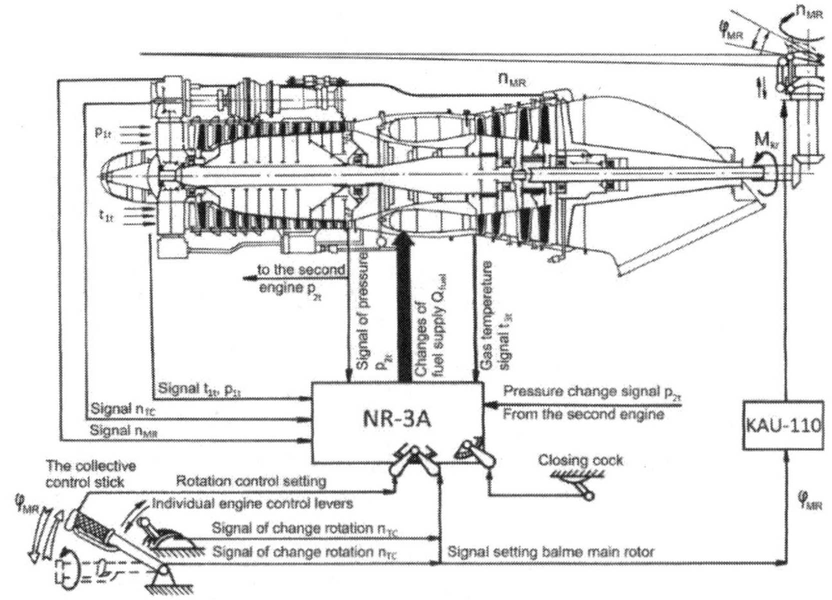

Fig. 1. Principle diagram of the collective control system with automatic maintenance of the rotor speed and control of the collective control of the helicopter rotor [4, 6].

The TV3-117 aircraft turbo-compressor engine is a turboshaft engine with an axial inlet, a twelve-stage axial compressor with rotary vane blades and two air bleed valves, an annular combustion chamber, a two-stage axial, cooled gas turbine compressor an outlet system with the outlet tube rotated at an angle of 60° to the right or left with respect to the engine [4].

Two TV3-117 3rd series turboshaft engines used to drive the Mi-24D and Mi-24DU helicopters, two TV3-117V engines of the Mi-24V drives the helicopter and two TV3-117MT engines drive the medium transport Mi-17 helicopter [5].

3 Force analysis when changing the angle of attack of the main rotor of the helicopter

When starting the engines and increasing the speed of the free turbine (when turning the correction handle to the right from idle to travel mode and when setting the collective control lever up), there is an excess torque (excess power) on the free turbine of the turboshaft engine, which is maintained until specified engine mode. With steady-state modes of operation of turboshaft engines, the automation of the drive unit ensures the maintenance of the set speed of the supporting rotor. However, the speed of the turbocharger rotor of the turboshaft engine changes in this case due to the load of the free gas turbine with the required torque (required power) as well as due to the change of the air temperature at the inlet to the engines t_{1t} (Fig. 2). When switching to reduced engine power (when reducing the setting angle of the carrier rotor blades and when setting the correction handle to the left to idle mode), the free gas turbine of the turboshaft engine generates insufficient torque (insufficient power) and the support rotor speed decreases. When moving the helicopter collective control lever while the turboshaft engines are operating in steady-state modes (travel, nominal, and take-off), the pilot controls the setting of the angle of adjustment of the supporting rotor blades by means of the collective control lever (by its continuous adjustment and complete rotation to the left) (Fig. 3). The fuel control system of the turboshaft engines adjusts the fuel supply to the main combustion chamber of the Q_{fuel} engines accordingly and the selected speed of the helicopter support rotor is maintained. When the pilot stops moving the collective control lever upwards (ready to do so), the controller will maintain a constant helicopter carrier rotor speed at three higher turbocharger rotor speeds, and thus at a torque compensation greater than before the collective control lever position change (e.g. in cruising mode of turboshaft engines). In the take-off mode, depending on the angle pattern, the settings of the helicopter carrier rotor blades, and thus depending on the required torque, are the equilibrium speeds of the helicopter carrier rotor slightly lower than in the cruising or nominal mode. Part of the increased required torque is compensated by the increase in the angle of flow of the gas through the pipeline to the blades of the free gas turbines of the turboshaft engines due to certain reductions in their speed compared to the speed in travel or nominal mode. The angle of attack to the flow of gas leading edge of the rotor blades of the gas turbine of free critical value, so that the gas flow separation from the surface of the rotor blade does not occur [7, 11, 12, 13].

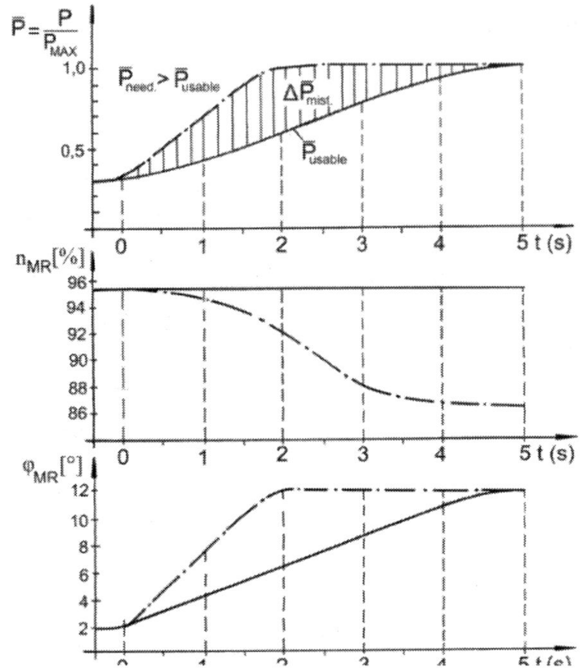

Fig. 2. Changes in relative power required $\overline{P}_{Req.}$, usable power $\overline{P}_{Usable.}$ and speed of the main rotor n_{MR} as a function of time at different speeds of increasing the setting angle of the main rotor blades φ_{MR} (full line corresponds to the set speed of the collective control sausage in accordance with the regulations for use and piloting technique, collective control lever of the helicopter support rotor).

A sharp change in the position of the collective control lever by the pilot is not recommended. Rapid movements of the collective control lever of the helicopter support rotor can tear off the gas stream from the surface of the blades of the free gas turbine of the turboshaft engine and consequently reduce its effective power and decrease the speed of the helicopter support rotor.

4 Unstable flow on a free gas turbine of a turboshaft helicopter engine

Figure 4 shows an unstable bypass of the rotor blades of one of the stages of a free gas turbine of a helicopter engine with a turbo shaft. The tearing of the gas flow on the back of the rotor blade of the free gas turbine of the turboshaft engine can occur when the angle of attack of the flowing gas at the leading edge of the rotor blade increases above the critical value (Fig. 5), which occurs:

- with a sharp increase in the absolute output velocity of the flow, the pipeline exiting the blades of the free gas turbine ($c_1 > c_{cal.}$), and at a constant speed of the free gas turbine of the turboshaft engine and a constant, calculated circumferential speed ($u_1 = u_{cal.}$);
- with a sharp reduce in the speed of the free gas turbine of the turboshaft engine, at which the peripheral speed decreases ($u_2 < u_{cal.}$), and a constant absolute output velocity of the gas flow from the guide blades of the free gas turbine ($c_2 = c_{cal}$);

- with increasing the absolute output velocity of the gas flow from the turbocharged engine free gas turbine blades ($c_3 = c_1$) and decreasing the turbocharged engine free gas turbine speed $u_3 = u_2 < u_{cal.}$).

Fig. 4. Free turbine of the turboshaft engine TV3-117 [8]

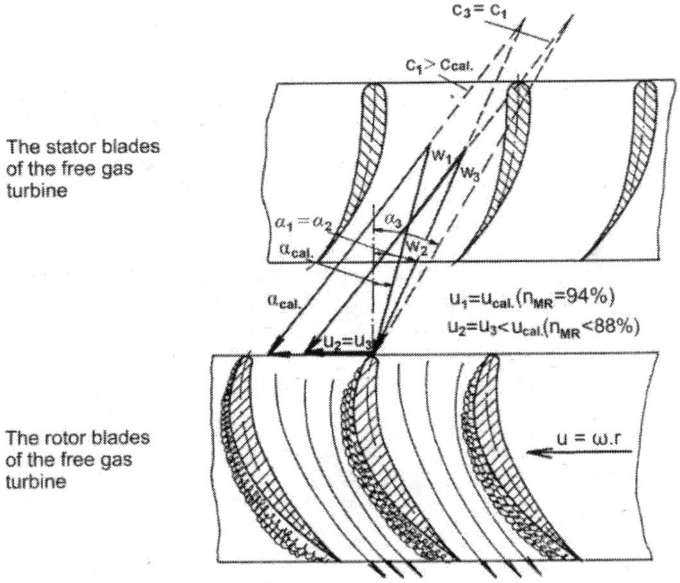

Fig. 3. Tearing of the gas flow at the back rotor blade of a free gas turbine, a turboshaft engine can occur when the angel's attack of the flowing gas at the leading edge for the rotor blade increases above a critical value [4, 6].

$c_{cal.}$ - absolute design gas velocity at the outlet of the turbocharged engine free gas turbine blades [m.s^{-1}];

w - relative velocity of the gas flow at the inlet to the rotor blades of the free gas turbine of the turboshaft engine [m.s^{-1}];

$u_{cal.}$ - calculating a rotor peripheral speed of the first step of free turbine turboshaft engine gas [ms^{-1}];

u_1 - peripheral speed of the first step of free turbine turboshaft engine gas for computing main rotor speed $n_{MR} = 94$ % [m.s^{-1}];

ω - angular speed of the rotor blades of a free gas turbine of a turboshaft engine at the design speed of the supporting rotor [rad];

r - radius of the rotor disk of the first stage of the free gas turbine of the turboshaft engine [m];

α - the angle of inclination of the gas flow at the leading edge of the rotor blades of the first stage of the free gas turbine of the turboshaft engine [°];

$\alpha_{cal.}$ - the calculated angle of inclination of the gas stream at the leading edge of the rotor blades of the first stage of the free gas turbine of the turboshaft engine at the calculated speed of the supporting rotor [°];

c_1 - the absolute flow velocity of the gas pipeline at the outlet of the deflector blades of the free gas turbine of the turboshaft engine, which corresponds to the speed of the main rotor $n_{MR} = 94$ % [m.s^{-1}];

α_1 - the angle of attack of the gas stream at the leading edge of the rotor blades of the first stage of the free gas turbine of the turboshaft engine, which corresponds to the speed of the main rotor $n_{MR} = 94$ % [m.s^{-1}];

w_1 - the relative velocity of the gas flow at the inlet to the rotor blades of the free gas turbine of the turboshaft engine, which corresponds to the speed of the supporting rotor $n_{MR} = 94$ % [m.s^{-1}];

u_2 - peripheral speed of the first step of free turbine turboshaft engine of the gas in the main rotor speed $n_{MR} < 88$ % [m.s^{-1}];

c_2 - the absolute flow velocity of the gas pipeline at the outlet of the deflection blades of the free gas turbine of the turboshaft engine, which corresponds to the speed of the main rotor $n_{MR} < 88$ % [m.s^{-1}];

α_2 - the angle of inclination of the gas flow at the leading edge of the rotor blades of the first stage of the free gas turbine of the turboshaft engine, which corresponds to the speed of the supporting rotor $n_{MR} = 94$ % [°].

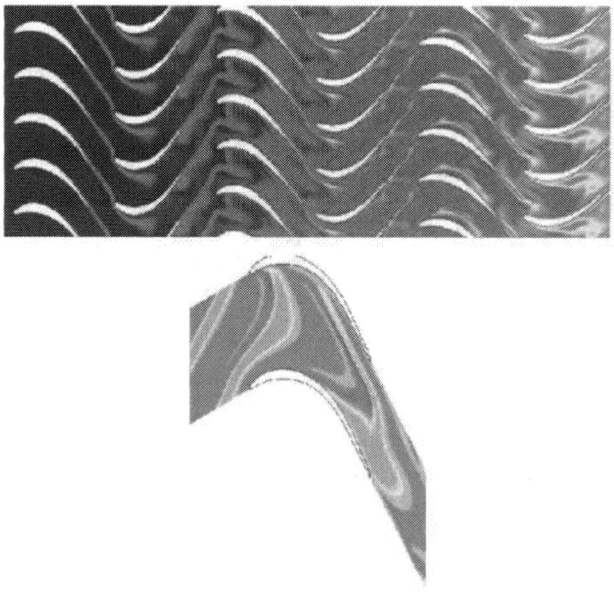

Fig. 4. Visualized character of gas flow in a multistage gas turbine and on the back of the stator blade of a free gas turbine of a turboshaft engine [9].

The Figure 6 shows a visualization of the tear-off of the jets at the gas turbine stages obtained under laboratory conditions [9, 10].

To tear off the gas flow can also be torn off in the rotor blade of the free gas turbine of the turboshaft engine (Fig. 7) with a considerable reduction of the flow angle of the flowing gas at the leading edge of the rotor blade below the critical value:

- with a sharp reduction in the absolute output velocity of the gas flow from the free gas turbine blades ($c_1 < c_{cal.}$) and a constant free gas turbine engine speed of constant, calculated peripheral speed ($u_1 = u_{cal.}$);
- while reducing the absolute output velocity of the gas stream from the free gas turbine blades ($c_2' = c_{cal.}$) and increasing the free gas turbine speed ($u_2' < u_{cal.}$).

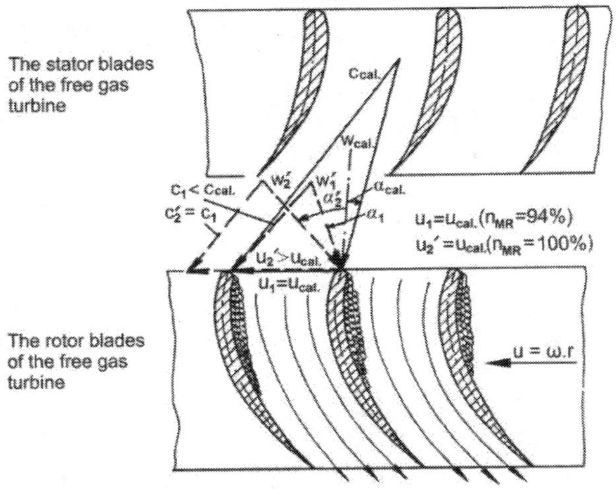

Fig. 5. Tearing of the gas flow at the blade rotor blade of a free gas turbine, a turboshaft engine can occur when the angel's attack of the flowing gas at the leading edge for the rotor blade increases above a critical value [4, 6].

Where:

u_2' - peripheral speed of the first step of free turbine turboshaft engine of the gas in the main rotor speed $n_{MR} = 100$ % [m.s^{-1}];

w_2' - the relative velocity of the gas flow at the inlet to the rotor blades of the free gas turbine of the turboshaft engine, which corresponds to the speed of the main rotor $n_{MR} = 100$ % [m.s^{-1}];

α_2' - the angle of an attack of the gas stream at the leading edge of the rotor blades of the first stage of the free gas turbine of the turboshaft engine, which corresponds to the speed of the main rotor $n_{MR} = 100$ % [m.s^{-1}].

In operation, the gas flow on the back of the rotor blades of a free gas turbine of a turboshaft engine is torn off most often when the carrier rotor speed drops below $n_{MR} =$ 88 %, regardless of the mode in which the turbo-compressors turboshaft (take-off, nominal, touring) operates.

In this case, the free gas turbine speed controller will increase the fuel supply to the engine's main combustion chamber to increase the free gas turbine power and adjust the support rotor speed to the desired value. Because at this point the angle of attack of the flowing gas at the leading edges of the rotor blades of the free gas turbine has reached a critical value, there is no increase in torque or power at the output shaft of the free gas turbine. In addition, assuming that the flow rate of gases from the blades of the free gas turbine of a turbocharged engine increases even more at this point, the flow angle of the flowing gas at the leading edge of the rotor blades will be even more supercritical so that the expected increase in torque due to increased speed flowing gas will not occur. This is due to the fact that by tearing off the gas stream, an increase in energy is absorbed, and increased gas flow rate is obtained, and the speed of the free gas turbine of the turboshaft engine decreases further. In operation, the gas turbine of the turboshaft engine may be selected to get into such a mode of a breakaway of the gas flow on the rotor blades free gas turbine if the collective control lever is set abruptly (determined as long as allowed) from the idle position to the position corresponding to take-off position. When the position of the carrier rotor blades changes abruptly to a large angle of attack, they increase just as fast to load the free gas turbine with increased torque, to a given mode. The speed of the helicopter's supporting rotor decreases rapidly and can fall below n_{MR} = 88 %. Although the turboshaft engine compressor will operate at maximum power during the rest of the engine operation, the angles of attack of the flowing gas at the leading edges of the rotor blades of the free gas turbine will be supercritical. If at this point, the pitch angle of the helicopter carrier rotor blades is further increased, then it will further complicate the existing situation. The helicopter's supporting rotor is overloaded and will not operate to develop the necessary thrust for the helicopter flight in the given helicopter flight mode. Frequent and long-term operation of a free gas turbine of a turboshaft engine in such or a similar mode contributes to a premature reduction in effective power and is not ruled out or damaged by the turboshaft engine. For this reason, the speed-changing position of the collective control lever must not exceed the specified value specified in the regulation for the use of the given type of helicopter.

5 Conclusion

As noted in the article, the nature of the flow on the rotor blades of a free gas turbine directly affects the efficiency of converting thermal energy into mechanical power free turbine turboshaft helicopter gas engine, which has a direct impact on the flight safety of the helicopter. For this reason, an understanding of these processes is of particular importance to helicopter pilots. Analogous processes also occur in industrial turboshaft engines with a free gas turbine, which drives electric generators in power plants or drives compressors in compressor stations of gas pipelines. These turbocharged engines experience the above processes due to a sharp change in the load of the electric generator or compressor.

References

1. Top. helicopter pad in press. Online: https://www.topky.sk/cl/100370/1688821/The reversal in the investigation of the fall of the vertical in the press everything is otherwise this should be the cause. 2018(sk).
2. Turning motor TV3-117, Manual of technical operation, PLN 078.00.5700, Book 1(RU).
3. Helicopter Mi-24V, Instructions for the helicopter crew, Part 1, Flight operation.9 (RU).
4. M. Hocko, Engine design TV3-117, VSŠL Košice, 2002, p. 169. (SK)
5. Using and piloting a helicopter Mi-17, Let-3-76. (SK)
6. Helicopter Mi-24V, Instructions for the helicopter crew, Part 1, Flight operation.(SK)
7. J. Růžek, Theory of aircraft engines I, VA AZ 1979, p. 373.(CZ)
8. M. Hocko, Turboshaft helicopter engines used in Czechoslovak, Czech and Slovak military aviation, LF TUKE Košice 2019, ISBN 978-80-553-3493-6, p. 220.(SK)

9. R. J. Howell, Unsteady Turbine Aerodynamics – from the Laboratory to the Engine, Whittle Laboratory Cambridge University Department of Engineering.
10. M. Hocko, Introduction to engine design TV3-117, VSŠL Košice, 2001, p. 120.(SK)
11. A.R. Samer, M. Hovanec, P. Korba, Y. J. Patel, J., V. Lukač, I. Vasilčin, B. Wysoczańská, M. Golisová, Enhancing the Aircraft maintenance management process for increasing safety Saleh ... [et al.] - 2022. In: 5th EAI International Conference on Management of Manufacturing Systems. - Cham (Switzerland) : Springer Nature s. 1-13 [online]. - ISBN 978-3-030-67240-9.
12. G. A. Misté, E. Benini Performance of a turboshaft engine for helicopter applications operating at variable shaft speed – 2012. In: Proceedings of the ASME 2012 Gas Turbine India Conference GTINDIA2012 - , Mumbai, Maharashtra, India.
13. W. Garre, T. Pflumm, M. Hajek, Enhamced efficiency and flight envelope by variable main rotor speed for different helicopter configurations - 2016. 42st European Rotorcraft Forum 2016.

Experimental steam turbine T10MW cold end cooling by water spraying

Michal Hoznedl[1*]

[1]Doosan Skoda Power s.r.o., Experimental Research of Flow, Tylova 1/57, 301 28, Pilsen, Czech Republic

Abstract. The paper deals with steam flow in experimental turbine T10MW, located in Škoda laboratory. The flow was examined for low or negative outputs of the turbine, i.e. for the so called last stages ventilation. The flow path of the turbine was in the Boiler Feed Pump Turbine (BFPT) version. It had all together 4 stages out of which two were last stages with the outlet to the condenser. In the area of each of the two outlets cooling nozzles were located with water for cooling the outlet steam flow and the area of last blades root cross-sections. Cooling of these areas is necessary due to the compression heat that occurs in the off design (ventilation) regimes. Various proportional amounts of cooling water and flowing steam were tested experimentally in constant pressure behind both last stages. Due to the fact that the flow path and the exhaust hood were fitted with many static pressure taps, thermometers and with the possibility of probing the temperature field along the outlet cross-section height, a number of results were achieved. These were mainly the turbine outputs, steam flows through the blades and cooling nozzles, determination of saturation limits in individual places at the outlet as well as temperature differences measured by the probe and stable thermometers. It was found out that the amount of cooling water was oversized for blade roots cooling, while the flow at the tip was cooled only minimally. The results are beneficial both in terms of further research of steam turbines in low regimes because this is how most newly produced machines are operated and for the designers of these machines.

Nomenclature

c	velocity	[m/s]
G	mass flow	[kg/hr]
h	enthalpy	[kJ/kg]
N	output	[kW]
p	pressure	[bar]
t	temperature	[°C]
u	circumferential velocity on mid blading diameter	[m/s]

* Corresponding author: michal.hoznedl@doosan.com

Abbreviations & subscripts

LSB	last stage blade	l	left side
L-0	3rd stage	m	middle
L-1	2nd stage from the inlet	mix	mixture
L-2	1st stage from the inlet	r	blade root, right side
rpm	revolutions per minute	s	steam
sat	at saturated line	t	blade tip, total
RB	rotor blades	x	axial direction
SB	stator blades	0	condition (plane) before stage
ST	stage	1	condition (plane) after stator blades of stage
cw	cooling water	2	condition (plane) after stage

1 Introduction

The flow research in steam turbines at low regimes is performed mainly using numerical simulations of the flow [1, 2]. In certain cases experimentally obtained results are available [3], when experiments were performed on model steam turbines. It is valuable to compare both methods, which ensures two-way data validation [4]. Due to the fact that ventilation regimes influence also the dynamic behaviour mainly of the last rotor blades, it is necessary to monitor the interaction of flow and dynamic load [5]. Similar results are rare, when experiments were performed on work (in the described case on 34 MW turbine in waste incinerator plant) [6]. Even in this case the flow parameters around the last stage are related to the amplitude of blade vibration defined by the BTT system (blade – tip – timing). It is often possible when researching low regimes in steam turbines to use the results and experience from the study of a similar phenomenon on axial compressors [7]. On these machines similar phenomena occur at first stages. Similar to turbine T10MW when measured in the BFPT version, in [8] a variable speed turbine was investigated. This paper is a significant follow-up to [9], which describes in detail research performed using a water brake (for high outputs) and a dynamometer (for low or negative outputs). The submitted paper investigates the influence of water flow through cooling nozzles located in the area of turbine last stages on the cooling of the outgoing steam flow. Similar works are unique and research is performed mainly using numerical simulations [10]. Also the phenomena related to the occurrence and movement of water films and water droplet of the liquid phase are numerically simulated [11, 12].

2 Experiment description

The experimental research on flow at the T10MW experimental steam turbine in configuration BFPT using an electric dynamometer was performed as work following experiments carried out with a water brake. The dynamometer is a device which is able to brake or actuate the turbine. It is also possible to set the regime in which the rotor or blades output is equal to zero. The dynamometer has a working speed in the range from -12 800 to +12 800 rpm. In this case the research on BFPT was carried out at nominal (design) speed of 8 000 rpm. The dynamometer is limited by the braking output on the level 900 kW @ 8 000 rpm. Thus, where it was possible, the results from the electric dynamometer were added to the results measured using the water brake. The water brake is a device for absorbing turbine output using water heated in the brake blades. The water brake was operated at a speed of 4000 to 5250 rpm, while the turbine output was up to 4 MW. Higher speed of the brake was not possible to reach for technical reasons. Results obtained using water brake are only supplementary for the purposes of this article.

During the measurement, the behaviour of LSB was observed at the turbine low load regime, achieved by the high backpressures or low LSB flow rates. The LSB regime falls from turbine mode to compressor mode during turbine ventilation. The medium is compressed, heated and dried. Temperature and pressure measurements in the tip area of both last stages was carried out together with the measurement of rotor output, amount of cooling water flowing through the jets and steam flowing through the flow path. Temperatures were measured at both last stage outlets using stable thermometers or temperature probe traversing. Due to the fact that both the probe and the stable thermometers were placed in axially and circumferentially different positions a spatially uneven arrangement of the flow was found.

During the evaluation a certain problem occurs due to different loads of last two stages. This difference is given by the pressure loss in the bypass channel and mainly by the different backpressure given by the obstacles in steam flow in direction to the condenser. As a result, the mass flow rate in both of LSBs (and also axial outlet velocity from LSB*)* was divided in ratio 50/50 from the total flow rate at the turbine inlet.

The positions of measuring points (static pressure taps and thermometers, or passages for probes) are shown in Fig. 1.

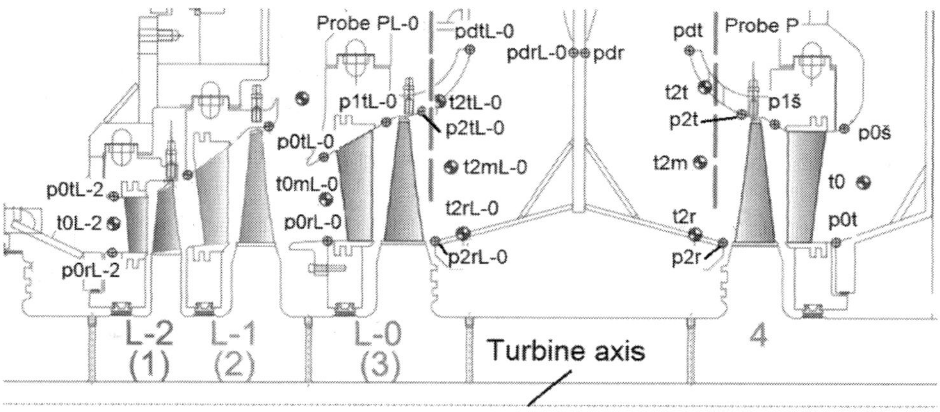

Fig. 1. Position of measuring points

At the same time Fig. 1 shows the section of the turbine. Steam enters the L-2 stage, where the inlet static pressure p_{0L-2} and inlet temperature t_{0L-2} are measured. After L-1 the steam flow is divided and about a half of the steam flows through six by-pass channels to the 4th stage, while the second half flows to the L-0 stage. Behind the outlet of both stages it is possible to probe or to measure temperatures at three heights of the outlet cross section. Due to wall pressures measuring it is also possible to define the state of steam (wet or superheated) in the area of root and tip section. All measuring points are doubled, static pressure taps and thermometers are located at two circumferentially different places.

While in Fig. 1 the position of measuring points is shown schematically, in Fig. 2 the positions of measuring ends of thermometers (circles) and the probe (dashed line) are drawn in a geometrical relations to LSB. It must be added that thermometers were placed 16° below the dividing plane, while the temperature probe moved 25° above the dividing plane at the L-0 stage and 15° above the dividing plane at the 4th stage. These positional differences are given by the design of passages and space constraints around the turbine. The longest direct distance between the thermometer t_{2tL-0} and the thermal probe position in the area of tip cross section is about 400 mm. Due to the fact that the flow in ventilation regimes is not circumferentially symmetrical the steam flows rather through the lower half of the cross section and thus large temperature differences can be expected between the temperatures detected by probing and those from stable thermometers.

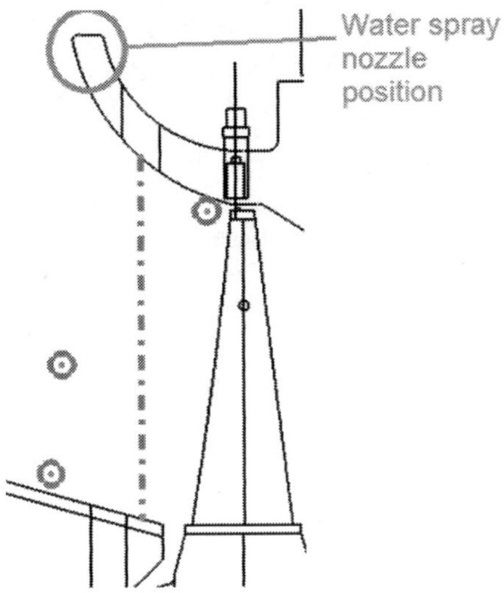

Fig. 2. Position of measuring points

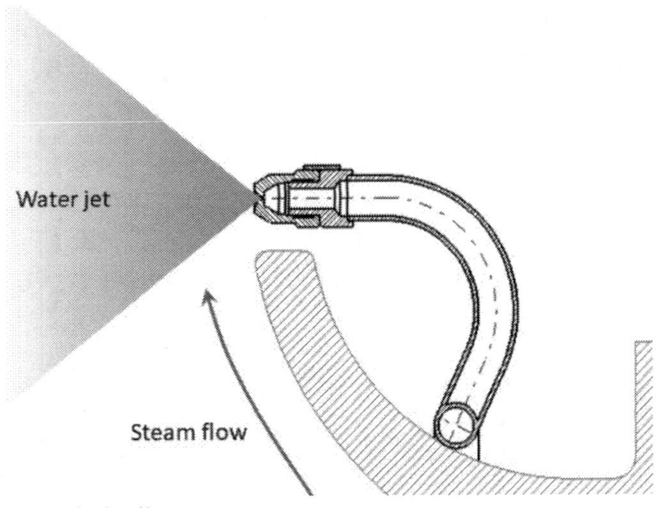

Fig. 3. Water spray nozzle detail

Measurement of static pressures using static pressure taps and probes was carried out using pressure scanners NetScanner 9022 and sensors NetScanner 9401 with the range of 0-1.06 bar. The combined standard uncertainty of measurement of the static pressure at the level 2σ under the pressure in the condenser about 5000 Pa is 110 Pa. The impulse piping was blown each 120 s in order to clean pressure pipes from condensed water. A total of 36 pressures were measured in the flow path.

Temperatures were measured using four-wire Pt100 class A resistance thermometers and measuring card NI 9216. Uncertainty of temperature measurements is ±0.3 °C. In the flow path 22 temperatures were measured. The flow of cooling water and steam to the turbine was determined using measuring orifices. The torque was determined using a torquemeter and at the same time also an electric dynamometer with measurement uncertainty of ±0.2 %.

In Fig. 2 the position of cooling nozzles placed at the end of the deflector and pointing in the axial direction is shown. In Fig. 3 there is a detail of the nozzle, the assumed cooling water spray cone and the indicated steam direction. The amount and velocity of water must be such as to overcome the steam flow from the rotor blades. At the same time the droplets must be small enough to evaporate before they reach the RB trailing edges. The nozzles at the outlet of the 3rd and 4th stage were circumferentially located after 90°, so behind each stage were 4 nozzles with the spray angle of 60°. The flow rate to both nozzles could be regulated by means of a simple ball valve. In addition to the nozzles behind both stages a third circuit was connected to cool the condenser. However, this circuit was disconnected during the described measurement. The water flow to all three circuits was measured with one orifice dimensioned for the maximum of 1000 kg/hr. Based on the manufacturer´s data the maximum flow to the total of 8 nozzles was determined to be 750 kg/hr for injected water pressure of 25 bar.

3 Measured regimes

A total of 13 regimes differing in rotor power were measured. This was given by the inlet pressure p_{0L-2} and temperature t_{0L-2} and pressure in the condenser p_c. The flow through the turbine which ranged between 9 000 kg/hr and 34 000 kg/hr depended on the inlet parameter. The turbine output measured by electric dynamometer after deducting mechanic losses was between – 400 kW (engine regime) and 650 kW (turbine regime). The designation and overview of individual regimes from the point of view of the condenser pressure p_c is in Fig. 4 a) and from the point of view of the flow, inlet pressure and condenser pressure is in Fig. 4 b).

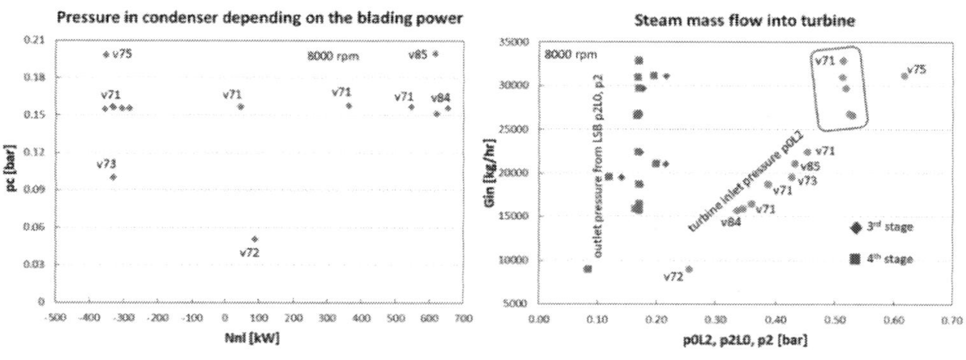

a) b)

Fig 4. a) Backpressure depending on the blading power; b) Steam mass flow into turbine

An overview of measured regimes including the flow through cooling nozzles related to steam flow through the last blades is shown in Tab. 1. Measurement of the influence of cooling water flow was carried out for variants v71a, b, c, d. In Tab. 2 other parameters of individual regimes are given. One of them is the ventilation coefficient c_{2x} / u, which indicates whether the last stage (not the whole rotor) is in ventilation, i.e. whether its output is equal to zero. Limit value of coefficient c_{2x} / u is for the given stage 0.181, which corresponds to the outlet axial velocity c_{2x} = 74.4 m/s. However, this figure does not say anything about the circumferential velocity that can be significant. The velocity vector can then approach the speed of sound. The value of cooling spray relative mass flow is a percentage of cooling water sprayed behind the stage to the steam flow through this stage provided that the steam and water are divided into both flows in the ratio of 50/50. In Tab. 1 the flow of cooling water through the circuits behind both last stages is given.

Table 1. Inlet and outlet parameters

Regime	**Inlet pressure p_{0L-2}**	**Inlet temp. t_{0L-2}**	**Pressure behind 3rd stage p_{2L-0}**	**Temp. behind 3rd stage t_{2L-0}**	**Pressure behind 4th stage p_2**	**Temp. behind 4th stage t_2**	**Condenser pressure p_c**
	[bar]	[°C]	[bar]	[°C]	[bar]	[°C]	[bar]
v73	0.4276	181.4	0.1412	65.2	0.1202	66.8	0.100
v75	0.6178	177.4	0.2160	98.6	0.1955	106.9	0.198
v85	0.4323	186.6	0.2146	144.3	0.1993	148.5	0.200
v84	0.3454	184.9	0.1677	124.9	0.1655	132.4	0.152
v71	0.5243	169.6	0.1708	78.7	0.1694	86.7	0.156
v71c	0.5187	174.9	0.1770	75.2	0.1705	77.4	0.157
v71	0.3877	182.5	0.1725	92.7	0.1711	107.5	0.158
v71	0.4545	178.9	0.1729	85.8	0.1701	97.7	0.157
v71	0.3596	186.2	0.1719	99.6	0.1706	116.6	0.157
v84	0.3359	189.7	0.1711	108.9	0.1704	131.7	0.156
v71d	0.5299	172.8	0.1687	78.1	0.1671	83.7	0.155
v71a	0.5145	167.9	0.1684	90.8	0.1704	99.9	0.156
v71b	0.5132	171.2	0.1684	95.0	0.1691	105.4	0.155

Table 2. General parameters

Regime	Inlet steam mass flow G_s	Spraying water mass flow G_{cw}	Cooling spray relative mass flow G_{cw}/G_s	Blading output N	Ventilation coefficient
	[kg/hr]	[kg/hr]		[kW]	[-]
v73	26 314	758	2-2.8 %	-331.6	0.091
v75	36 347	750		-352.4	0.102
v85	30 841	704		617.4	0.077
v84	26 403	701		619.8	0.069
v71	34 223	794		-330.2	0.100
v71c	32 942	410	1.24 %	-333.2	0.107
v71	28 471	813	2-2.8 %	364.2	0.073
v71	31 985	800		45.8	0.085
v71	26 399	794		545.9	0.066
v84	26 518	776		654.8	0.065
v71d	33 501	810	2.42 %	-356.9	0.100
v71a	33 282	45	0.14 %	-304.6	0.128
v71b	32 517	172	0.53 %	-283.9	0.123

4 Experiment results

Due to different velocity of the turbine operating with a dynamometer and a water brake it is necessary to normalize the velocity by changing the flow, i.e. by the axial outlet velocity. In Fig. 4 temperatures are shown behind the 3rd and the 4th stage at the root, tip and in the middle of the channel length. The x-axis shows the ventilation coefficient $c_{2xL\text{-}0}$ / u for the 3rd stage, or c_{2x} / u for the 4th stage. Cooling using nozzles took place only for measuring with dynamometer, i.e. for $c_{2x} / u < 0.13$ and $c_{2xL\text{-}0} / u < 0.12$. During measuring with water brake the cooling nozzles were switched off. Despite the maximum cooling water flow in most regimes (except for regimes v71a, b, c) there was no cooling of the steam flow near the rotor blades tip and with decreasing ventilation coefficient, on the contrary, the temperature behind the tip further increased. The steam flow at the blades root and near the middle of the outlet channel was constantly cooled regardless of the ventilation coefficient value.

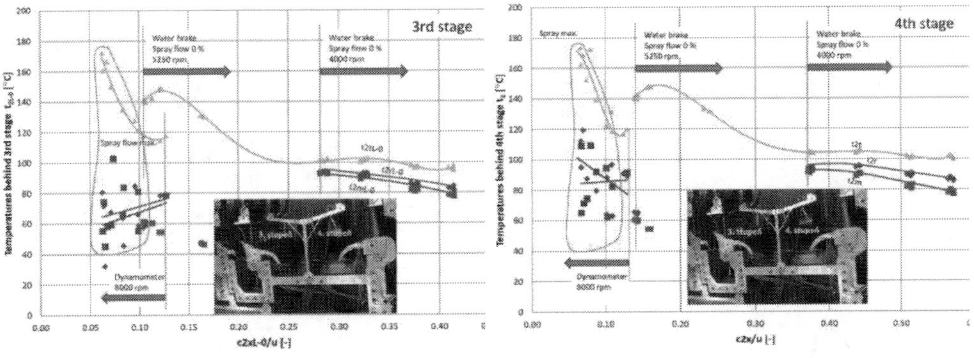

3rd stage) 4th stage)

Fig. 5. Temperatures measured behind both last stages

Contours of velocity vectors in the area of both last stages and the outlet diffuser obtained using CFD calculations of flow together with the cooling water cones are shown in Fig. 6 for parameters corresponding with regime v71a. The area of high velocity near the outer limiting wall of the diffuser is evident and below there is a huge area with ventilation vortex. From the measurement results it can be concluded that water droplets sprayed from nozzles pass the area with high velocity, get to the ventilation vortex, cool the thermometers at the root and in the middle of the channel, but, at the same time, probably do not cause erosion of the RB trailing edges at the root.

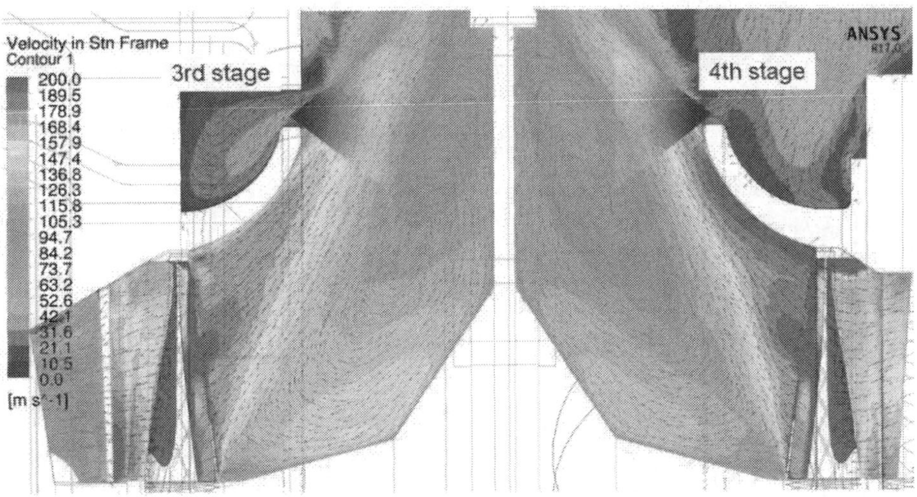

Fig 6. Velocity vector contours

The quality of steam cooling in the area of root and tip limiting wall of both last stages can be determined also from measured temperatures and temperatures at the saturation limit. These temperatures are calculated on the basis of the pressures measured by wall pressure taps. The steam can then be maximally cooled just to the saturation temperature. If the measured temperature is higher than the saturation temperature for the given pressure then water cooling is not sufficient. In Fig. 7 courses of measured temperatures and saturation temperatures are shown for both stages, for regimes measured using water brake (c_{2x} / u a c_{2xL-0} / u > 0.13) and using dynamometer. For most regimes measured with dynamometer the flow of cooling water was maximal. Nevertheless, the steam flow behind the RB tip could never be cooled and the temperature further increased with deteriorating low operation. Although the temperature in the root cross section area also did not reach the saturation limit, the steam was able to cool significantly. However, it is not possible to determine whether the cooling was caused by cooling water or by the influence of the ventilation vortex, see Fig. 6.

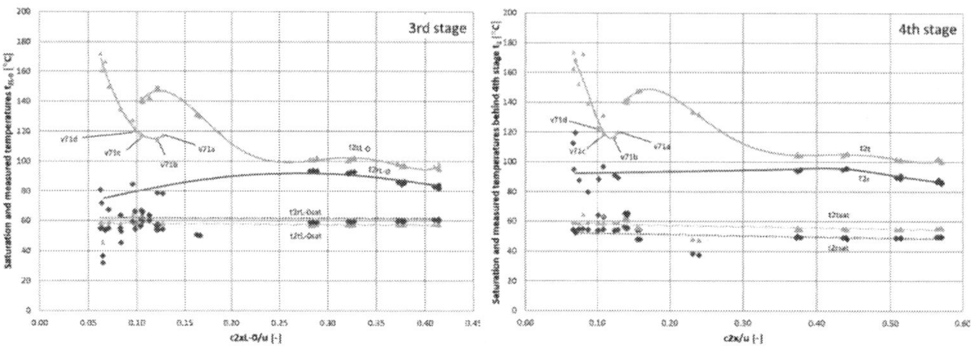

3rd stage) 4th stage)

Fig. 7. Saturation and measured temperatures behind both last stages

For regimes v71a, v71b, v71c and v71d it is possible to write energy evaluation for steam, cooling water and the final mixture, here e.g. for the 4th stage:

$$G_{2sL-0} \cdot h_{2sL-0} + G_{2cwL-0} \cdot h_{2cwL-0} = G_{2mixL-0} \cdot h_{2mixL-0} \quad (1)$$

Cooling water enthalpy h_{cw} can be defined using the IAPWS_97 tables as a function of water pressure (29 bar) and water temperature (32 °C), h_{2cwL-0} = 136.729 kJ/kg. Enthalpy of the water and steam mixture behind the 3rd stage blades $h_{2mixL-0}$ can be defined from measurement using mean pressure p_{2L-0} and temperature t_{2L-0}. Temperature t_{2L-0} is determined as a mean value from temperatures t_{2rL-0}, t_{2mL-0} and t_{2tL-0}, pressure is determined as the mean value of pressures p_{2rL-0} and p_{2tL-0}.

Now the enthalpy of steam behind the blades can be calculated:

$$h_{2sL-0} = \frac{G_{2mixL-0} \cdot h_{2mixL-0} - G_{2cwL-0} \cdot h_{2cwL-0}}{G_{2sL-0}} \quad (2)$$

At this stage of calculation, data already exist for the calculation of theoretical steam temperature t_{2sL-0} for various injected amounts, but without the influence of injecting, as a function $t_{2sL-0} = f(p_{2L-0}, h_{2sL-0})$. The calculated steam temperature should be the same for all 4 evaluated regimes, but in fact oscillates between 90 and 108 °C. This is caused mainly by different steam flows (26 500 up to 32 800 kg/h) and partly also by different pressures in the condenser (0.168 up to 0.177 bar).

In the next step it is necessary to calculate also the enthalpy of the mixture, but assuming complete evaporation of cooling water. The enthalpy of the mixture will be determined by measured pressure behind the blades and the steam dryness x = 1.

$$\frac{G_{2sL-0} \cdot h_{2sL-0} + G_{2cwL-0} \cdot h_{2cwL-0}}{G_{2mixL-0}} = h_{2mixL-0} = f(p_{2L-0}, x = 1) \quad (3)$$

Iteratively the flow rate $G_{2cwL\text{-}0}$ can be determined such that the saturation temperature for pressure $p_{2L\text{-}0}$ corresponds with the saturation temperature of the mixture, given by the enthalpy $h_{2mixL\text{-}0}$.. Both temperatures must then be equal at the end of the iterative calculation. The determined flow $G_{2cwL\text{-}0}$ can then be substituted to the equation (3) and the relative flow of water to the cooling nozzles can be determined. For the 3rd stage the relative flow to reach saturation limit of steam in the whole cross section is $G_{2sytL\text{-}0} = G_{2cwL\text{-}0} / G_{2sL\text{-}0} = 2.74$ % and for the 4th stage it is $G_{2cw} / G_{2s} = 3.43$ % (the steam flow through each of both last stages is 16 640 kg/h). The temperature of steam flow in plane 2 that equals to the saturated steam temperature assuming total evaporation of water can be referred to as $t_{2mixL\text{-}0\text{-}minimal}$.

Temperatures $t_{2sL\text{-}0}$ and $t_{2mixL\text{-}0\text{-}minimal}$ give the range in which the measured temperature $t_{2L\text{-}0}$ should be found. This is confirmed also in Fig. 8. Also the courses of temperatures $t_{2rL\text{-}0}$, $t_{2mL\text{-}0}$ and $t_{2tL\text{-}0}$ are shown in this figure. While the temperature behind the RB tip is still high, temperatures in the middle and at the root of the channel decrease and already at a relative flow rate of 1.2 % for the 3rd stage it cools to saturation temperature for a given pressure $t_{2L\text{-}0sat}$. On the other hand, temperature at the tip never decreases due to cooling water spraying. It corresponds with the measured temperature $t_{2L\text{-}0}$ that is for relative flow 1.2 % still at the minimal possible value, while for a higher value of relative flow it increases due to the fact that the steam flow behind the tip cannot be cooled by the given spray configuration. Thus, it can be said that about 50 % more cooling water is sprayed to the steam flow than would be necessary to cool the mixture to the saturation temperature. The unevaporated water can cause erosion of trailing edges at the rotor blades root. Similar conclusions can be drawn also for the 4th stage, see Fig. 8 b).

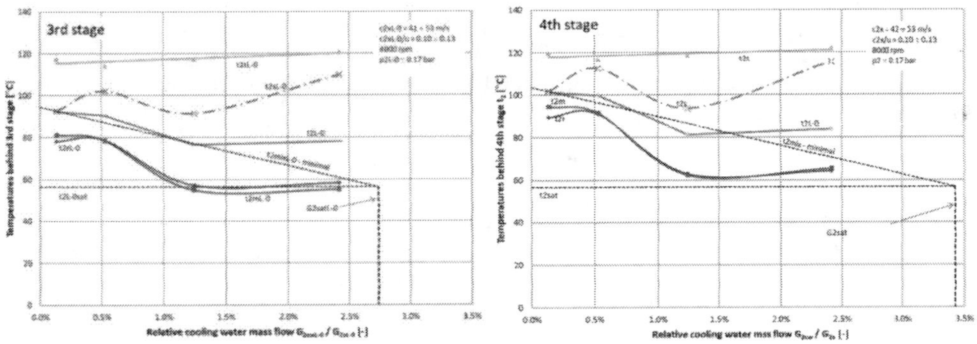

3rd stage) 4th stage)

Fig. 8. Saturation and measured temperatures behind both last stages

5 Conclusions

Experimental research of water steam flow in the 4-stage flow path was carried out on an experimental steam turbine T10MW in BFPT design. Energy of the rotor was absorbed by water brake or by electric dynamometer. In certain operations, on the other hand, the rotor was driven by the dynamometer, and the turbine output was negative. The main emphasis in the research was placed on the outlet steam flow behaviour for nominal and mainly for ventilation regimes of the turbine. The steam flow in ventilation regimes is supplied with heat from the last stages that operate in the compressor regime. Steam at the last rotor blades outlet had to be therefore cooled by external water spray.

Measurements without the spray influence revealed that the highest steam temperature is in the area behind the rotor blade tips for both the nominal and ventilation regimes. As the steam flow through the blades decreases, the temperature behind the rotor blade tip increases significantly. The steam temperature at this point can reach up to 180 °C, while steam temperature at the area of root and middle blade diameter remains unchained, about 100 °C. The temperature of steam saturation at the root and the tip is lower than the measured temperature. The steam is superheated throughout the cross section.

In the next part of the experiments the influence of the amount of cooling water related to the steam flow through the blades was determined. Gradually, the amounts of 0, 0.14, 0.53, 1.24 and 2.42 % of relative cooling water mass flow related to the steam flow through the last stage were injected to the exhaust hood area. For relative flow of 1.24 % and higher the outlet steam flow was cooled near the root and middle blade diameter to the saturation limit. The steam at the tip could not be cooled even at maximum flow of cooling water. Maximum water flow was designed with regard to the assumed cooling of the whole outlet cross section to the saturation limit. However, the entire cross section can never be cooled, as the cooling water droplets can never reach the fast flowing steam. Needlessly, about twice as much water is sprayed to the area near the root and middle of the blades as would be necessary to reach the saturation temperature. Unevaporated droplets can then cause erosion of trailing edges at the root of rotor blades.

The presented work was financially supported by the Technology Agency of the Czech Republic; project TK01020029 Efficiency Increasing of Turbine Wet Steam Last Stages, by National Energy Centre TN01000007 and by Technology Agency of the Czech Republic; project FW03010515 Ultra flexible operation regimes of steam turbines.

References

1. TANUMA, T. et al. *Unsteady flow effects on steam turbine last stage blades at very low load operating conditions*. Proc. ASME TurboExpo, 2018. GT2018-76498, https://doi.org/10.1115/GT2018-76498 .
2. MAMBRO, A. et al. *CFD modelling of steam turbine last stage blades at low load using multiple mixing plane approach*. Proc. ASME TurboExpo, 2020. GT2020-14667, https://doi.org/10.1115/GT2020-14667 .
3. Haller, B. *Alleviation of rotating pressure oscillations in the last LP turbine stage during low-load conditions.* Proc. ASME TurboExpo, 2016. GT2016-56088, https://doi.org/10.1115/GT2018-76498 .
4. SIGG, R., HEINZ, C., CASEY, M. *Numerical and experimental investigation of a low-pressure steam turbine during windage.*. Proceedings of the Institution of Mechanical Engineers, Part A: Journal of Power and Energy. Volume: 223 issue: 6, page(s): 697-708, https://doi.org/10.1243/09576509JPE826 .
5. SHIBUKAWA, N. et al. *A Correlation Between Vibration Stresses and Flow Features of Steam Turbine Long Blades in Low Load Conditions*. Proc. ASME TurboExpo 2011. GT2011-46368, https://doi.org/10.1115/GT2011-46368 .
6. HOZNEDL, M. et al. *Experimental and Numerical Study of Flow and Dynamics on LSB at 34 MW Steam Turbine.* Proc. ASME TurboExpo, 2020. GT2020-14280, https://doi.org/10.1115/GT2020-14280 .
7. GARNIER, V. H. et al. *Rotating waves as a stall inception indication in axial compressors* Transactions of the ASME. Vol.113, pp. 290-302, April 1991.
8. SCHUBERT, M., TUSCHE, J. *Development of a Robust LP Blade Family for Variable Speed Applications.* Proc. ASME TurboExpo 2018. GT2018-76539, https://doi.org/10.1115/GT2018-76539 .

9. TUPÝ, D., HOZNEDL, M. *Experimental research on ventilation at T10MW experimental steam turbine.* Proc. AIP Conference Proceedings 2323, 060004 (2021); https://doi.org/10.1063/5.0041416 .
10. XU, M et al. Effect of water spray in exhaust passage of steam turbine on flow field of the last stage during windage. In: International Journal of Heat and Mass Transfer 161:120296, https://doi.org/10.1016/j.ijheatmasstransfer.2020.120296 .
11. TISCHENKO, V.A, ALEKSEEV, R.A. *Numerically Simulating the Formation and Motion of Water Films and Erosion-Hazardous Droplets in Flow-Through Parts of Steam Turbines.* In: Thermal Engineering, 2019, Vol 66, No. 11, pp. 830-838. https://doi.org/10.1134/S0040601519110053 .
12. GRIBIN, V.G. et al. *Experimental study of the features of the motion of liquid-phase particles in the interblade channel of the nozzle array of a steam turbine.* In: Power Technol Eng 51, 82–88 (2017). https://doi.org/10.1007/s10749-017-0788-x

An experimental study of the CO_2 hydrates production for the long term storage

Ondřej Bartoš [1*], and *Matěj* Hrnčíř[1]

[1]CTU in Prague, dep. Energy Engineering, Technická 4, Prague 6, 166 07, Czechia

Abstract. An aim of the paper is to show recent data obtained from a new experimental set-up build for the production of the CO_2 gas hydrates. The purpose of the experimental set-up is to analyse the practical aspects of the transformation gaseous CO_2 to the hydrates. The deserving effort to decrease impacts of the global warming is containing the more questionable attempt to capture the CO_2 produced within the electricity production and to avoid a releasing to the atmosphere. The storage in the form of the gas hydrates present an alternative way to more known technologies involved in the projects of CCS (Carbon Capture and Storage). The production of the gas hydrate is observed in the set-up with simultaneously acquired data of state condition close to the phase boundary. The presented work has two goals, first is the estimation of the transformation efficiency of the CO_2 to the hydrates in compare with the theory and second goal is obtaining of the data for new CO_2 hydrates production set-up with liquid circulation and possibility to separate pure hydrate. The experimental analysis of the gas hydrates production process can help to estimate the practical aspects of the hydrates production for a possibility of CO_2 storage in this form.

1 Introduction

As concerns about greenhouse gases such as CO_2 are still not sufficiently solved, it has been proposed to isolate its hydrates form at great ocean depths [1]. Regarding the standard condition, one can expect that 1m3 of hydrate contains approximately 170 nm3 of CO2 with depending on the formation process. Investigating the properties of CO_2 hydrate and reaction kinetics both in the laboratory and in the marine environment is important for a potential industrial way to long lasting storage of the CO2 without danger of the sudden release of the gas to the atmosphere. Therefore, a reliable, reproducible and effective method for the synthesis of CO_2 hydrate should be found. The purpose of this work should be performed the observation and measurement of the state condition of the hydrates formation in laboratory conditions, both in fresh water and in saline [2]. The thermodynamics steady state condition of the CO2 hydrates formation is at significantly higher pressure than atmosphere. From this reason a measurement set-up to study hydrates formation was developed and manufactured. The hydrates are formed on the phase boundary and this results that not all supplying gas is transformed to the hydrate. This paper present the data from the measurement of the formation process with the effort to establish the efficiency of the gas transformation. The

* Corresponding author: ondrej.bartos@fs.cvut.cz

measurement set-up was equipped with the precise measurement of the temperature and gas flow. Obtained experience will be used for the development of the more advance facility which will be close to the real industry.

Hydrates are the crystalline products (clathrate) of water interactions with inorganic and organic substances, and are characterized by a strict structure for the given type of substance. All hydrophobic gases and volatile organic compounds have the ability to form hydrates since their molecular size will be 3.8 - 9.2 Å {10-10m} (Ar, N_2, O_2, CH_4, C_2H_4, C_2H_6, C_3H_8, iso-C_4H_{10}, Cl_2, CS_2) and hydrophilic compounds (CO_2, SO_2), ethylene oxide (C_2H_4O), tetrahydrofuran (THF, C_4H_8O), acetone. By their shape, the hydrates resemble snow or loose ice, but under normal atmospheric conditions, hydrates are rarely encountered [3].

The hydrates storage on the deep ocean bottom requires a detailed study of the deep ocean condition and properties of the hydrates namely the density. The density of the hydrates is very close to the density of the sea water on the bottom. The density of the hydrate is higher than the density of the sea water but the hydrate produced in the laboratory contains small bubbles and its density is lower than the water. This is one of many reason for study of the production process in the laboratory.

2 Measurement setup

The aim of the experiment was to create carbon dioxide hydrate (CO_2 6 H_2O) in laboratory conditions and confirm that CO_2 can be produced for the further storage. The carbon dioxide is drawn to the nozzle under the water level and released as the bubbles through the cold distilled, drinking or salty water.

The set-up (Fig. 1) contains pressure vessel (reactor) made from cast iron and is equipped with two 15 mm thick borosilicate glass in both front and rear to observe the formation of hydrates. Stainless steel flanges tightened with four M12 screws are located at the top and bottom. In the lower flange there is a pressure gauge and a hole for cooling water hoses. In the top flange there is a blow-off valve, carbon dioxide inlet port and a thermometer-positioning tube that measures the inner temperature of the vessel. On the sides of the pressure vessel there are welded tubes with holes for the hoses through which there is a device inside which the gas can be released from the bottom through the individual bubbles. Under this device, there is a copper pipe coil, which is also supplied with cooling water, to cool the vessel more efficiently. In our case, an operational temperature is above 1 ° C to avoid the water freezing. The interior of the pressure vessel was coated with the zinc paint which was proved as the best protection against the corrosion. Cooling system is connected to the thermal bath to maintain constant temperature.

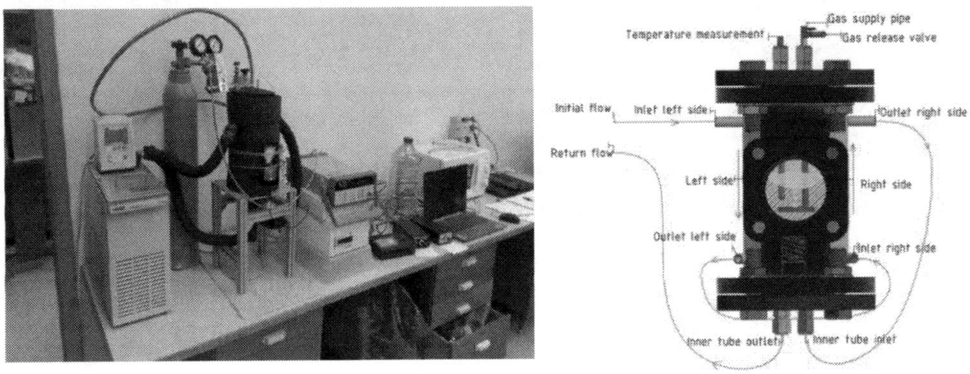

(a) (b)

Fig. 1. Entire set-up (a), Schema of the reactor connection (b).

The instrumentation of the experiment consists of the classical probes. RTD precisely calibrated for the measurement of the reactor internal temperature and the inlet and outlet of the cooling water together with the its flow by the vortex flowmeter. Pressure is the measured by standard transducer and mass flow of the CO2 is measured by the mass flow meter. Data are acquired by the datalogger. Entire reactor and connecting pipes are isolated by the PP foam.

3 Acquired data

The production of the CO2 hydrate in the set-up was initialized when the temperature in the reactor interior decreased under the equilibrium temperature corresponding with the pressure [4]. See Fig 2a. The hydrate nucleation starts very slowly at the beginning and sub-cooling of few degrees Celsius is necessary. On the Fig. 2b are presented the capture images of the hydrates. The measurement set-up was design only for a demonstration of the hydrates formation, but after gaining some experimental skills the precise measurement of the energy fluxes was performed. The aim was to compare the heat released while the hydrates are produced with the CO2 mass flow to the reactor. The result would be efficiency of the transformation. The direct determination of the formed hydrate is not possible due to the high pressure in the reactor, only non-direct determination methods are available.

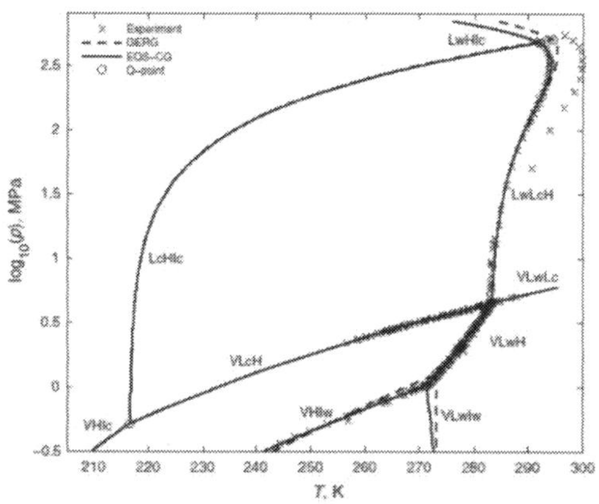

(a)

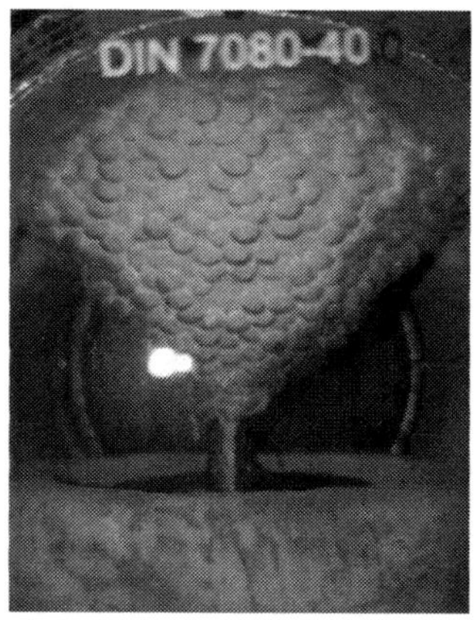

(b)

Fig. 2. CO_2-Water-Hydrate phase diagram [4] (a), Hydrate in the reactor in a typical "bubble" structure (b).

The measurement of the temperature inside of the reactor and temperature of the cooling water at the inlet and outlet is shown on the Fig. 3. The mass flow of the cooling water was nearly constant app. 1.68 L/min. The start of the production was app. at 14:34 and the end 14:44 when the pressure inside the reactor was decreased. Several measurement were performed but in the paper only one measurement is shown.

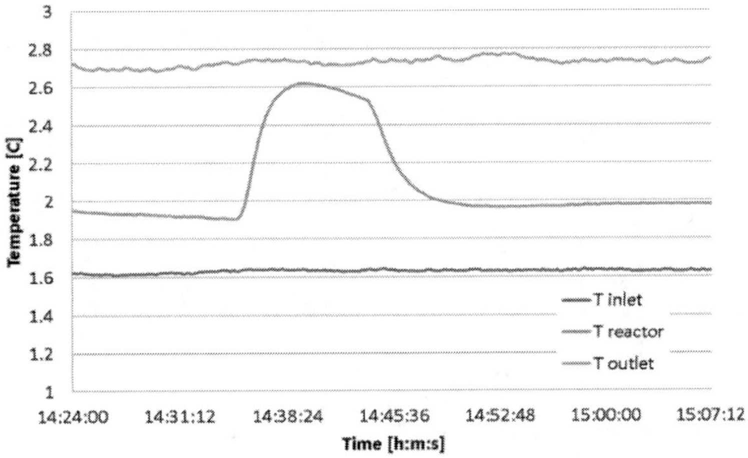

Fig. 3. The temperature inside the rector when the hydrate was produced

4 Data processing and results

The main goal for the data processing was to determine the amount of the hydrate in the reactor. The direct measurement is impossible due to the high pressure inside the vessel. From this reason some non-direct method were tested. The first way how to determine the mass of the hydrate formed in the reactor is from the heat balance. The hydrate formation energy is known app. Hf = 65kJ/mol. Several measurements were averaged and the average formation rate 0.05 mol/min. The second way is to measure direst flow of the gas to the reactor at the inlet and outlet. The difference should be the hydrate formation rate. Unfortunately this method is not possible to use in this simple manner. Two mass-flow meters were used on the inlet and the outlet of the reactor, but both gauges work at the different pressure with different gas than the gas for the calibration. The entire measurement chain contains many uncertainties which cause the high inaccuracy of the measurement. Even the solubility of the CO2 and accumulation in the reactor were involved.

Another method for the determination of the transformation ate is modelling the process in the Simulink environment. For the simulation the simple equation of the energy conservation was used (1).

$$m_w c_p \frac{dT}{dt} = H_f \frac{dm_g}{dt} - C_2(T - T_{in}) \tag{1}$$

where T temperature in the reactor, m_w is mass of the water in the reactor, c_p specific heat capacity, H_f hydrate formation enthalpy, T_{in} temperature of the cooling water, m_g mass of the transformed CO_2, C_2 unknown constant which respect heat transfer from the reactor.

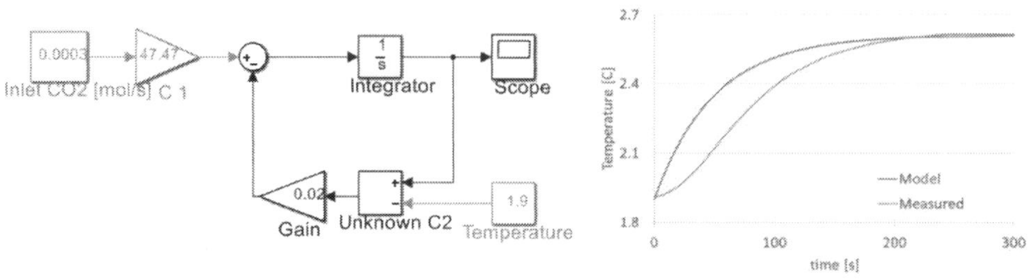

(a) (b)

Fig. 4. The temperature inside the rector when the hydrate was produced

On the Fig. 4 a and b are shown the wire diagram and progress of the measured and the modelled temperature profile. The eq. 1 describes 1st order system, but measured profile seems to be rather more 2nd order system, but after the more advance simulation we found that this is sufficient description, because main reason for the second order system is the delay in the temperature measurement. The RTD probe is in the pipe which protect the probe against the high pressure. The final transformation rate was determined 0.02 mol/min.

5 Conclusion

The current experiment proves that suggested method is possible to use for the CO2 hydrates production, but bigger set-up for production testing would be better. The direst measurement of the transformation efficiency or measurement of the formed hydrate amount is hardly achievable, but the combination of the energy flux measurement and numerical simulation one can get a promising result.

The obtained knowledge presented in this paper will be used for the construction of the bigger unit for the hydrate production with focusing on the industrial use. Main goal will be determined the energy required for the hydrates production. This will be crucial information if the storage in form of hydrates should be studied more.

The more general effect of this work is gaining knowledges around the gas hydrates, for example methane hydrates are presented on the sea bottom and in the permafrost in the north areas of the Earth and in the future the releasing of the methane to the atmosphere due to the global warming and make a greenhouse effect even worse.

We gratefully acknowledge the support by the Grant Agency of the Czech Technical University in Prague, Grant No. SGS 21/154.

References

1. "Gas hydrates for deep ocean storage CO2", Aker Kvaerner, (2003)
2. O. Bartoš, S. Chystiakov and P. Hýbek, "A study of CO_2 hydrates for deep ocean storage" , ERIN, The 13th International Conference for Young Researchers and PhD Students April 9th-11th, (2019), Blansko - Češkovice, Czech Republic
3. E.D. Sloan, C.A. Koh, "Clathrate Hydrates of Natural Gases", third ed., CRC Press, Taylor & Francis group, New York, (2008)
4. A. Jägera, V. Vinš a J. Gernert , "Phase equilibria with hydrate formation in H2O + CO2 mixtures modeled with" Fluid Phase Equilibria, (2013)

MATEC Web of Conferences **345**, 00012 (2021) https://doi.org/10.1051/matecconf/202134500012
Power System Engineering 2021

The effect of recuperator on the efficiency of ORC and TFC with very dry working fluid

Aram Mohammed Ahmed[1, 2], and *Attila R.* Imre[1, 3*]

[1]Budapest University of Technology and Economics, Faculty of Mechanical Engineering, Department of Energy Engineering, Műegyetem rkp. 3, H-1111 Budapest, Hungary
[2]Northern Technical University - Technical College of Kirkuk, 36001 Kirkuk, Iraq
[3]Department of Thermohydraulics, Centre for Energy Research, POB. 49, H-1525 Budapest, Hungary

Abstract. Organic Rankine Cycles (ORC) and Trilateral Flash Cycles (TFC) are very similar power cycles; ideally, they have a reversible adiabatic (isentropic) compression, an isobaric heating, an isentropic expansion and an isobaric cooling. The main difference is that for ORC, the heating includes the full evaporation of the working fluid (prior expansion); therefore, the expansion starts in a saturated or dry vapour state, while for TFC, the heating terminates upon reaching the saturated liquid states. Therefore, for TFT, expansion liquid/vapour state (in bubbly liquid or in vapour dispersed with droplets), requiring a special two-phase expander. Being ORC a more "complete" cycle, one would expect that its thermodynamic efficiency is always higher than for a TFC, between the same temperatures and using the same working fluids. Surprisingly, it was shown that for very dry working fluids, the efficiency of TFC can exceed the efficiency of basic (i.e. recuperator- and superheater-free) ORC, choosing sufficiently high (but still subcritical) maximal cycle temperature. Therefore in these cases, TFC (having a simpler heat exchange unit for heating) can be a better choice than ORC. The presence of a recuperator can influence the situation; by recovering the proper percentage of the remaining heat (after the expansion), the efficiency of ORC can reach and even pass the efficiency of TFC.

1 Introduction

The utilization of low-temperature heat sources for energy conversion required novel technologies; one of the novelty is the use of alternative working fluids (like organic materials or carbon dioxide) for Rankine cycles [1]. Using these methods, one can produce electricity from heat sources unavailable for this purpose before, like low-temperature geothermal heat, unfocused solar heat or low-temperature industrial waste heat. Most of these heat sources are existing ones; without proper utilization, they would only dissipate their heat into the environment; sometimes, this is "only" the waste of sources, but in several cases, it can have a direct negative environmental impact. Therefore the application of ORC- (or some similar technology-) based power plants can be used to reach the goals for a greener World.

* Corresponding author: imreattila@energia.bme.hu

The idealized basic form of the Organic Rankine cycle (just like the Steam Rankine cycle) has four steps, two isobaric heat exchange and two adiabatic (isentropic) expansion/compression. This can be seen in Figure 1/a, using a so-called wet or ACZ-type working fluid, like water, carbon dioxide, propane, etc. [2]. The type of the working fluid (namely the shape of the T-s diagram has very strong effect on the process [3,4,5]. For wet working fluids, in the first step, the working fluid is in a saturated, cold liquid state (point 1), where the temperature is usually equal to the environmental one; it is compressed then to a compressed liquid state with a little bit higher temperature (point 2). In an ideal cycle, the compression would be isentropic; in a real system, there is a small entropy production, and the entropy of point 2 would be higher than the entropy of point 1; but these non-idealities (together with the non-idealities of the other steps) are neglected in this study. The next step is isobaric heating; for basic ORC, it has two sub-steps. First, the liquid heating (from point 2 to point 3*), then the evaporation (from 3* to 3) happen. Technically, this can be done by two different heat exchanger (a liquid heater and an evaporator) or is one having two parts. In point 3*, the working fluid is in a high-temperature saturated *liquid* state, while in point 3*, it is in a high-temperature saturated *vapour* state. In a basic ORC, heating terminates here (i.e. no superheating to reach dry vapour state). The expansion also starts from point 3* (a high-temperature saturated vapour state) and terminates at point 4 (a low-temperature, two-phase state; in this case, it is wet vapour, i.e. vapour with liquid droplets). Then a second heat exchanger (a condenser) would fully condenses this vapour till the initial state (point 1, saturated cold liquid) state is reached.

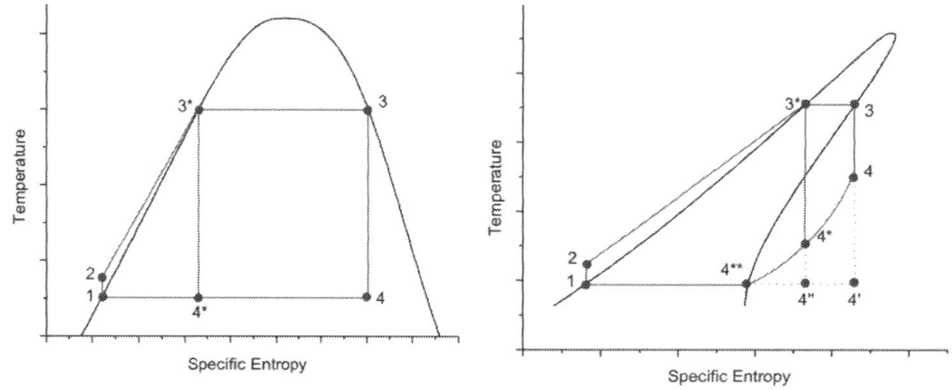

(a) (b)

Fig. 1. Comparison of the recuperator- and superheater-free ORC and TFC cycles (see text).

In Figure 1/a, an alternative technology can also be seen; the so-called (organic) Trilateral Flash Cycle (TFC) [6]. In TFC, heating terminates at point 3* (hot saturated liquid state); expansion starts there and terminates in the two-phase region, at point 4*, which is – at least in this case, concerning the position of the point – is a very wet vapour state with a lot of bubbles. Similarly to ORC, a condenser condenses the vapour from this state to the initial cold saturated liquid state (point 1).

ORC is usually considered superior to TFC for several reasons:

- efficiency is higher (see later);
- due to the evaporation (between points 3* to 3), the heat input is higher and therefore, power output is also higher (even with equal efficiency);
- for the realization of a TFC cycle, a very special expander is needed because of the very high fraction of droplets (especially in the early stage of the expansion); this can be, for example, a volumetric expander [7,8].

Concerning the efficiency, one can realize very easily, that by using wet (ACZ-type) working fluid, the efficiency of ORC is higher than the efficiency of TFC. From Figure 1/a, it can be seen that the ORC (i.e. the whole 1→2→3*→3→4→4*→1 cycle) is virtually the combination of a TFC (1→2→3*→4*→1) and a strange pseudo-Carnot cycle, running entirely in the two-phase region (4*→3*→3→4→4*). Although one cannot simply sum the efficiencies of the TFC and the pseudo-Carnot cycle, still - being the efficiency of Carnot-cycles is the highest possible one among all others, between the same maximal and minimal cycle temperatures – it is plausible to assume that the addition of this highly efficient pseudo-Carnot cycle to the TFC to complete it to ORC would increase the efficiency. It was shown previously that in this case, using the same temperature ranges, the efficiency (using wet working fluid) of ORC is always above the TFC, except at the critical point (the top of the *T-s* dome), where points 3 and 3* would merge, making the TFC and ORC identical. This has a surprising consequence, namely that the efficiency of ORC (with a fixed minimal cycle temperature) has to reach a maximum at high temperatures (usually a very few Kelvin below the critical point) then it will decrease and smoothly reach the continuously increasing efficiency of TFC at the critical point [9,10]. An example for a typical wet working fluid (CO_2) can be seen on Figure 2/a; the efficiencies of ORC (magenta) and TFC (blue) cycles are with minimal cycle temperature fixed at 217 K and maximal one varied between 217 K and the critical temperature. A well-recognizable maximum for ORC efficiency of around 300 K (used as maximal cycle temperature) can be seen; it means that increasing the maximal cycle temperature above this value is counterproductive. For TFC, the efficiency smoothly increases along with the whole temperature range [9].

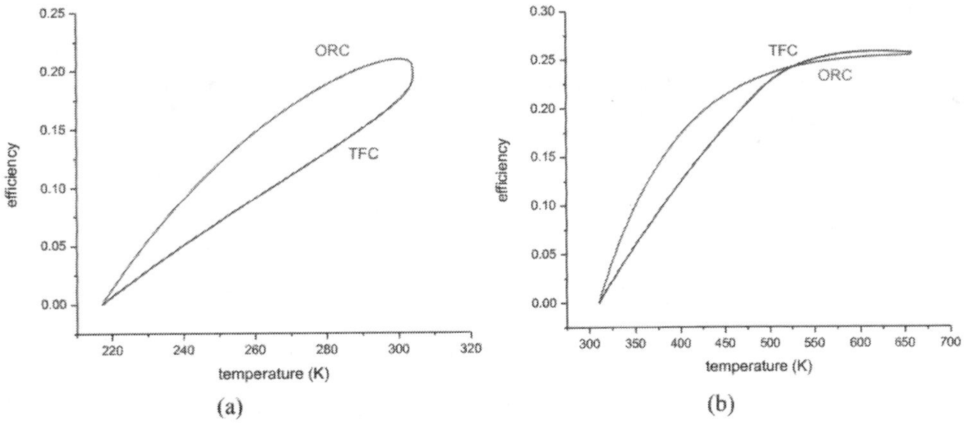

Fig. 2. The thermodynamic efficiencies of the ORC and TFC cycles using wet (a: CO_2) and very dry (b: dodecane) working fluid, using the basic layout, without superheating and heat recuperation. Minimal cycle temperatures are fixed (a: 217 K; b: 310 K), while the maximal cycle temperatures gradually increased up to the critical temperature of the given fluid.

In the case of dry working fluids without a recuperator, one can lose some (even significant) part of the heat without converting it to work. The total lost part (as can be seen in Figures 1a and 1b) are almost triangular ones, 4-4'-4** for ORC and 4*-4''-4** for TFC. This loss will cause some loss of efficiency. As it can be seen in Figure 1b, sometimes the loss for ORC can be significantly bigger than for TFC, causing higher efficiency loss. For moderately dry working fluids [9], it can cause a smaller and smaller efficiency surplus for ORC. In extremely dry fluids (like dodecane, a class ANZCM –fluid [2] the loss for ORC can be so high that choosing high maximal cycle temperature (while keeping the low one in the original value) can shift ORC efficiency below TFC one, as can be seen in Figure 2b for dodecane with minimal cycle temperature at 310 K and maximal one above this value, up to the critical point. In this case, the maximum for ORC efficiency disappears, while the efficiency curve for TFC can show a small, insignificant maximum [9].

Although the minimal temperatures chosen for this demonstration for carbon dioxide (as 217 K) and dodecane (as 310 K) are quite low, they are not unrealistic. 217 K can be easily reached in cryogenic cycles (for example using liquid methane, LNG as heat sink), see for example in Ref. [11], while 310 K (almost 33 Celsius) can be obtained by a good air cooling. It is important to remark, that in both cases, the shape of the curves would be the same by choosing higher minimal temperature, only the loops would be smaller and therefore the effect would be less remarkable.

2 The effect of heat recovery

As it was already mentioned, some part of the input heat can be lost when dry working fluids are used. Some part of the lost heat can be recovered by using a recuperative or regenerative method. By using this part for liquid pre-heating, the efficiencies can be increased. Theoretically, this part is the part of the former triangular regions located at temperatures above the temperature of point 2; technically, one should consider the loss of an extra "few K-wide" region to provide proper heat flow in the recuperator or regenerator.

In Figure 3, one can see the cycle for dodecane (like in Figure 2b), without recovering any heat (rec=0); recovering half of the theoretically recoverable one (rec=0.5) and recovering all of the theoretically possible one (rec=1). One should remember that even for rec=1; the part below the temperature of point 2 is still lost. In Figure 3, the recuperated and un-recuperated cases are shown at two different temperatures. In Figure 3/a, the case with minimal temperature fixed at 310 K can be seen (like for Figure 2/b), while in Figure 3/b, the curves represent the case with minimal temperature fixed at 272 K. Technically, this second temperature is less relevant (although it can be reached even in mild cryogenic cases), this case demonstrates that both effects (the crossing of the two efficiency lines with low recuperation ration and the disappearance of the crossing point at higher recuperation ratio) exist on various temperatures, i.e. it is a general phenomenon.

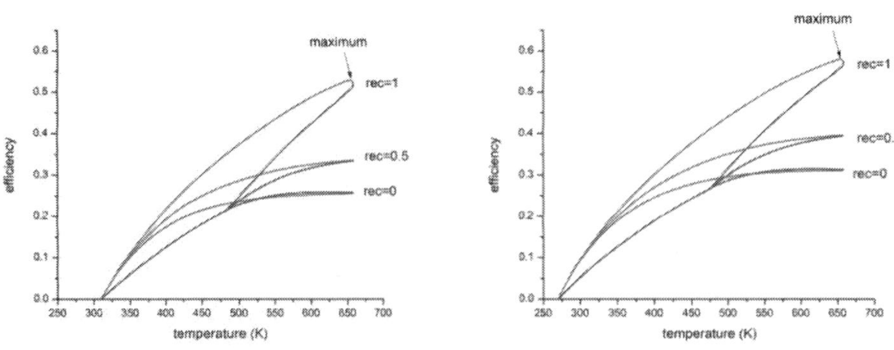

(a) (b)

Fig. 3. The effect of lost heat recovery and the use of it for liquid pre-heating in ORC and TFC with extremely dry working fluid (dodecane), on two different minimal temperatures.

Heat recovery is clearly advantageous for ORC as well as TCT efficiencies. One can see that by recovering a proper amount of heat, ORC will be more efficient than TFC, although one have to consider the cost of appropriate heat regenerators versus the positive effect of efficiency increase (compared to recovery-free ORC). It might happen that the extra earning caused by the efficiency increase (connected with power output increase) is negligible; in this case the use of recuperator-free system can be advantageous, and among them, TFC-type system can surpass ORC-type one in efficiency (as well as in power output)

3 Discussion

The advantages and disadvantages for ORC-and TFC based systems with and without recuperator are summarized here:

Advantages using simple TFC over simple ORC with very dry working fluid

- Simple initial heating unit (only liquid heater); i.e. lower installation and maintenance cost
- Higher efficiency, due to the lower heat lost after expansion

Disadvantages using simple TFC over simple ORC with very dry working fluid

- The need for a special (wetness resistant) expander; i.e. higher installation and maintenance costs.

Advantages using recuperated ORC over simple TFC with very dry working fluid

- Possibility to reach efficiency above the efficiency of simple TFC

Disadvantages using recuperated ORC over simple TFC with very dry working fluid

- Need of complex initial heating unit (liquid heater plus evaporator); i.e. higher installation and maintenance costs.
- Need of recuperative heat exchanger to recover part of the leftover heat after expansion; i.e. higher installation and maintenance costs.

Advantages using recuperated ORC over recuperated TFC with very dry working fluid

- Possibility to reach efficiency above the efficiency of recuperated TFC (the potential increase is bigger for ORC than for TFC)

Disadvantages using recuperated ORC over recuperated TFC with very dry working fluid

- Need of complex initial heating unit (liquid heater plus evaporator); i.e. higher installation and maintenance costs.
- Although both systems need a recuperative heat exchanger to recover part of the leftover heat after expansion, this heat exchanger has to be bigger for ORC, i.e. higher installation and maintenance costs (although the difference here will be only minor one).

4 Conclusions

The choice of working fluid is a multi-dimensional optimization process involving various (chemical, environmental, economical, etc.) criteria. Under some circumstances – for example, due to some compatibility problem - the number of potential working fluids can be pretty low; therefore, the utilization of a given heat source has to be solved with a given working fluid. In that case, one has to tune other parameters, like the type of cycle (ORC or TFC) or the system layout (with or without recuperator). Using our results, one might choose the proper (most effective) system to turn heat to work (and power), when the use of extremely dry working fluid (like dodecane) would be advantageous for some other reason (like temperature levels, etc.).

Part of the research reported in this paper and carried out at BME has been supported by the NRDI Fund (TKP2020 NC, Grant No. BME-NC) based on the charter of bolster issued by the NRDI Office under the auspices of the Ministry for Innovation and Technology.

References

1. Macchi, E. and Astolfi, M., Organic Rankine Cycle (ORC) Power Systems: Technologies and Applications (Elsevier-Woodhead Publishing: Duxford, UK, 2016).
2. Györke, G., Deiters, U.K., Groniewsky, A., Lassu, I. and Imre, A.R., Energy, **145**, 288–300 (2018).
3. Garrido, J.M., Quinteros-Lama, H., Mejía, A., Wisniak, J. and Segura, S. Energy **45**, 888–899 (2012).
4. White, J.A. and Velasco, S. Ind. Eng. Chem. Res. **58**, 1038–1043 (2019).
5. Zhang, X., Zhang, Y. and Wang, J., Energy **201**, 117722 (2020).
6. Fischer, J. Energy **36** 6208 (2011).
7. Kolasiński, P. Energy Conversion and Management **231**, 113859 (2021).
8. Richter, L., AIP Conf. Proc., **1889**, 020032 (2017)
9. Ahmed, A.M., Kondor, L. and Imre, A.R., Energies **14** 307 (2021).
10. Ahmed, A.M. and Imre, A.R., Pollack Periodica, **16**, 73 (2021).
11. Daniarta, S. and Imre A.R., Periodica Polytechnica Mechanical Engineering, 64, 432 (2020).

Validation of large eddy simulation of flow behind a circular cylinder

Martin Isoz[1,2,*], *Tomáš* Hlavatý[1,2], *Václav* Uruba[1,3], and *Pavel* Procházka[1]

[1]Institute of Thermomechanics of the Czech Academy of Sciences, v.v.i., Dolejškova 1402/5, Prague 182 00, Czech Republic

[2]University of Chemistry and Technology, Prague, Department of Mathematics, Technická 5, Prague 166 28, Czech Republic

[3]University of West Bohemia, Pilsen, Department of Power System Engineering, Univerzitní, Pilsen 301 00, Czech Republic

Abstract. The flow in the wake behind a circular cylinder in a cross-flow at Reynolds number of 4815 was studied both experimentally and via mathematical modeling. The mathematical model was performed as a Large Eddy Simulation (LES), while the experiments were carried out using the time-resolved variant of the Particle Image Velocimetry (PIV) method. Both the simulation and experiment took into account the dynamical aspects of the studied phenomenon, which enabled a detailed validation of the mathematical model. The overall statistical properties of the simulated flow were validated via comparing the time-averaged measured and computed velocity and vorticity fields. To validate the dynamical behavior, the velocity spectra were examined first. Next, the Proper Orthogonal Decomposition (POD) of the spatio-temporal velocity data was performed on both the experimental and numerical data and a comparison of the obtained energetic modes was carried out. All the performed validations have shown a satisfactory agreement between the simulation and the experiment.

1 Introduction

Two strong trends have been observed in fluid dynamics during the past decade. First, the increasing available computing power have enabled applications of high-fidelity modeling approaches, such as the large eddy simulation (LES), even to industrial-scale problems [1–3]. Second, improving measurement techniques, e.g. the particle image velocimetry (PIV), provide more and more detailed data on flow dynamics. Nevertheless, it seems that even the dynamic LE simulations are commonly validated against static data, see e.g. [2, 4–6]. Furthermore, if any analysis of the system dynamics is performed, it is usually limited to the comparison of turbulence spectra as in [7, 8]. Such a situation follows from the fact that most existing verification and validation methods for fluid flow simulations are derived for the computationally cheaper and better established Reynolds-averaged Navier-Stokes equations (RANS) [9].

In the present paper, we aim to promote a discussion on LES data validation. Especially with respect to the flow dynamics and validation of the simulated turbulent structures. A

*e-mail: isozm@it.cas.cz

geometrically simple set-up of a wide and thin circular cylinder in a cross-flow is used in both experiment and LE simulation. The experiment is designed in order to provide detailed data on flow dynamics and is similar as published in [10, 11]. Furthermore, the simulation is prepared with a level of detail corresponding to the industrial LES, i.e. the whole problem geometry (experimental test section) is not modeled using the computationally expensive LES. Instead, we employed the $k-\omega$ shear stress transport (SST) detached eddy simulation (DES) model based on the work [12], which is a hybrid LES-RANS model.

The selected modeling approach provides high-resolution data in the zone of interest corresponding to the measured region. Both experimental and numerical data are, similarly to [7, 8] analyzed with respect to (i) time-averaged quantities, and (ii) spectra of velocity components at given locations. Furthermore, the measured and modeled flow fields are studied via the means of proper orthogonal decomposition (POD) [13–16], where we focus on the most energetic coherent flow structures. Note that POD analysis of both experimental [10, 11, 17] and simulation [18–20] results is present in the literature. However, the possibility of comparing measured and computed POD results in order to validate flow simulation seems not to be well studied.

2 Problem geometry and experimental setup

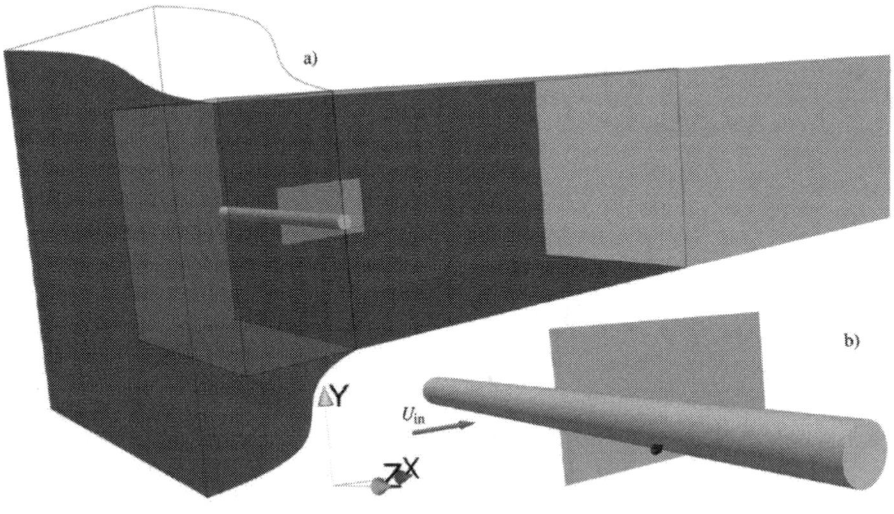

Fig. 1. (a) Sketch of the experimental set up geometry and of the corresponding computational domain. Geometrical discrepancy between the two is highlighted in red. The plane of measurement used in experiment is shown in turquoise. (b) Detail of the cylinder and plane of measurement.

The studied problem comprises the cross-flow around a circular cylinder. The experimental data were measured at a blow-down facility at the Institute of Thermomechanics of the Czech Academy of Sciences with a closed test section of $250 \times 250\,\mathrm{mm}^2$. The cylinder model of diameter $d = 15\,\mathrm{mm}$ and height $h = 250\,\mathrm{mm}$ (cross-section blockage of about 6 %) was placed at the test section center and oriented transversely with respect to the main flow direction. The situation is schematically depicted in Figure 1a.

The test section inlet nozzle provides a flow field with a low turbulence of intensity $I \approx 0.2\,\%$, a good regularity with departures below 1 %, and mean velocity $U_{\mathrm{in}} = 5\,\mathrm{m\,s^{-1}}$ resulting in the Reynolds number $\mathrm{Re} = U_{\mathrm{in}} d/\nu = 4815$, where ν is the fluid (air) kinematic viscosity. The inlet nozzle provides a well specified flow at the contact with the cylinder model. However, it causes a geometrical discrepancy between the experimental and simulation setup. In particular, the nozzle itself cannot be modeled as the conditions at the nozzle inlet are not known. At the same time, the region in front of the cylinder needs to be included in the computational domain. To extend the computational domain in front of the cylinder while not having to resolve the nozzle inlet, the test section was prolonged in the upstream direction. However, the test section walls in front of the cylinder were considered as impermeable but not posing any tangential resistance to the flow. The situation is depicted in Figure 1a, where the special computational boundary (*virtual walls*) is shown in red.

The experimental velocity fields were obtained using the Particle Image Velocimetry (PIV) method. The measurement apparatus consists of a laser and CMOS camera by the Dantec company. The laser is the New Wave Pegasus, Nd:YLF double head with wavelength of 527 nm, maximal frequency 10 kHz, shot energy of 10 mJ (for 1 kHz) and the corresponding power of 10 W per one head. The camera Phantom V611 has the resolution of 1 280 x 800 pixels, is able to acquire double snaps with frequency up to 3000 Hz (at full resolution), and it uses an internal memory 8 GB. The data were acquired and post-processed in the DynamicStudio software [21]. To visualize the flow, the SAFEX particles, oil droplets 1 μm in diameter, have been used.

The PIV measurements were performed in the plane perpendicular to the cylinder axis and parallel to the main flow direction, i.e. in the $x-y$ plane as indicated in Figure 1. The evaluated velocity field comprised 159×99 vectors and was acquired at the sampling frequency of 2 kHz. One measurement consisted of 4000 double-snapshots representing 2 s of the real-time flow. More details on the experiment may be found in [10, 11].

3 Mathematical model

3.1 Flow governing equations

The flow in the problem geometry shown in Figure 1 is assumed to be governed by Navier-Stokes equations for an incompressible Newtonian fluid of the standard form

$$\begin{gathered}\frac{\partial \boldsymbol{u}}{\partial t} + \nabla \cdot (\boldsymbol{u} \otimes \boldsymbol{u}) - \nabla \cdot \boldsymbol{T} = -\nabla \tilde{p} \\ \nabla \cdot \boldsymbol{u} = 0\end{gathered} \tag{1}$$

where $\boldsymbol{u}$ is the fluid velocity, $\tilde{p} = p/\rho$ is the kinematic pressure, where p is the pressure and ρ the fluid density. Finally, $\boldsymbol{T}$ is the viscous stress tensor.

In general, the turbulent flow was solved within the large eddy simulation framework in which $\boldsymbol{u}$ and $\tilde{p}$ are replaced by their filtered counterparts $\overline{\boldsymbol{u}}$ and $\overline{\boldsymbol{p}}$ and the stress tensor $\boldsymbol{T}$ is decomposed as

$$\boldsymbol{T} = \overline{\boldsymbol{\sigma}} + \boldsymbol{\tau}, \tag{2}$$

where $\overline{\boldsymbol{\sigma}}$ is the filtered, or resolved scale, strain rate tensor and $\boldsymbol{\tau}$ is the unknown subgrid scale (SGS) stress tensor that represents the effects of SGS motion on the resolved LES fields [1].

Furthermore, in order to correctly and efficiently account for the computational domain being relatively large compared to the region of interest containing the plane of measurement, we use a hybrid LES-RANS turbulence model based on $k-\omega$ SST approach introduced by Strelets [12]. This particular approach known as the detached eddy simulation (DES) turbulence model automatically adapts between LES and RANS computation in dependence on a DES length scale $\tilde{l}$, which is determined by the local cell size Δ and turbulent kinetic energy (TKE, k) as described in [12, 22]. The flow governing equations were solved utilizing the open-source CFD library OpenFOAM [23], which is based on the finite volume method.

3.2 Computational mesh and boundary conditions

The computational mesh was constructed such that the region of interest (RoI: $x \in [-0.5d, 7d], y \in [-2d, 2d], z = \in [-2d, 2d]$) as well as the domain parts directly affecting the flow in RoI comprise a structured hex-dominant grid with a resolution sufficient to fully simulate all the relevant flow scales. On the other hand, the main purpose of the computational domain outside of RoI is to provide physically correct boundary conditions at the RoI boundaries. Thus, the mesh in these regions is coarsened and the flow there is simulated using the $k - \omega$ SST unsteady RANS model.

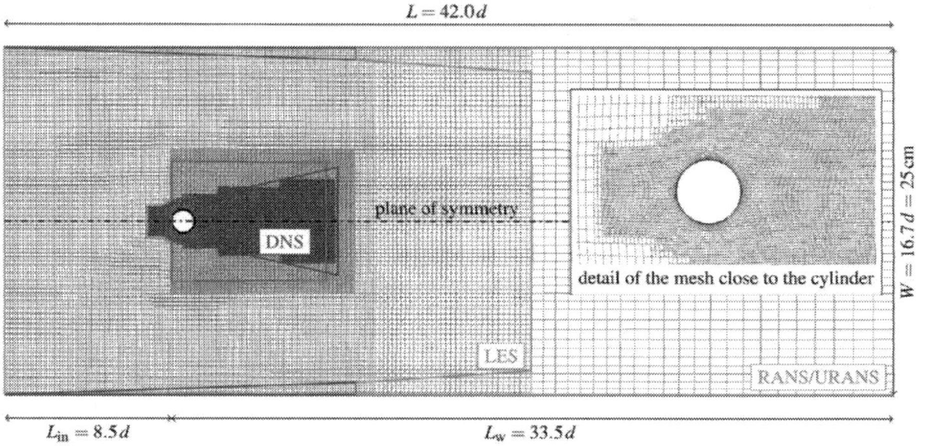

Fig. 2. Cut through the computational mesh in the plane of measurement. Regions of activity for different turbulence models are highlighted. The region of interest is outlined in red.

The overall mesh structure is shown in Figure 2, where all the distances are scaled by the cylinder diameter d. The center of the coordinate system is located in the cylinder center with the z axis aligned with the cylinder principal axis, and the x and y axes oriented in the streamwise and spanwise directions, respectively.

Note the high mesh resolution directly in front of the cylinder and the grid refinement towards the test section walls, see the blue regions in Figure 2. The high-resolution mesh in front of the obstacle is necessary to correctly resolve the pressure field, which affects the flow separation. The regions close to the test section walls needed to be fully resolved in order to account for the wall effects that, for the studied flow conditions of Re = 4815, affect the wake behind the cylinder already in the region of interest.

Table 1: Applied boundary conditions. The *virtual walls* are shown in red in Figure 1.

Boundary	Conditions	Boundary	Conditions	Boundary	Conditions
inlet	$\boldsymbol{u} = (U_{\text{in}}, 0, 0)^{\text{T}}$ $\boldsymbol{n} \cdot \nabla p = 0$ $k = k_0 \leftarrow I = 0.2\%$ $\omega = \omega_0 \leftarrow I = 0.2\%$	outlet	$\boldsymbol{n} \cdot \boldsymbol{u} = (0,0,0)^{\text{T}}$ $p = 0$ $\boldsymbol{n} \cdot \nabla k = 0$ $\boldsymbol{n} \nabla \omega = 0$	cylinder	$\boldsymbol{u} = (0,0,0)^{\text{T}} \leftarrow$ *no-slip* $\boldsymbol{n} \cdot \nabla p = 0$ $k = 0$ $\omega = 0$
virtual walls	$\boldsymbol{u} \leftarrow$ *slip* $\boldsymbol{n} \cdot \nabla p = 0$ $\boldsymbol{n} \cdot \nabla k = 0$ $\boldsymbol{n} \cdot \nabla \omega = 0$	walls	$\boldsymbol{u} = (0,0,0)^{\text{T}} \leftarrow$ *no-slip* $\boldsymbol{n} \cdot \nabla p = 0$ $k \leftarrow$ wall function $\omega \leftarrow$ wall function		

The used flow boundary conditions were mostly standard and are summarized in Table **??**. However, it was necessary to resolve the geometric discrepancy between the experiment and simulation, shown in Figure 1. The flow field at the inlet is well defined with $U_{\text{in}} = 5\,\text{m}\,\text{s}^{-1}$ and $I = 0.2\,\%$. Still, in the simulation, the inlet needed to be moved upstream, see the red (*virtual*) walls in Figure 1a. To keep the flow uniformity at the contact with the cylinder, the *slip* boundary condition,

$$\boldsymbol{u} \cdot \boldsymbol{n} = 0 \quad \text{and} \quad \boldsymbol{t}_{\text{w},t} = \boldsymbol{t}_{\text{w}} - (\boldsymbol{n} \cdot \boldsymbol{t}_{\text{w}})\,\boldsymbol{n} = 0\,, \tag{3}$$

where $\boldsymbol{t}_{\text{w}}$ is the wall stress vector, is applied at the *virtual walls*. Finally, note that although the wall functions are used for the turbulence variables at the test section walls, the mesh resolution is such that they are effectively turned-off for most of the test-section and affect the flow solution only downstream from the region of interest.

4 Proper Orthogonal Decomposition

The proper orthogonal decomposition (POD) is a method suitable to (i) identify high-energy coherent structures in spatio-temporal data, and (ii) obtain a low-dimensional approximation of a high-dimensional process [24–26]. As such, it is often applied in the fields of turbulence analysis [10, 27] and modeling [28, 29] or in model order reduction [26, 30]. The main advantages of POD are its robustness and the fact that it is a purely algebraic matrix-based method.

Both the experiment and simulation generate data through sampling of the studied system at given times,

$$\boldsymbol{S} = \left\{\boldsymbol{y}_j := \boldsymbol{y}(t_j)\right\}_{j=1}^{n}\,, \quad t_j \in (0,T]\,, \tag{4}$$

where $\boldsymbol{S}$ is a set of system *snapshots* and $\boldsymbol{y}_j$ is the vector of data at time t_j.

If all the system snapshots have the same dimension, i.e. $\boldsymbol{y}_j \in \mathbb{R}^m\,, j = 1,\ldots,n$, the set $\boldsymbol{S}$ may be rearranged into the matrix of snapshots

$$Y = [\boldsymbol{y}_1,\ldots,\boldsymbol{y}_n] \in \mathbb{R}^{m\times n}\,, \quad \text{rank}\,(Y) = n \tag{5}$$

where we assumed that the number of system snapshots n is lower than the data dimension m, which is commonly true for both experimental and numerical data.

The standard POD algorithms are based on the singular value decomposition (SVD) of the matrix Y.

$$\begin{gathered} Y = \Psi\Sigma\Phi^{\text{T}}\,, \Psi \in \mathbb{R}^{m\times m}, \Phi \in \mathbb{R}^{n\times n}, \Sigma = \text{diag}\,(\sigma_1,\ldots,\sigma_m) \in \mathbb{R}^{m\times n} \\ \Psi\Psi^{\text{T}} = E_m\,, \Phi\Phi^{T} = E_n\,, \sigma_1 \geq \sigma_2 \geq \cdots \geq \sigma_n > 0 = \sigma_{n+1} = \cdots = \sigma_m \end{gathered} \tag{6}$$

where E_m and E_n are identity matrices in $\mathbb{R}^{m\times m}$ and $\mathbb{R}^{n\times n}$, respectively, and the first n columns of the matrix Ψ form an orthonormal basis of the column space of the matrix Y, i.e.

$$\boldsymbol{y}_j = \Psi_n \boldsymbol{\eta}_j\,, \quad H = [\boldsymbol{\eta}_1,\ldots,\boldsymbol{\eta}_n] := \Psi_n^{\text{T}} Y\,, \Psi_n = [\boldsymbol{\psi}_1,\ldots,\boldsymbol{\psi}_n] \in \mathbb{R}^{m\times n}\,. \tag{7}$$

In (7), each column $\boldsymbol{\psi}_j$, $j = 1,\ldots,n$ of the matrix Ψ_n represents a temporally coherent spatial structure, *topos*. Furthermore, at any time t_j, the corresponding data vector $\boldsymbol{y}_j$ may be reconstructed as a linear combination of toposes, utilizing the coefficients $\boldsymbol{\eta}_j$. Thus, each row j of the matrix $H \in \mathbb{R}^{n\times n}$ represents the temporal evolution of the contribution of the topos $\boldsymbol{\psi}_j$ to the reconstruction of the vector $\boldsymbol{y}_j$, *chronos*.

Finally, let us note that given the SVD properties, the toposes, or POD modes, are sorted in descending order by the amount of variance in the original data they represent. In particular, if the data $\boldsymbol{y}_j$, $j = 1,\ldots,n$ correspond to the flow velocity field fluctuations $\boldsymbol{u}' = \boldsymbol{u} - \text{mean}\,\boldsymbol{u}$, the POD modes are ordered by the amount of total kinetic energy they hold. The importance of each topos $\boldsymbol{\psi}_j$ is proportional to the corresponding singular value σ_j.

Computational remark:

The standard approach to POD analysis is to first compute the full matrix Ψ_n and to subsequently truncate as $\Psi_\ell = [\psi_1, \dots, \psi_\ell] \in \mathbb{R}^{m\times\ell}$ based on a cut-off energy level or error in the Frobenius norm in the Y matrix reconstruction [26, 31]. However, such a task usually requires the matrix Y to be loaded into the computer memory, which may prove a problem when processing a fully three-dimensional LES simulation data [32].

Require: matrix of snapshots Y, POD rank ℓ, number of power iterations r
1: Generate Gaussian random $P \in \mathbb{R}^{n\times\ell}$
2: Compute $\tilde{Y} := YP \in \mathbb{R}^{m\times\ell}$
3: **for** $k = 1$ **to** r **do**
4: update $\tilde{Y} := Y(Y^{\mathrm{T}}\tilde{Y})$
5: **end for**
6: Decompose $\tilde{Y} = QR \to Q \in \mathbb{R}^{m\times\ell}$
7: Compute $B = Q^{\mathrm{T}}Y \in \mathbb{R}^{\ell\times n}$
8: Compute $\tilde{\Psi}_\ell, \Sigma_\ell, \Phi_\ell = \mathrm{svd}(B)$
return $\Psi_\ell = Q\tilde{\Psi}_\ell, \Sigma_\ell, \Phi_\ell$

Algorithm 1: Out-of-memory randomized POD modes computation

In order to limit the memory demands of POD modes computation, we employed an out-of-memory randomized SVD algorithm stemming from the works [33–35]. Utilizing this approach, the number of modes of interest $\ell \ll n = \mathrm{rank}(Y)$ needs to be known a-priori. The POD modes computation is based on a generation of a Gaussian random matrix $P \in \mathbb{R}^{n\times\ell}$ and summarized in Algorithm 1. Note that all the computations involving the original matrix Y, i.e. the lines 2, 4 and 7 of Algorithm 1 may be decomposed into a series of outer vector products and the matrix Y does not have to be fully loaded into the computer memory.

5 Results and discussion

5.1 Mean quantities

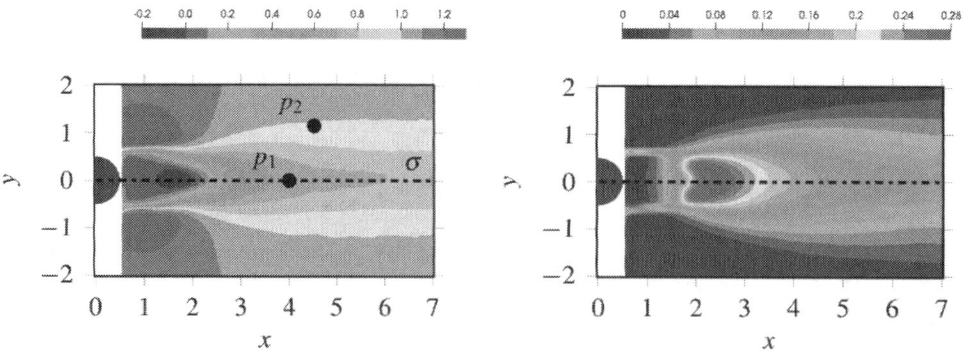

Fig. 4. Time mean fields of measured streamwise velocity component (left) and Turbulent Kinetic Energy (right). Reference points p_1 and p_2 for turbulence spectra analysis are shown in the right figure.

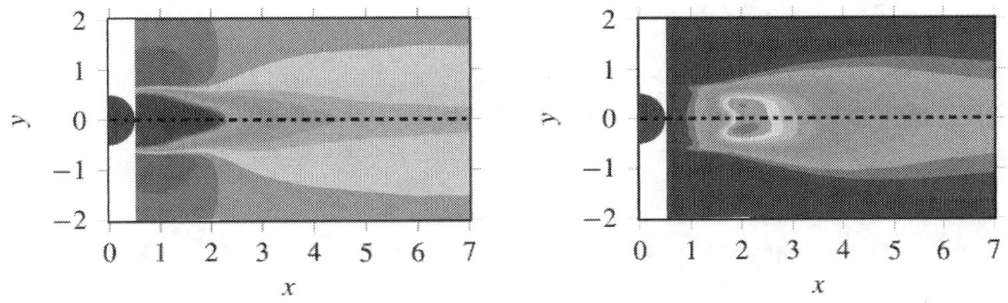

Fig. 6. Time mean fields of computed streamwise velocity component (left) and Turbulent Kinetic Energy (right). The scales are the same as in Figure 4.

The first results to be compared are the mean or time-averaged measured and computed quantities. In the following, we concentrate on the streamwise velocity component U and the turbulent kinetic energy (TKE, k) on the plane of measurement (PoM: $x \in [0.5d, 7d]$, $y \in [-2d, 2d]$, $z = 0$) and its computational counterpart. Note that, all the spatial coordinates are scaled by the cylinder diameter d and U is scaled by U_{in}.

The PIV data are shown in Figure 4, while the simulation results are depicted in Figure 6. Corresponding data in the figures are always colored utilizing the same scales. There is a significant qualitative agreement between the experiment and simulation. Both the measured and computed velocity field comprise two recirculation zones in the near-wake directly behind the cylinder and ending approximately at $x = 2$. Furthermore, the simulated and measured wake dimensions are roughly equal, as indicated by the yellow color in the left hand side of Figures 4 and 6. However, compared to the PIV time-averaged streamwise velocity field, the simulation estimates a slightly mode pronounced recirculation zone and a slightly bigger wake, albeit with a less apparent velocity deficiency.

A similar information as from the comparison of U may be gathered from the TKE distribution. Namely, in both the experiment and simulation, there is a region of low turbulence kinetic energy right behind the cylinder and a local maximum close to the test-section plane of symmetry at $x \approx 2.5$. Still, the simulated TKE is lower than the measured one and the maximum is moved by $\Delta x \approx 0.2$ towards the cylinder. Moreover, the computed low-TKE region behind the cylinder is smaller and not bounded by high-TKE bands, cf. the right hand sides of Figures 4 and 6 at $x \approx 1$.

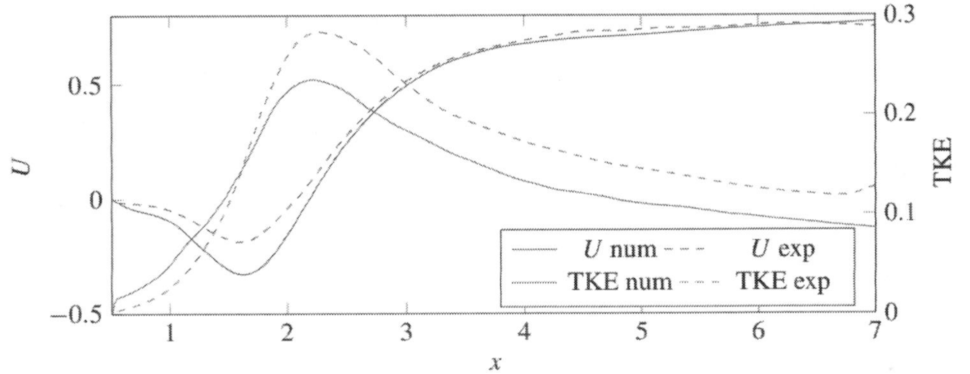

Fig. 7. Quantitative comparison of measured and computed streamwise velocity and turbulence kinetic energy on the line σ.

Finally, to evaluate the quantitative agreement between the measured and computed time-averaged fields, we plot the streamwise velocity component U and the turbulent kinetic energy k along the line $\sigma : x \in [0.5, 7], y = 0, z = 0$ shown in the left hand side of Figure 4. The local measured mean velocity minimum is placed at $x \doteq 1.57$ and has the value of $\min U^{\sigma}_{\text{exp}} \doteq -0.18$. The computed local velocity minimum is $\min U^{\sigma}_{\text{sim}} \doteq -0.33$ and it is placed at $x \doteq 1.61$. With respect to the ranges of measured data, the relative difference between the experiment and simulation is $\Delta U^{\sigma,r}_{\text{min}} = 16\,\%$ and $\Delta x^r = 0.6\,\%$. Still, the agreement between the measured and computed U is actually the worse in $\min U^{\sigma}$. The average relative error in U on σ is $1.2\,\%$.

The situation with the turbulence kinetic energy is similar to the streamwise velocity component. Overall, the simulation under-estimates the amount of TKE in the system by approximately $12\,\%$. At the same time, for $x \in [0.5, 1.5]$, the computed TKE seems to be actually over-estimated by roughly $27\,\%$, which is observable even via a comparison of the qualitative results in Figures 4 and 6,respectively. Also, similarly to the case of U, the biggest discrepancy in the data is encountered for the local TKE maximum observed at $x \doteq 2.28$ in experiment and at $x \doteq 2.22$ in simulation (relative error of $0.9\,\%$) with the respective TKE values of 0.28 and 0.24, respectively (relative error of $14\,\%$).

Despite the quantitative discrepancies in the data, all the qualitative trends in the time-averaged variables are in agreement. Furthermore, the quantitative discrepancies between the experiment and LES are similar or lower than the ones reported for example in [2, 8] or in [19].

5.2 Turbulence spectra

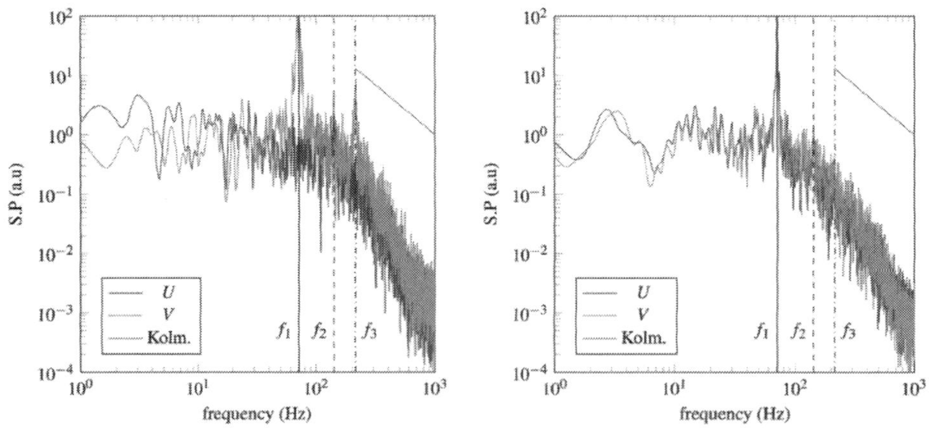

Fig. 8. Measured spectra of streamwise and spanwise velocity components in reference points p_1 (left) and p_2 (right).

The initial comparison of measured and computed flow dynamics is performed utilizing the spectra of streamwise and spanwise velocity components evaluated in the points p_1 and p_2 depicted in the left hand side of Figure 4 and located at $(4, 0, 0)$ and $(4.53, 1.15, 0)$, respectively. The measured and computed spectra are shown in Figures 8 and 9, respectively. Similarly to [11], all the results were processed using the standard method with the Hamming window.

In both Figure 8 and 9, we added vertical lines representing the frequencies of the three most important harmonics. All the shown vertical lines, i.e. for both the experiment and simulation, are placed at the same positions. Namely, the fundamental harmonic $f_1 = 71\,\text{Hz}$

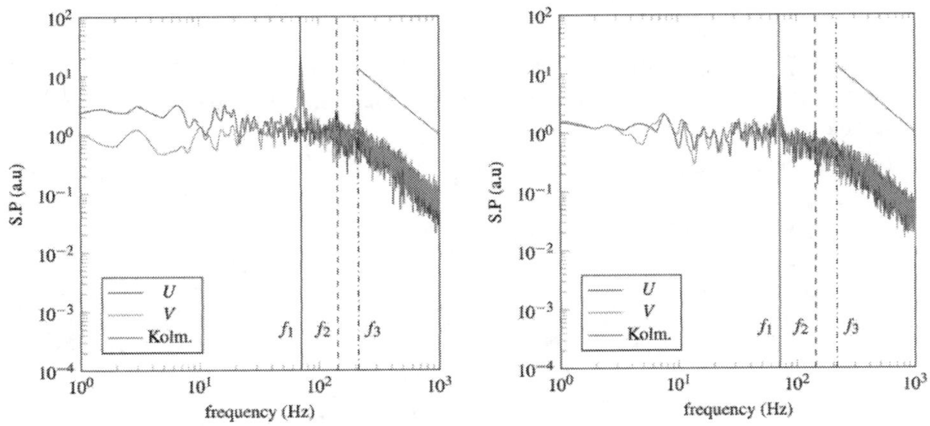

Fig. 9. Computed spectra of streamwise and spanwise velocity components in reference points p_1 (left) and p_2 (right).

corresponds to the Strouhal number $\mathrm{Sr} = f_1\, d/U_{\mathrm{in}} = 0.21 \approx 0.2$, which is a typical value of Sr in the broad range of Reynolds numbers starting with $\mathrm{Re} \gtrsim 200$, see e.g. [36] and references therein.

The higher harmonics are located at $f_2 = 142\,\mathrm{Hz}$ and $f_3 = 213\,\mathrm{Hz}$, respectively. All the three harmonics may be identified in both the measured and computed spectra. Still, the measured spectra are in general more pronounced. For example, after an identical postprocessing, the maximum value of the peak V-spectrum measured in p_1 connected to f_1 is about 100, while the same maximum in the computed data is approximately 20. Furthermore, all the spectra contain the turbulent part as follows from the decrease in the inertial part of the turbulence spectrum in agreement with the Kolmogorov hypothesis ($-5/3$ law). However, the decrease in the measured spectra for frequencies higher than f_3 seems a little bit steeper than the $-5/3$ law suggests. On the other hand, at the decrease in the computed spectra in the same region follows the Kolmogorov hypothesis almost exactly.

Finally, both the simulation and experiment provide considerably different spectra in p_1 and p_2 with the spectra corresponding to the same point being qualitatively similar. For example, the peak at f_1 is missing in both the U-spectra in point p_1 and it is present in all the remaining data, i.e. in the V-spectra in p_1 and in U and V-spectra in p_2. Furthermore, the U-spectra in point p_1, which is located on the σ line and inside the cylinder wake show identifiable peaks at f_2 and the V-spectra at the same location demonstrate peaks at f_1 and f_3. The spectra in p_2, placed close to the wake border, exhibit a distinct peak at the fundamental frequency only.

5.3 High-energy POD modes

The final part of the present contribution is devoted to the POD analysis of both experimental data and simulation results. The main comparison is performed on the plane of measurement (PoM) and utilizing the streamwise and spanwise velocity components. However, the section is concluded by a presentation of fully three-dimensional POD results in the whole region of interest (RoI) as obtained from the simulation. The main idea of utilizing the POD analysis for dynamic simulation validation is based on the following. First, the topological structure of the POD modes *toposes* should be similar between the modes representing similar fraction of the system total kinetic energy. Second, the power spectra of *chronoses* corresponding to similar *toposes* should exhibit the same peaks or lack of them.

In both the experiment and simulation, 4000 system velocity snapshots sampled at the frequency of 2 kHz were decomposed using the POD. The experimental data were processed via a standard POD procedure implemented in the DynamicStudio software [21]. The simulation data were processed utilizing the out-of-memory randomized POD, which is summarized in Algorithm 1. Although the out-of-memory approach is not necessary for the PoM data processing, it enables the full 3D POD based on the RoI data.

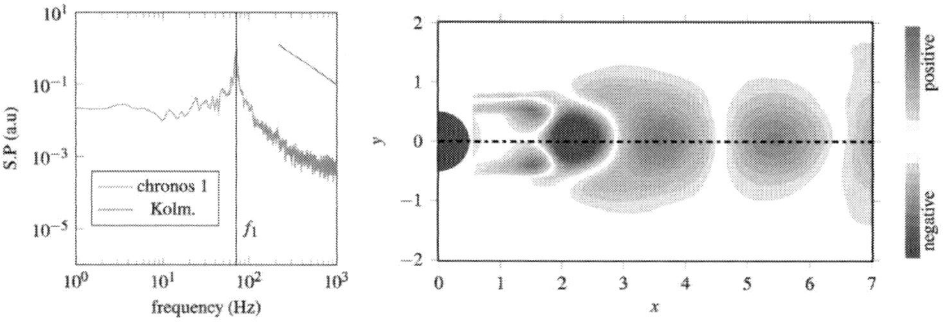

Fig. 10. POD mode 1 as obtained from experimental data, spectrum of *chronos* (left) and *topos* (right).

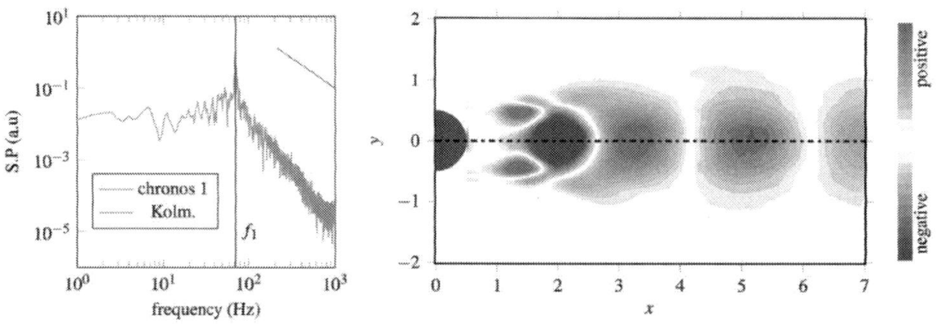

Fig. 11. POD mode 1 as obtained from simulation, spectrum of *chronos* (left) and *topos* (right).

The analysis was limited to the two most energetic POD modes, which combined contain more than 60 % of the system total kinetic energy for both the experimental and numerical results on PoM. In particular, the experimental modes 1 and 2 account for 35 % and 33 % of the total kinetic energy, while their simulation counterparts represent 25 % and 24 % respectively. Such a situation indicates a quasi-periodic flow, the periodic part of which is covered by the POD modes 1 and 2. The higher modes than contain the periodic flow perturbations leading to the observed chaotic turbulent flow.

The most energy-rich modes for experiment and simulation are given in Figures 10 and 11, respectively. In both the figures, the mode *topos* is visualized in the right hand side and the power spectrum of the mode *chronos* is given on the left. The *topos* is represented by the vorticity distribution. Comparing the *toposes* of the POD modes 1 based on experiment and simulation, it is possible to identify strongly similar vortex structures symmetrical with respect to the line σ. Furthermore, the local vorticity maxima and minima are placed at similar positions. Still, the computed vortices in the POD mode 1 are slightly closer to the cylinder than the corresponding experimental results, which is to be expected based on the analysis of the time-averaged flow fields.

Focusing on the *chronoses* of the POD mode 1, in both the measured and computed power spectra, there is a distinct peak at $f_1 = 71\,\text{Hz}$. Thus, both the real and simulated flow fields exhibit a strong underlying periodicity partially defined by the POD mode 2 oscillating at the system fundamental harmonics. In other words, the POD-based reconstruction of the original flow field as defined by the equation (7) is a superposition method where the time-stable *toposes* are superposed utilizing the *chronoses*. Thus, a strong harmonics in a POD mode *chronos* suggests an almost periodic component of the trajectory of the system $\boldsymbol{\eta}(t) = \Psi_n^{\mathrm{T}}\boldsymbol{u}(t)$, cf. equation (7). Consequently, approximating the original variable $\boldsymbol{u}'$ as $\boldsymbol{u}'_\ell := \Psi_\ell^{\mathrm{T}}\boldsymbol{u}(t)$, where ℓ is the number of modes exhibiting strong harmonics in the *chronos* leads to an (almost) periodic flow field approximation.

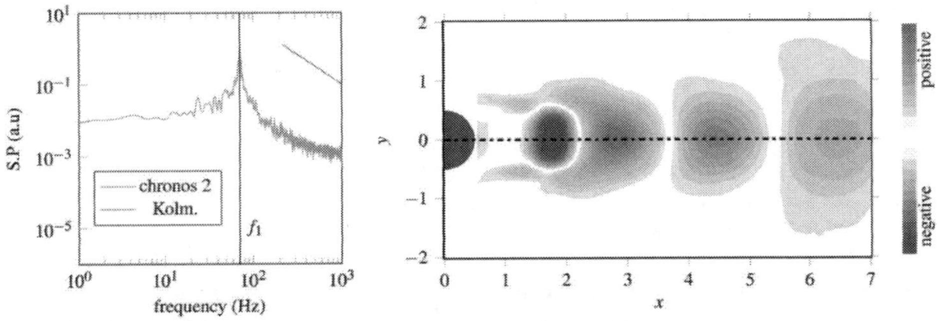

Fig. 12. POD mode 2 as obtained from experimental data, spectrum of *chronos* (left) and *topos* (right).

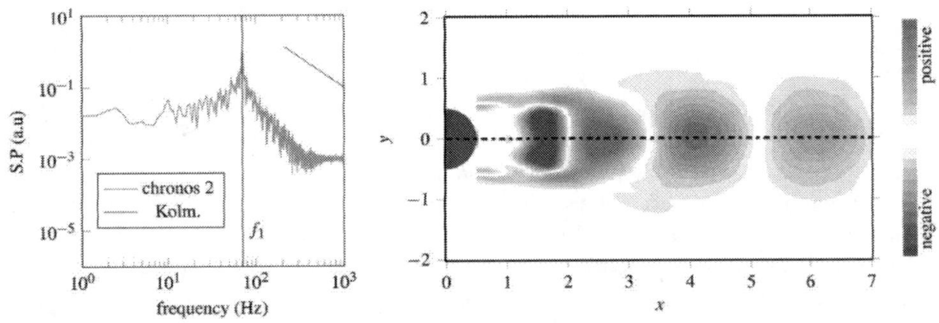

Fig. 13. POD mode 2 as obtained from simulation, spectrum of *chronos* (left) and *topos* (right).

The situation is almost the same with the POD mode 2. Once more, there is a qualitative agreement between the experiment and simulation results, comprising the similar topology of the modes *toposes* and presence of a strong fundamental harmonics in the *chronoses* spectra. The biggest discrepancy in the structure of *toposes* may be identified in the region of negative vorticity close to $x = 1.8$. In the experiment, there is a single local minimum placed on the line σ, while in the simulation, there are two local minima located below and above σ. A similar situation may be observed even comparing the right hand sides of Figures 4 and 6, where the experiment suggests a single maximum of the turbulence kinetic energy located on σ at $x = 2.28$ and the simulation result shows two distinct maxima located at a similar streamwise position, but symmetrically below and above σ.

Fig. 15. Visualization of *toposes* of fully three-dimensional POD modes 1 (left) and 2 (right) computed from simulation results.

The fully three-dimensional *toposes* of POD modes 1 and 2 in the region of interest as obtained from the simulation data are visualized in Figure 15. The dominant vortex structures are identified via contours of the Q-criterion and colored by the vorticity component oriented in accordace with the cylinder axis (z axis in the used coordinate system), i.e. the coloring is the same as in Figures 10–13. Combined, the modes 1 and 2 form an (almost) periodic base of the chaotic turbulent flow. Moreover, the identified structures are almost uniform in the z axis direction, suggesting that the most energy-rich part of the flow is close to the 2D structure. Note that the energetic prevalence of the modes 1 and 2 usually decreases with (i) the increasing flow Reynolds number and flow chaoticity, and (ii) changes in the system geometry promoting three-dimensional flow. For example, at $\mathrm{Re} = 100$ and 2D flow approximation of a flow behind a circular cylinder, the modes 1 and 2 account for more than 90 % of the flow total kinetic energy [27].

6 Conclusion

A validation procedure for a large eddy simulation of the flow around a circular cylinder at $\mathrm{Re} = 4815$ was presented. The validation steps include comparison of (i) time-averaged flow quantities, (ii) turbulence spectra, and (iii) POD analysis results, between the experiment and simulation. It might be assumed that a simulation returning results similar to the experiment for all the three steps reflects the real flow dynamics. However, the first step, i.e. a similarity of time-averaged fields, is only a necessary condition for the success of the simulation validation. At the same time, it is the only step that is easily quantified and for which exist standard validation procedures; albeit, developed for the RANS-based turbulence modeling. As for the other steps, arguably, a model that produces similar both turbulence spectra and POD modes as the validation experiment contains similar coherent structures that are evolving in time in a similar manner and can provide a representative insight into the real flow dynamics. Unfortunately, how to evaluate the "similarity" between experimental and simulation POD analysis is still an open question. As an example, let us mention that it is not clear how many simulated POD modes need to correspond to the experimental data in order for the model to accurately catch all the important flow characteristics. In the subsequent work, we plan to extend the POD analysis towards higher modes, to include the data measured on a cross-oriented plane of measurement (i.e. in the $y - z$ plane, see [10]) into the validation dataset, and to devise some guidelines for using POD for transient simulations validation.

The work was supported by the Grant project with *No. GA19-02288J* of the *Czech Science Foundation*, within institutional support *RVO:61388998* and by the Centre of Excellence for Nonlinear Dynamic Behaviour of Advanced Materials in Engineering *CZ.02.1.01/0.0/0.0/15_003/0000493* (Excellent Research Teams) in the framework of Operational Programme Research, Development and Education.

References

1. Z. Yiang, Chin. J. Aeron. **28**, 11 (2015)
2. X. Wen, K. Luo, Y. Luo, H.I. Kassem, H. Jin, J. Fan, Appl. En. **183**, 1086 (2016)
3. R. Lohner, C. Othmer, M. Mrosek, A. Figueroa, A. Degro, Comp. Fluids **214**, 104771 (2021)
4. M. Fatica, H.J. Kaltenbach, R. Mittal, Cent. Turb. Res. Ann. Res. Briefs **10**, 23 (1997)
5. E.A. Sewall, D.K. Tafti, A.B. Graham, K.A. Thole, Int. J. Heat Fluid Flow **27**, 243 (2006)
6. T.E. Tokyay, S.G. Constantinescu, J. Hydraul. Eng. **132**, 1303 (2006)
7. H. Jie, Y.Z. Liu, J. Vis. **19**, 61 (2016)
8. I. Gonzalez-Neria, A. Alonzo-Garcia, S.A. Martinez-Delgadillo, V.X. Mendoza-Escamilla, J.A. Yanez-Varela, P.G. Verdin, Chem. Eng. J. **374**, 1138 (2019)
9. T. Xing, J. Hydrodyn. **27**, 163 (2015)
10. V. Uruba, P. Prochazka, V. Skala, *On the 3D dynamics of the wake behind a circular cylinder*, in *Proceedings of the conference Topical Problems of Fluid Mechanics*, edited by T. Bodnar, D. Simurda (IT CAS, 2020), p. 105
11. V. Uruba, P. Prochazka, *POD Spectrum of the Wake behind a Circular Cylinder*, in *Proceedings of the MATEC Web of Conferences* (EDP Sciences, 2020), p. 328
12. M. Strelets, *Detached eddy simulation of massively separated flows* (AIAA, 2001), pp. 2001–0879
13. K. Pearson, Phil. Mag. **2**, 559 (1901)
14. H. Hoetelling, Psychometrica **1**, 27 (1935)
15. K. Karhunen, Ann. Aca. Sci. Fennicae Se. A. I. Math.-Phys. **34**, 7 (1946)
16. M. Loeve, Revue **48**, 195 (1946)
17. W. Mickiewicz, PAN Arch. Ac. **40**, 475 (2015)
18. G. Janiga, Chem. Eng. Sci. **201**, 132 (2019)
19. Y.S. Mustafa, H.C. Lim, AIP Adv. **11**, 040511 (2021)
20. V.S. Fraga, G. Yin, M.C. Ong, Ships Offsh. Struct. p. 1977939 (2021)
21. Dantec Dynamics, *Dynamicstudio* (2021), `https://www.dantecdynamics.com/components/dynamicstudio/`
22. F.R. Menter, M. Kuntz, R. Langtry, *Ten years of experience with the SST turbulence model* (Begell House, 2003), pp. 625–632, 1st edn.
23. OpenCFD, *OpenFOAM: The Open Source CFD Toolbox. User Guide Version 1.4, OpenCFD Limited* (Reading UK, 2007)
24. M. Samani, Ph.D. thesis, University of Saskatchewan (2014)
25. M. Isoz, Ph.D. thesis, University of Chemistry and Technology (2018)
26. M. Isoz, Ocean Eng. **186**, 106083 (2019)
27. K. Taira, M.S. Hemati, S.L. Brunton, Y. Sun, K. Duraisamy, S. Bagheri, S.T.M. Dawson, C.A. Yeh, AIAA J. **58**, 2737 (2020)
28. T.B. Gatski, M.N. Glauser, *Proper orthogonal decomposition based turbulence modeling*, in *Proceedings of the conference ICASE/LaRC Workshop on Transition and Turbulence* (NASA, 1991), p. 22

29. V. Shinde, SIAM J. Sci. Comput. p. 2737 (2020)
30. S. Chaturantabut, D.C. Sorensen, SIAM J. Sci. Comput. **32**, 2737 (2010)
31. S. Volkwein, *Proper orthogonal decomposition: theory and reduced-order modelling*, 1st edn. (Universtity of Konstanz, Konstanz, Germany, 2013)
32. K. Kabir, A. Haidar, S. Tomov, A. Bouteiller, J. Dongarra, *A Framework for Out of Memory SVD Algorithms*, in *High Performance Computing*, edited by J.M. Kunkel, R. Yokota, P. Balaji, D. Keyes (Springer International Publishing, 2017), p. 158
33. V. Rokhlin, A. Szlam, M. Tygert, SIAM J. Matr. An. Appl. **31**, 1100 (2009)
34. N. Halko, P.G. Martisson, J.A. Tropp, SIAM Review **53**, 217 (2011)
35. M. Gu, SIAM J. Sci. Comput. **33**, 2580 (2015)
36. J.H. Lienhard, *Synopsis of lift, drag, and vortex frequency data for rigid circular cylinders* (Pullman, Wash: Technical Extension Service, Washington State University, 1966)

Developing of experimental combustion boiler with stationary fluidized bed and multifuels combustion

Marek Jadlovec[1]*, and *Stanislav* Honus[1]

[1]VSB - Technical University of Ostrava, 17. listopadu 2172/15, Ostrava-Poruba, Czech Republic

Abstract. Due to the outdated higher requirements for reducing emissions of harmful substances in the combustion of fossil fuels and waste in particular, more and more research is needed, which will lead to the provision of various alternatives to influence the emerging pollutants into the atmosphere. This study deals with the development of experimental combustion units with a stationary fluidized bed with regulated parameters of primary and secondary air supply, performed fuel supply and flue gas parameters. Combustion is carried out in a bubbling stationary fluidized bed with a height of 100 mm and a diameter of 140 mm. The fluidized bed consists of ST 54 glass sand with a mean grain size of 0.22 mm. The basic purpose of the study is to describe the construction of the experimental unit with verification of its functionality. On the experimental equipment was realized experiments in terms of combustion various types of fuels, whether they are fossil fuels (coal), alternative fuels as straw pellets, sludge from wastewater treatment plants, sludge from paper mills or their combinations. Determination of temperature and pressure along the entire length of the unit is a device equipped with thermocouples and pressure sensors.

1 Introduction

Fluidized bed combustion is one of the three most basic types of solid fuel combustion. In practice, fluidized bed combustion units are experimental with output from single digit kW to dezens kW up to heating plants with output in hundreds kW. [1] This study uses an experimental unit with a maximum heat output of 20 kW. The main goal of the design of the experimental fluidized unit was to design it so that it is possible to change the combustion parameters and at the same time the type of fuel almost immediately. [2] This is the main advantage of the whole stand and that is its versatility. To achieve the required versatility, the unit must be equipped with the correct source of primary and secondary combustion air, as well as a sufficiently large smoke fan for proper exhaust of combustion products. Fluidized bed combustion is realized at temperatures up to 900 ° C. [3] This fact has a positive effect on the formation of NO_x compounds. Another advantage is the possibility of dosing the sorbent directly into the fluidized bed to reduce SO_x emissions. However, this article focuses mainly on the combustion of alternative fuels that could replace fossil fuels. [4], [5], [6], [7]

* Corresponding author: marek.jadlovec@vsb.cz

The disadvantage of combustion such fuels in a stationary fluidized bed is the greater amount of ash that the fuels carry. This fact leads to the degradation of the fluidized bed after a certain time and thus to a change in the character of combustion.

This article briefly describes all the limitations and the way to deal with the transformation of the boiler to another type of fuel. Thermocouples, pressure sensors, flow regulators and flue gas analyzers are implemented in the boiler to assess the quality of this transformation. [8], [9]

2 Experimental setup

2.1 Experimental device and analyses equipment

The experimental unit is located on the premises of VŠB - TUO. As you can see in Figure 1, the dominant element of the whole device is the primary air blower with performance 57 $m^3{}_N$/hod with pressure ratio 1, 7 and with controllability in the range 0 – 100 %. A primary combustion air flow meter is included in the pipeline to measure the flow. Another technological device is the preheating of primary air with an output of 4 kW, which can preheat the combustion air up to a temperature of 650 ° C. The preheated air is distributed on a "grate" in this case a silicon filter disc with a diameter of 90 mm, a thickness of 6 mm and a porosity of 16 micrometers. This is followed by a fluidized bed, which consists of glass sand ST 54 with a mean grain size of 0,22 mm. A preview of the fluidized bed is shown in Figure 2, with its steady state height being 100 mm. [10] The fluidized bed is surrounded by resistance heaters with a total output of 8,8 kW, which have the task of heating the combustion chamber to the required temperature. Above the fluidized bed is a secondary air inlet, which is initiated by a LEISTER MISTRAL hot air blower with performance 24 $m^3{}_N$/hod. This is followed by the flue gas tract itself, which is equipped with a cyclone separator, water cooler and especially a smoke fan with performance 300 $m^3{}_N$/hod.

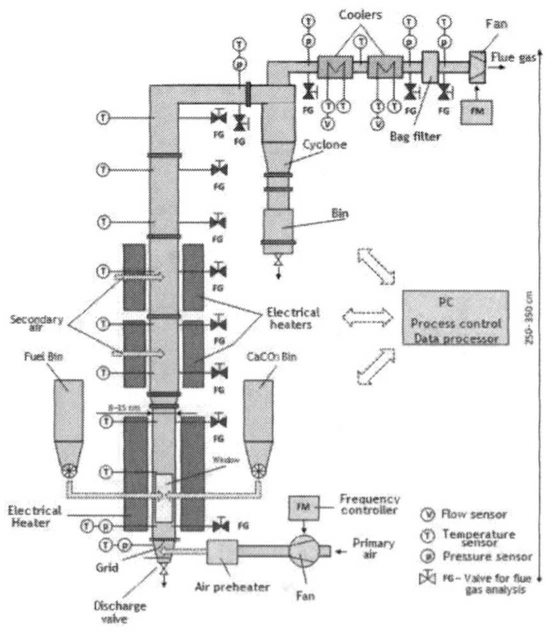

Fig. 1. Schematic view on experimental unit

The chapter itself is fuel economy. For the transport of various materials, the device is equipped with two screw feeders with a pitch of 6 mm and 20 mm as you can see on Figure 3 (a), (b). The screw feeder is connected to the electric motor by means of a Hardy coupling. The transport capacity of the feeder with a pitch of 6 mm is up to 1,75 kg/hod and for a pitch of 20 mm is up to 5,5 kg/hod. There are two possible systems of connecting the feeder, either state 1 Figure 4 (a), when the screw is led directly into the fluidized bed and state 2 Figure 4 (b), when the fuel is supplied indirectly into the fluidized bed by free fall through the connecting pipe. State 1 is more advantageous for fuels with a higher moisture content, where the fuel dries before entering the fluidized layer, and condition 2, for fuels with a higher heating value and lower moisture, where there is a risk that burns could occur during the dosing itself.

An interpretation of four temperatures T_1, T_2, T_3, T_4 appears in the article. The temperature T_1 is measured at a height of 100 mm above the filter disc, while T_2, T_3, T_4 at a height of 300, 500 and 700 mm.

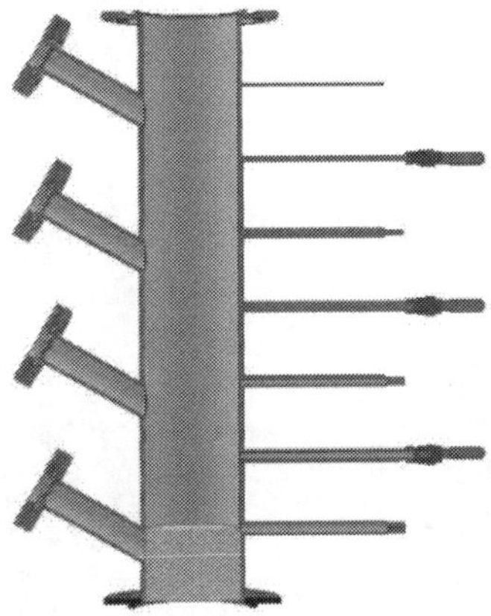

Fig. 2. Cross section of experimental unit with a view on fluidized bed

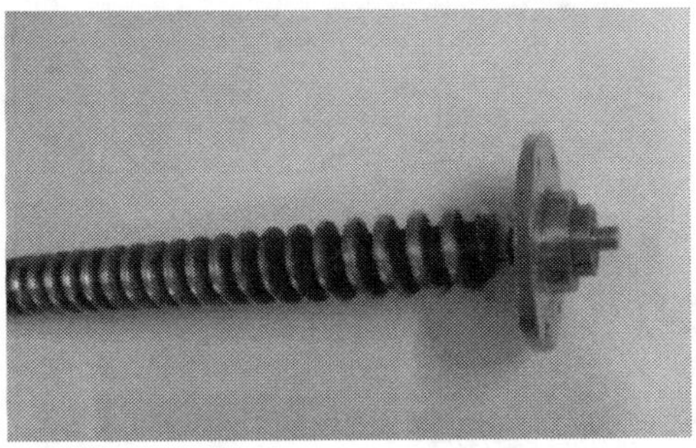

(a)

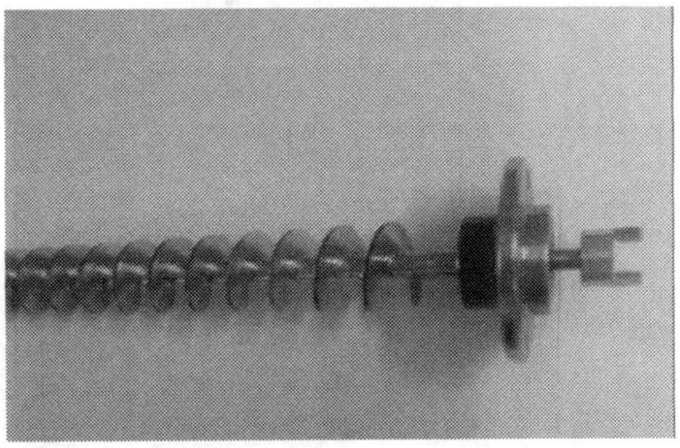

(b)

Fig. 3. Two types of screw conveyor (a) with a pitch 6 mm (b) with a pitch 20 mm

Quantitative and qualitative evaluation of flue gases is performed by Gasmet DX4000 portable FTIR gas analyzer, see Figure 5 (c), which is able to analyze CO_2, CO, N2O, NO, NO_2, SO_2, NH_3, HCl, HF, CH_4, H_2O and O_2. For additive analysis and control, a Testo 350 analyzer is available for the determination of CO, NO, NO_2, SO_2, H_2O, see Figure 5 (b). DURAG HM-1400 TRX Figure 5 (a) is available for Hg analysis in flue gas. [11], [12]

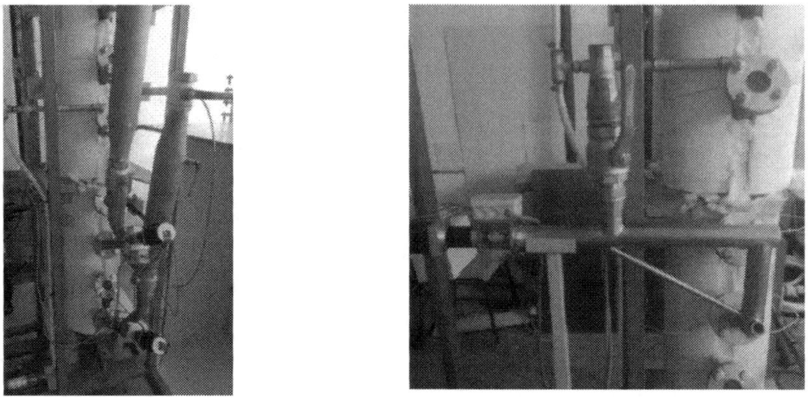

(a) (b)

Fig. 4. Two types of fuel conveyor connection (a) directly to the fluidized bed (b) outside the fluidized bed

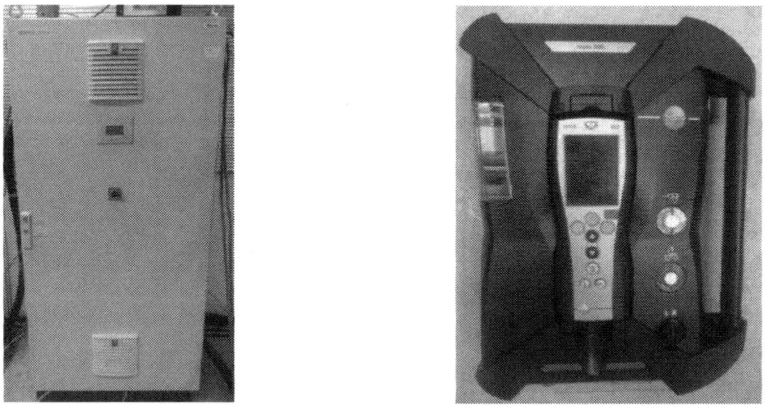

(a) (b)

(c)

Fig. 5. Analyses equipment (a) DURAG HM-1400 TRX (b) Testo 350 (c) Gasmet DX4000 portable FTIR gas analyzer

2.2 Fuels Definition

As mentioned in the paragraphs above, one of the most important features of the experimental unit is its versatility. This ability has been demonstrated in the combustion of multiple fuels. So far, six different types of fuels with different heating values and compositions have been burned in the combustion unit. These were straw pellets, sludge from wastewater treatment plants, sludge from paper mills their combination, lignite, and coal. Straw pellets can be seen in Figure 6. Figure 7 (a), (b), (c) shows a sludge from wastewater plants, a mixture of sludge from wastewater plants with a sludge from paper mills to a ratio of 4:1 and 2:1. Figure 8 (a) is lignite and Figure 8 (b) coal. [13]. Table 3 shows the elemental composition of the fuels used together with ash, moisture and heating value.

Fig. 6. Straw pellets

(a) (b)

(c)

Fig. 7. (a) Sludge wastewater plants (b) Mix of sludge wastewater plants and paper mills sludge with ratio 4:1 (c) Mix of sludge wastewater plants and paper mills sludge with ratio 2:1

(a) (b)

Fig. 8. (a) Lignite (b) Coal

Table 1. Ultimate analysis of fuels

Parametr	Unit	Straw pellets	Sewage sludge	Paper mill sludge	Lignite	Coal
Carbon	[wt. %]	42,14 ± 0,21	29,07 ± 0,15	43,6 ± 0,22	65,8 ± 0,33	75,81 ± 0,38
Hydrogen	[wt. %]	5,14 ± 0,10	4,04 ± 0,08	4,2 ± 0,08	5,1 ± 0,10	4,62 ± 0,09
Sulphur	[wt. %]	0,12 ± 0,004	0,69 ± 0,021	1,3 ± 0,039	0,69 ± 0,021	0,33 ± 0,010
Nitrogen	[wt. %]	0,55 ± 0,017	2,93 ± 0,088	2,1 ± 0,063	1,03 ± 0,031	1,38 ± 0,041
Oxygen	[wt. %]	32,28	16,53	18,1	14,63	9,68
Moisture	[wt. %]	8,2 ± 0,16	21,8 ± 0,44	22,3 ± 0,45	10,59 ± 0,21	2,4 ± 0,05
Ash	[wt. %]	11,2 ± 0,11	24,94 ± 0,25	8,4 ± 0,08	2,14 ± 0,02	5,78 ± 0,06
Heating value	[MJ/kg]	15,53 ± 0,54	9,83 ± 0,34	12,82 ± 0,45	26,8 ± 0,94	30,53 ± 1,07

3 Results

3.1 Calculation fluidization velocity

Based on the correlation according to Broadhurst and Becker, the fluidization velocity is calculated. Table 2 defines the sand and air properties for the calculation. [14]

Table 2. Properties of air and sand

Parametr	Unit	Air	Sand
Diameter	[m]	x	0,00022
Density	[kg/m^3]	2624	0,327
Temperature	[K]	1073,15	x
Viscosity	[Pa.s]	4,26E-05	x
Pressure	[kPa]	101,1	x

$$Ar = \frac{D_p^3 \rho_f (\rho_p - \rho_f) g}{\mu^2} \tag{1}$$

$$Re_{mf}^2 = \frac{Ar^{1,85}}{37,7 Ar^{0,85} + 242000 \left(\frac{\rho_p}{\rho_f}\right)^{0,13}} \tag{2}$$

$$U_{mf} = \frac{\mu Re_{mf}}{D_p \rho_f} \tag{3}$$

Where Ar is Archimedes number, Dp is particle diameter, ρ_p is particle density, ρ_f is fluid density, g is gravitational constant, μ is dynamic viscosity, Re is Reynolds number, U_{mf} is minimum fluidization velocity.

The results are recorded in Table 3.

TABLE 3. The resulting correlation values

Parametr	Unit	Value
Ar	[-]	49,275
Re_{mf}	[-]	0,042
U_{mf}	[m/s]	0,025

The actual velocity of the primary air at the entrance to the fluidization layer is 0,6 – 0,9 m/s, which is up to 30 times faster than the calculated minimum fluidization velocity. However, even at this speed, there is no flight of particles, due to the huge pressure drop on the filter disc and in the sand layer itself.

3.2 Combustion Regime

As you can see in Table 1, the elemental composition, the moisture ash content, and the related heating value is different for each fuel. The high diversity of heating value affects the combustion regime and therefore it is necessary to approach each fuel individually. It follows that an external heat source must be handled sensitively to heat the fluidized bed. For example, in the case of fuels with a heating value below 10, it is a problem to achieve autothermal combustion, so it is necessary to have the external heating switched on continuously or to use co-combustion with another fuel.

The graphs below show the course of temperatures along the height of the fluidized bed and the concentration of emission compounds in the flue gases of individual fuels. Concentrations were monitored CO, NO_x (sum of NO, N_2O a NO_2), HCl, SO_2, CO_2, O_2 and total mercury. All experiments took place for 30 minutes after the onset of the combustion process.

Straw pellets are among the heatest fuels and therefore the temperature in the fluidized bed sometimes exceeded up to 900 ° C. Figure 9 (a) shows that homogeneity was not sufficiently guaranteed, and unstable combustion occurred, indicating a slightly increasing CO and 100 mg/m^3 limit. One of the causes of combustion instability was the larger size of the individual pellets. Their length was up to 40 mm, which could lead to incorrect fuel dosing.

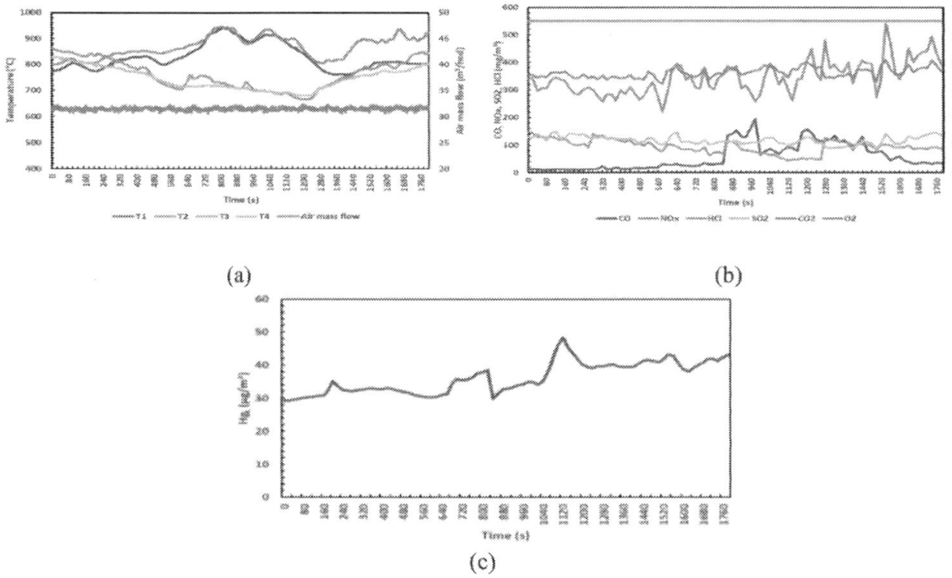

Fig. 9. Straw pellets combustion (a) The course of temperatures in the combustion chamber and air mass flow (b) Concentration of emission compounds of flue gases (c) Mercury concentration of flue gases

Sewage sludge is the fuel with the lowest heating value and therefore, as you can see from Figure 10 (a), the combustion temperature reached a maximum of 750 °C, up to the upper floors of the combustion chamber. It follows that external heating had to be active throughout the combustion process. The fact that the temperature was higher in the upper floors of the combustion chamber means that it was in the upper floors that the generated CO burned down. Sewage sludge also has a higher mercury content, which is clearly seen in Figure 10 (c), where the maximum value of total mercury was 142 $\mu g/m^3{}_N$.

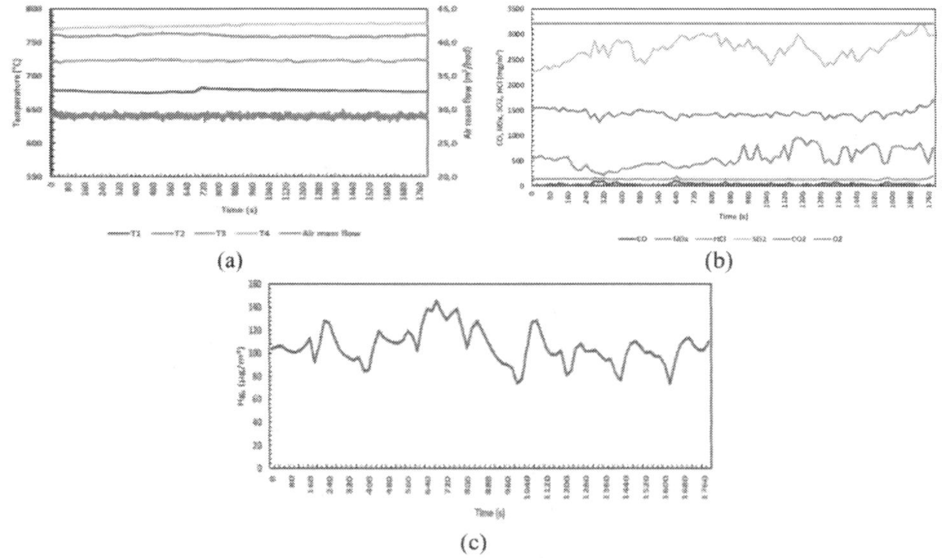

Fig. 10. Sewage sludge combustion (a) The course of temperatures in the combustion chamber and air mass flow (b) Concentration of emission compounds of flue gases (c) Mercury concentration of flue gases

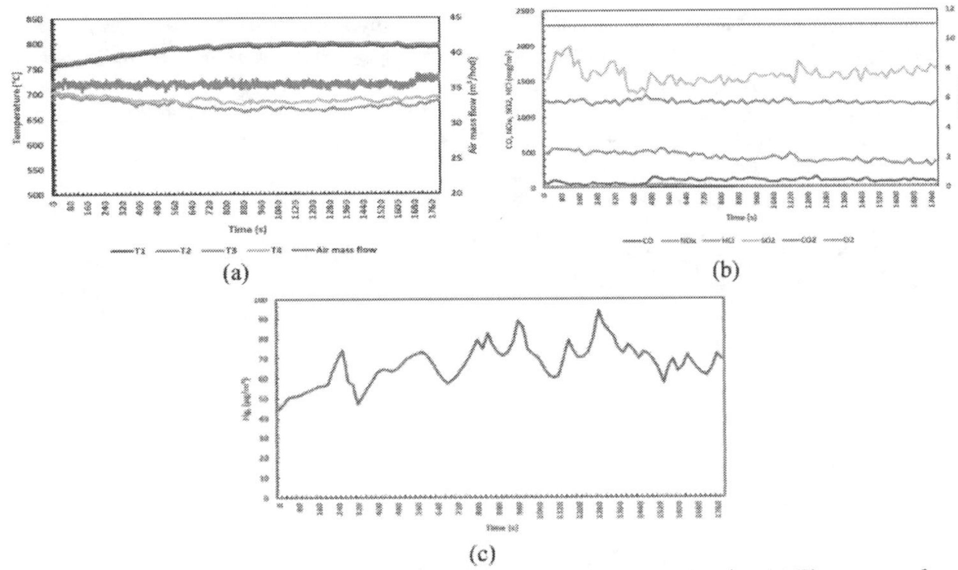

Fig. 11. Mix of sewage sludge and paper mills sludge with ratio 4:1 combustion (a) The course of temperatures in the combustion chamber and air mass flow (b) Concentration of emission compounds of flue gases (c) Mercury concentration of flue gases

This first mix with a ratio of 4:1 already has a higher heating value than pure sewage sludge, so the combustion takes place at a slightly higher temperature. Also, the ash content of this fuel is lower than that of a pure sewage sludge, which is why the fluidized bed is not so clogged with ash and thus degraded. The NO_x value was slightly lower than the previous fuel, while the SO_2 value was relatively significantly lower.

The second of the 2:1 mix has the lowest mercury content, as shown in Figure 12 (c). At the same time, it has the highest calorific value of all sludge incinerated, so it can be said that it is the most suitable for incineration. The combustion temperature is kept almost constant below 1000 ° C, which is accurate for fluidized bed combustion.

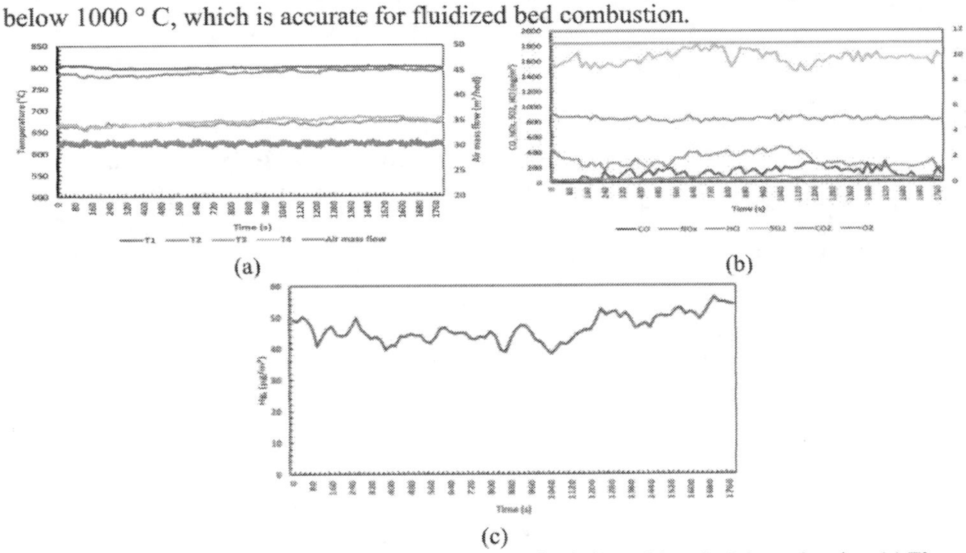

Fig. 12. Mix of sludge wastewater plants and paper mills sludge with ratio 2:1 combustion (a) The course of temperatures in the combustion chamber and air mass flow (b) Concentration of emission compounds of flue gases (c) Mercury concentration of flue gases

Lignite is one of two types of coal that are combusted in the experimental unit. Thanks to its favorable granulometry (0,71 – 2 mm) and high heating value, it is a fuel with quality properties, as can be seen in Figure 13. Thanks to this granulometry, it is possible to use a screw conveyor with a lower pitch. The combustion temperature is almost constant, at the limit of 1000 ° C as well as pollutant emissions.

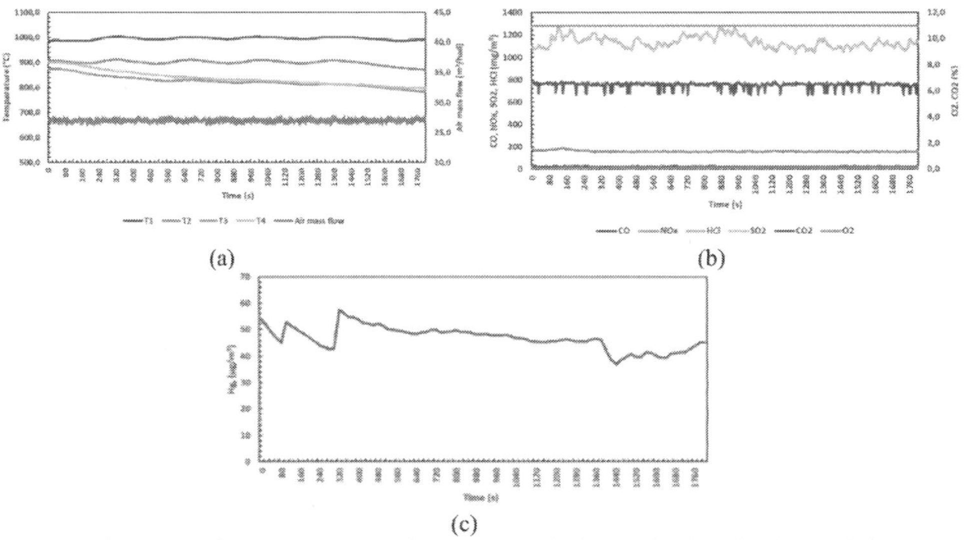

Fig. 13. Lignite combustion (a) The course of temperatures in the combustion chamber and air mass flow (b) Concentration of emission compounds of flue gases (c) Mercury concentration of flue gases

Coal is a fuel with the highest heating value, for which it is very difficult to maintain constant combustion. As you can see in Figure 14 (a), the primary combustion air flow had to be increased three times during combustion. This increase is due to the high consumption of oxygen, when the oxygen value decreased very rapidly and there was a possibility of a rapid increase in CO.

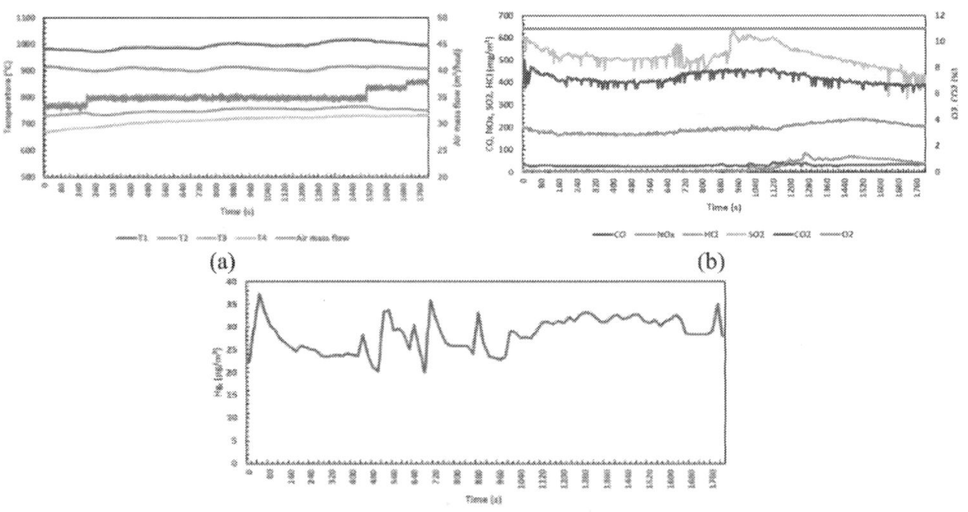

Fig. 14. Coal combustion (a) The course of temperatures in the combustion chamber and air mass flow (b) Concentration of emission compounds of flue gases (c) Mercury concentration of flue gases

Table 4 summarizes the average combustion temperatures, air mass flow and concentration of emission compounds for each fuel.

TABLE 4. Conclusion of average combustion parameters for each fuel

Parametr	Unit	Straw pellets	Sewage sludge	Mix 4:1	Mix 2:1	Lignite	Coal
Temperature	[°C]	762,8	718,9	788,2	793,5	947,2	950,6
Air mass flow	[m^3_N/hod]	31,4	29,1	37,5	30,5	27,1	35,0
CO	[mg/m^3_N]	56,4	41,9	92,7	126,3	23,0	30,0
NO_x	[mg/m^3_N]	345,0	554,4	435,8	295,8	158,2	194,5
HCl	[mg/m^3_N]	94,6	141,9	17,8	58,9	9,2	24,3
SO_2	[mg/m^3_N]	117,5	2721,6	1586,7	1637,1	1134,3	516,1
CO_2	[%]	7,5	5,3	5,7	5,0	6,4	7,2
Hg	[µg/m^3_N]	36,1	105,9	52,0	46,5	46,8	28,4

4 conclusion

This article focused on the combustion of alternative and convection fuels in an experimental combustion unit with a stationary fluidized bed. For the correct design of the unit, calculations were performed to determine the flight speed. Six different types of fuel with variable calorific values in the range of 9,83 – 30,53 MJ/kg were tested on the unit. In all cases, the ability of equipment with certain limitations to burn fuel was demonstrated, both in the form of pellets and in bulk form. All measured values were interpreted in Figure 9 - 14. All average measured values were processed and interpreted in Table 4 for better clarity. Due to the reduction of the measurement deviation, only the so-called Mercury total was measured, as the sum of the elemental and oxidized forms of mercury. In such a small unit with a not very stable combustion, large inaccuracies occurred during the measurement of individual forms of mercury, which led to invalid dates. It has been found that the top two external electrical heaters need to be switched on throughout the combustion due to the afterburning of the CO produced. The biggest problem for the unit was more ash fuels, as over time the fluidized bed filled with ash, which led to its degradation.

This study was supported by the student project No. SP2021/110

References

1. YANG, Wen-ching (ed.). Handbook of fluidization and fluid-particle systems. CRC press, (2003).
2. OKA, Simeon. Fluidized bed combustion. CRC press, (2003).
3. ANTHONY, E. J. Fluidized bed combustion of alternative solid fuels; status, successes and problems of the technology. Progress in Energy and Combustion Science, (1995), 21.3: 239-268.
4. FANG, M., et al. Experimental study on rice husk combustion in a circulating fluidized bed. Fuel processing technology, (2004), 85.11: 1273-1282.

5. WERTHER, J.; OGADA, Ti. Sewage sludge combustion. Progress in energy and combustion science, (1999), 25.1: 55-116.
6. MAGDZIARZ, Aneta, et al. Properties of ash generated during sewage sludge combustion: A multifaceted analysis. Energy, (2016), 113: 85-94.
7. KIJO-KLECZKOWSKA, Agnieszka, et al. Mechanisms and kinetics of granulated sewage sludge combustion. Waste Management, (2015), 46: 459-471.
8. HAGMAN, Henrik; BACKMAN, Rainer; BOSTROM, D. Co-combustion of animal waste, peat, waste wood, forest residues, and industrial sludge in a 50 MWth circulating fluidized-bed boiler: ash transformation, ash/deposit characteristics, and boiler failures. Energy & fuels, (2013), 27.10: 5617-5627.
9. RAJCZYK, Rafał, et al. Co-combustion of municipal sewage sludge and hard coal on fluidized bed boiler WF-6. Archives of Environmental Protection, (2014), 40.3.
10. KOORNNEEF, Joris; JUNGINGER, Martin; FAAIJ, André. Development of fluidized bed combustion—An overview of trends, performance and cost. Progress in energy and combustion science, (2007), 33.1: 19-55.
11. BAK, Jimmy; CLAUSEN, Sønnik. FTIR emission spectroscopy methods and procedures for real time quantitative gas analysis in industrial environments. Measurement Science and Technology, (2001), 13.2: 150.
12. ERXLEBEN, Holger; RUZICKA, Jaromir. Atomic absorption spectroscopy for mercury, automated by sequential injection and miniaturized in lab-on-valve system. Analytical chemistry, (2005), 77.16: 5124-5128.
13. STASIAK, Mateusz, et al. Mechanical and combustion properties of sawdust—Straw pellets blended in different proportions. Fuel Processing Technology, (2017), 156: 366-375.
14. BROADHURST, T. E.; BECKER, H. A. Onset of fluidization and slugging in beds of uniform particles. AIChE Journal, (1975), 21.2: 238-247.

Designing a top cooling system for an electromagnetic calorimeter

Matěj Jeřábek[1*], *Michal* Volf[1], and *Daniel* Duda[1]

[1]University of West Bohemia, Faculty of Mechanical Engineering, Department of Power System Engineering, Univerzitní 22, Pilsen, Czech Republic

Abstract. The article describes a numerical simulation of flow in the cooling system of an electromagnetic calorimeter by analysing the temperature and pressure fields. Two fundamentally different approaches were used to analyse the pressure field - analytical 1D calculation and numerical 3D flow simulation. The article contains a detailed evaluation and description of individual analyses using the commercial software ANSYS 2020 R1.

1 Introduction

What is matter, where does it come from and what is its structure? These are questions that could be answered in experiments at the new Facility for Antiproton and Ion Research (FAIR), which will be part of the existing Helmholtz Centre for Heavy Ion Research (GSI). The complex includes, among other things, the HESR (High Energy Storage Ring) particle accelerator, which houses the PANDA (AntiProton ANnihilation at DArmstadt) detector.

The purpose of this detector is to conduct physical research on topics related to the strong interaction, exotic states and the structure of hadrons [1]. But why? The purpose is related to a phrase we may all have heard before, namely that the whole is not merely the sum of its parts. For some physical cases this theorem certainly does not hold, but for the proton this theorem takes on its true value, since the sum of the masses of the valence quarks in the proton is less than 2% of its total mass. The rest of the mass is the result of the kinetic and binding energy between the quarks due to the dynamics of the strong nuclear interaction [2].

In order to collect all the necessary information from antiproton-proton collisions, an EMC (Electro Magnetic Calorimeter) will be built as a part of the PANDA detector. The EMC will be able to provide accurate trajectory reconstruction, energy and momentum measurements and very efficient identification of the charged particles. The electromagnetic calorimeter works on the principle of scintillation, i.e., the radiation emitted by the particle collisions reacts with a scintillator that produces a series of flashes of different intensities [3]. These flashes are captured by electronics and transformed into a digital signal. The EMC uses hundreds of cube-shaped scintillators made of $PbWO_4$, whose scintillation properties are strongly dependent on temperature. Since the light yield from the scintillator crystals increases with decreasing temperature, it is essential to keep these crystals at very low temperatures. In addition, the noise produced by the amplification and subtraction technique

* Corresponding author: jerabeekm@gmail.com

also decreases with decreasing temperature. Along with the flash intensity, the decay time also varies with temperature [3]. Thus, efforts are made to keep the temperature of the crystals as stable as possible.

This paper briefly describes the design of a cooling system that would allow low crystal temperatures to be maintained, and homogeneous distribution and stability of the temperature over all the crystals while the detector is in operation. Furthermore, the paper will describe the analysis of the effect of the pressure drop of the whole cooling system, from which the design and requirements for the main pump of the system are then derived. Two approaches have been used for the pressure drop analysis, namely an analytical 1D calculation and then a numerical 3D simulation, where the entire geometry of the cooling system is used for the numerical simulation, but without other components occurring in the calorimeter (crystals, segments, etc.). Only a part of the calorimeter geometry was selected for the temperature field analyses, but it was modelled in great detail.

2 Description of the panda detector

As already mentioned, HESR is one of the accelerators located in the newly built FAIR complex. Its main components are the PANDA detector and the Koala detector which is used to measure the differential cross section of antiproton-proton elastic scattering [2]. The PANDA detector is composed of several large systems, which can be seen in Fig. 1 [2]. However, in this article we will focus only on the calorimeter of which the proposed cooling system will be a part.

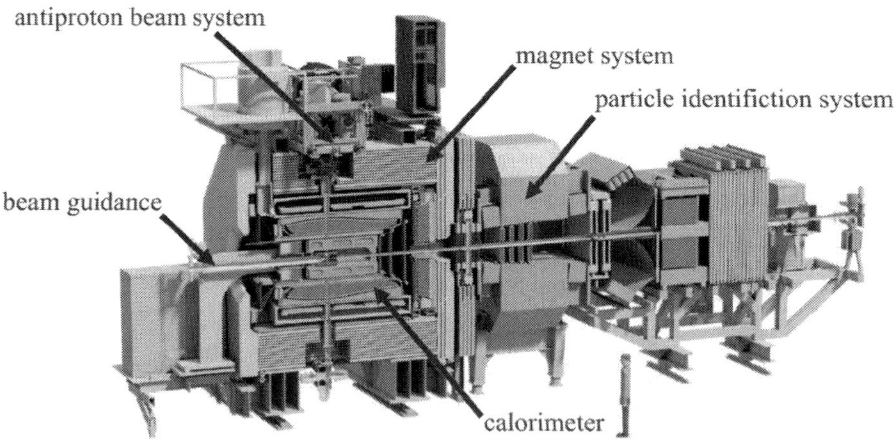

Fig. 1. PANDA detector model [2]

2.1 Description of the electromagnetic calorimeter

The calorimeter itself, see Fig. 2a, is made up of 16 segments assembled in the shape of a cylinder. Each segment contains 710 scintillation detectors (blue crystals) rotated to a common point where the particle collisions take place. In Fig. 2a we can also see the crystals located at the back of the calorimeter. However, the cooling system for these (green) crystals needs to be designed separately, as different requirements apply to it.

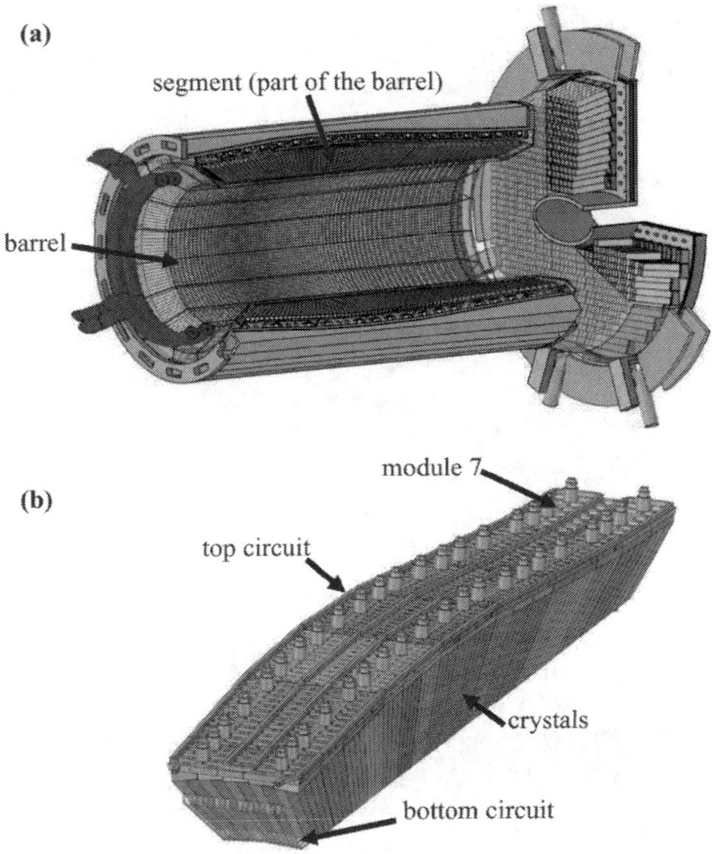

Fig. 2. Top (a): Electromagnetic calorimeter model [4]. Bottom (b): Segment detail

In Fig. 2b we can see the segment without covers and measuring devices, which is divided into seven colour-coded modules. It should be pointed out that the individual modules differ in the inclination of the crystals and also in their number. For subsequent analyses, it is also important to mention that the last module, number 7, has revolving tubes of cooling circuits and therefore it can be assumed that there will be a slightly more intense heat transfer than the other modules.

In addition, a preliminary design of the cooling circuits can be seen in Fig. 2b. It can be seen that each segment will have an upper and a lower cooling circuit.

3 Cooling system overview

The primary task of the cooling system is to maintain a very low, stable temperature and homogeneous temperature distribution over all the crystals.

The cooling capacity requirements of the system are determined by the heat sources present in the calorimeter segments. These sources are the measuring or sensing electronics chips, which can be seen in Fig. 3a, and the bus cables, see Fig. 3b. As an additional source of heat, it is necessary to consider the heat that enters the measuring compartment via conduction from the surroundings.

As mentioned, the segments have two separate cooling circuits. Therefore, when designing the cooling system, two separate cooling systems will have to be designed. The upper refrigeration system will consist of the upper refrigeration circuit, as well as the distribution piping that will take care of the refrigerant supply and discharge from the upper refrigeration circuits of each segment. The lower cooling system will consist of the same components as the upper cooling system, except that it will supply and discharge refrigerant from the lower cooling circuits. The second main task is to design the cooling systems depending on the total pressure drop, which will ultimately influence the selection of the main pump of the system. The cooling system is schematically shown in Fig. 3c.

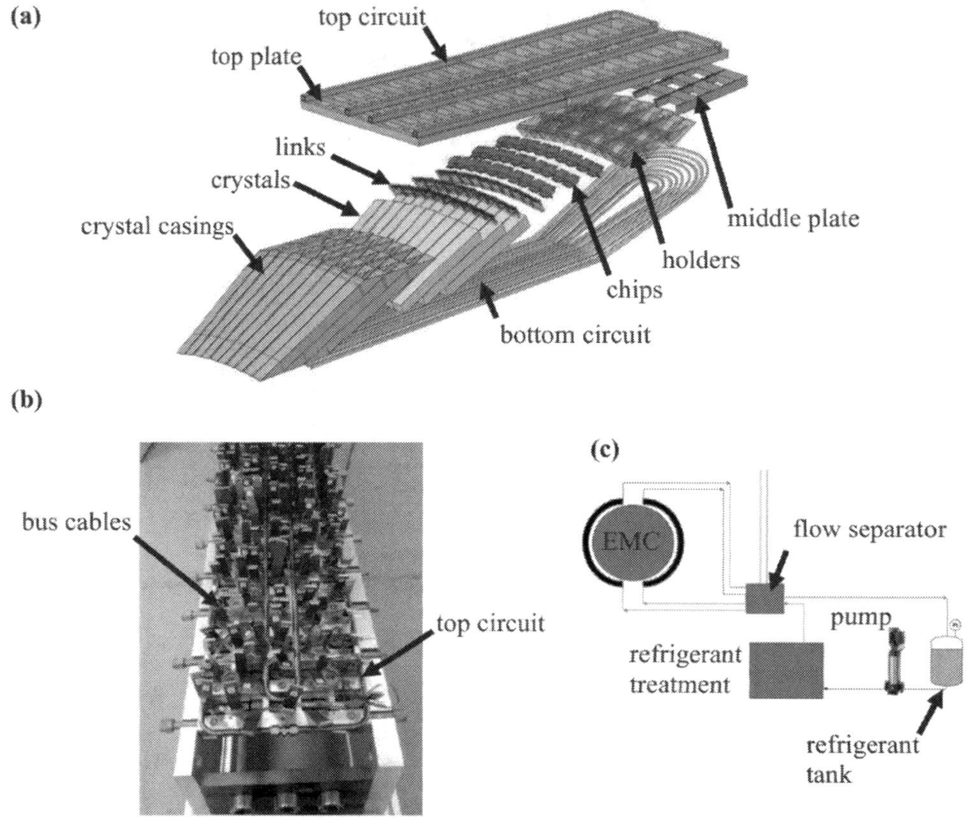

Fig. 3. Top (a): Parts of the module 7. Bottom (b): Preassembled slice [3]. Bottom (c): Cooling system schematics [5]

3.1 Cooling system requirements

In terms of temperature field distribution, the requirements are as follows. The operating range of temperatures in which the crystals must move while the calorimeter is running is in the region -30 °C to -20 °C. The stability and homogeneity of the temperature field of the crystals is then determined by the requirement that the maximum temperature difference of all the crystals must be within 1 °C.

In terms of pressure drop, the requirements for the cooling systems are specified in such a way that the pressure drop of the two cooling systems, i.e., the cooling system as a whole, should not exceed 1 bar at the required coolant flow rate of 2.78 kg/s. This condition follows from the requirement for the pump performance and operating costs. A summary of the initial requirements for the whole cooling system can be found in Table 1.

Table 1. Summary of requirements [5]

t [°C]	Δt [°C]	Δp [bar]	m [kg/s]
-30 to -20	≤1	≤1	≤2.78

3.2 Cooling system working medium

In order to meet the above requirements, both systems will operate with a 40/60 mix of water and methanol as the coolant. This type of refrigerant is specified by the client, however other mixtures such as water and ethanol could also be considered. The following limiting criteria apply to all types of refrigerants. The refrigerant must maintain a liquid phase even at the very low operating temperatures mentioned. Furthermore, the viscosity of the refrigerant also plays a role in the selection of the refrigerant, as this will ultimately affect the resulting pressure drop of the system, i.e., as the viscosity of the medium increases, the pressure drop increases. Finally, health risks, price, flammability, etc. also play a role in the selection of the refrigerant. The material properties of the refrigerant used are shown in Table 2.

Table 2. Material properties of the refrigerant [5]

Medium	c_p [J/(kg·K)]	ρ [kg/ m³]	μ [Pa·s]	λ [W/m·K]	M [kg/mol]
water/methanol (40/60)	3151	930	$7.7e^{-3}$	0.341	26.5

Where c_p is the specific heat capacity of the medium, ρ is the density of the medium, μ is the dynamic viscosity, λ is the thermal conductivity coefficient and M is the molar mass.

3.3 Cooling system design methodology

The requirements meant that we approached the design of the cooling system from two perspectives. First of all, the design was based on a numerical simulation performed by researchers at UWB [3], where the bottom circuit was included in the initial simulation of the temperature field and the heat transfer between the individual cooling circuits and crystals was evaluated using this mathematical model.

Secondly, the paper deals with the design of the distribution piping and the implication of this piping in the preliminary design of the cooling circuits, so as to create a complete cooling system that will ensure the transport of the medium to the cooling circuits and then its drainage back to the tank. The system was analysed using two different computational procedures. Firstly, a 3D simulation was performed using CFD (Computational Fluid Dynamics) on a realistic pipe. However, due to the difficulty of performing the simulation (long calculation time, computational complexity, etc.), this simulation was not very suitable for analysing the effect of the changes in the design of the system (number of pipes, change of cross-section, change of dimensions, etc.) on the magnitude of the pressure loss. Therefore, a fast and sufficiently accurate 1D computational tool was developed that would allow a rapid analysis of the effect of changing the system geometry on the pressure drop. This 1D tool will be verified by performing a detailed CFD calculation as experimental data is not available.

4 Cooling system design in terms of temperature field

The computational domain for the numerical simulation of the temperature field was primarily based on the study above [3]. At the outset, it should be added that the geometry of the barrel is quite complicated and simulating the entire barrel would be too time consuming, so we tried to greatly simplify the domain for the analysis. Since each of the segments has its own cooling system, i.e., there is only one cooling system with parallel tubes for each section, it is sufficient to simulate the cooling process in only one section i.e., for one segment instead of the whole barrel. As already mentioned, the segment is composed of seven modules. However, simulating all the modules would again prohibitively increase the computational complexity of the entire simulation. Therefore, we tried to simulate only a part of the geometry in such a way that all the phenomena related to the physical description of the problem are captured and the results are then applicable to the whole geometry. This ultimately allows us to significantly reduce computation time and computational requirements.

In the study [3], it was decided to simulate only the last two modules. This is based on the assumption that module 6 is very similar in design and topology to the other modules, and that module 7 has a reversal of the coolant flow and is therefore expected to have a different heat transfer compared to the other modules.

Taking into account the limitations, assumptions, and outputs of the study [3], a new computational domain was constructed to complement the default one with a bottom cooling circuit. The implementation of the new domain into the default domain can be seen in Fig. 4.

The geometry of the preliminary design of the lower circuit was used to prepare the complementary domain. Thus, 9 circular tubes with an outer diameter of 6 mm and an inner diameter of 4 mm were used. Since the heat transfer had to be considered for the temperature field analysis, the geometry of the tubes was kept in its entirety and was additionally supplemented with a domain representing the refrigerant, i.e., the inverse volume of the tubes, and a domain representing the ambient air.

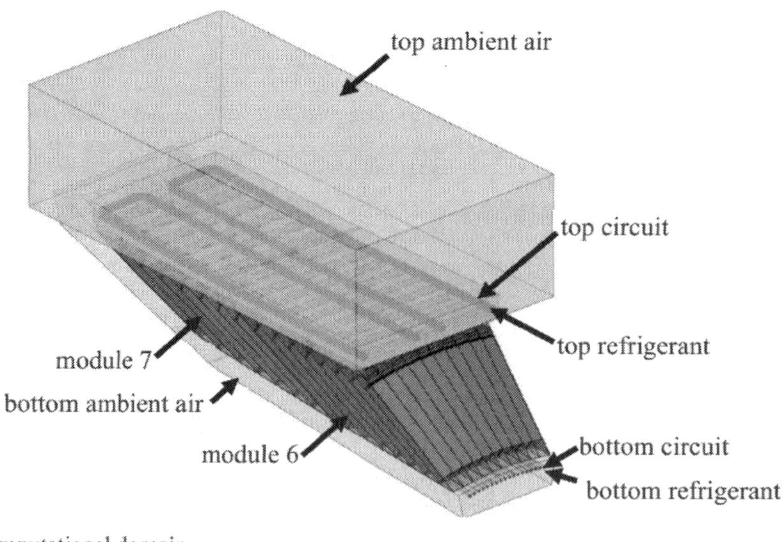

Fig. 4. Computational domain

4.1 Simulation setup (temperature field)

When setting up the numerical simulation, the following inputs were entered into the calculation: the reference pressure of the refrigerant domain was set to 0 bar, so it was not necessary to distinguish between relative and absolute pressure when defining the pressure boundary conditions at the outlet. The turbulence model was chosen to simulate the shedding in the tubes of the circuits by SST k-"ω" ("Shear Stress Transport") with a turbulence intensity at the inlet of 5%. Gravity was also included in the simulation since the vertical differences in the tubes, especially in the lower circuit, are not negligible. Initial and boundary conditions are also needed to define the CFD simulation. Since this was a stationary simulation, only boundary conditions were used. The inlet coolant boundary condition was set to a temperature of -30 °C and a mass flow rate of 0.12 kg/s. At the outlet of the medium, the outlet pressure condition was set to 1.012 bar, corresponding to the outlet to the atmosphere. The medium used is the mixture of water and methanol (40/60) mentioned earlier, whose material properties are given in Table 2. The heat source in the domain is considered to be the chips of the measuring electronics, whose thermal power is estimated to be 150 mW [3]. This heat flux will be transferred to the inner walls of the top plate where the processors are expected to be located. The heat flux produced by the bus cables is considered to be 200 mW [3]. The individual components located in the domain can be seen in Fig 3a. As can be seen (Fig. 3a), the top circuit tubes are located up above the top plate, not around the chips, which are considered as a heat source. Thus, heat transfer is mainly by conduction through a number of components, and thus not only a suitable contact between them must be ensured, but also a lower temperature of the cooling medium to create a sufficient temperature gradient.

A related issue is the need to ensure turbulent flow in the otherwise long and straight sections of tubing where, if laminar flow were to occur, there would be a decrease in heat transfer, thereby increasing the temperature of the refrigerant at the wall. This would ultimately reduce the cooling effect and heat dissipation from the domain.

Finally, the ambient air temperature above the computational domain of the modules is set to +25 °C. This boundary condition simulates the heat transfer that enters the calorimeter via conduction from the environment. The ambient air temperature below the computational domain of the modules is then set to -25 °C, simulating the steady state condition that the calorimeter will be in during operation. In other words, the steady state can be understood to mean that in the middle, i.e., in the empty space between the segments, the calorimeter will be cool relative to the crystals. The air in the two domains representing the ambient air is considered as an ideal gas. The reference pressure of the ambient air domains is set to 1 atm. Since the ambient air domain has no inlet or outlet, the pressure level must be defined over the reference pressure.

4.2 Results (temperature field)

The results of the stationary CFD simulations of the temperature field show that the addition of the extra domain of the lower cooling circuit has a very positive effect on the resulting temperature field distribution in the crystals. The temperature contours within the computational domain constructed for the numerical temperature field simulations can be seen in Fig. 5.

While in the initial study [3] there was an affected region which meant that the requirement for a homogeneous temperature distribution over all crystals was not met when evaluating the results, this region is no longer present in the completed domain. This change is a major improvement in the temperature field results. From Fig. 5 it can be seen that the temperature difference between all the crystals is now 0.3 °C, and so it can be concluded that the requirement for a maximum temperature difference across all the crystals of up to 1 °C is satisfied. It can also be seen from Fig. 5 that the refrigerant in the upper circuit heats up significantly more than the refrigerant in the lower circuit. This is due to the fact that all the heat sources in the segment are located in the upper part. This implies that the lower circuit may have a lower mass flow rate than the upper circuit. Furthermore, it can be concluded that the preliminary design of the lower circuit is satisfactory for the required crystal cooling.

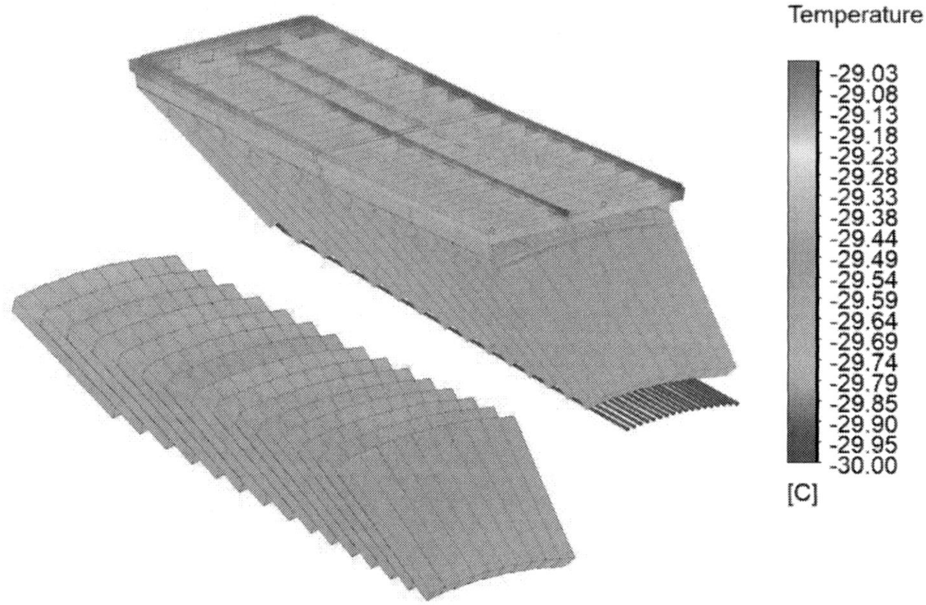

Fig. 5. Temperature field distribution

5 Cooling system design in terms of pressure loss

The preparation of the computational domain is different for each of the two design approaches mentioned above. For the pressure loss analysis, the complete domain of the cooling system is considered, i.e., the cooling circuits on all the segments, including the supply and discharge distribution piping. The distribution piping was modelled for the upper and lower circuits. The implementation of the geometry of the distribution piping and individual circuits in the EMC model can be seen in Fig. 6 (a). The systems themselves can be seen in Fig. 6 (b) and (c).

(a)

(b)

(c)

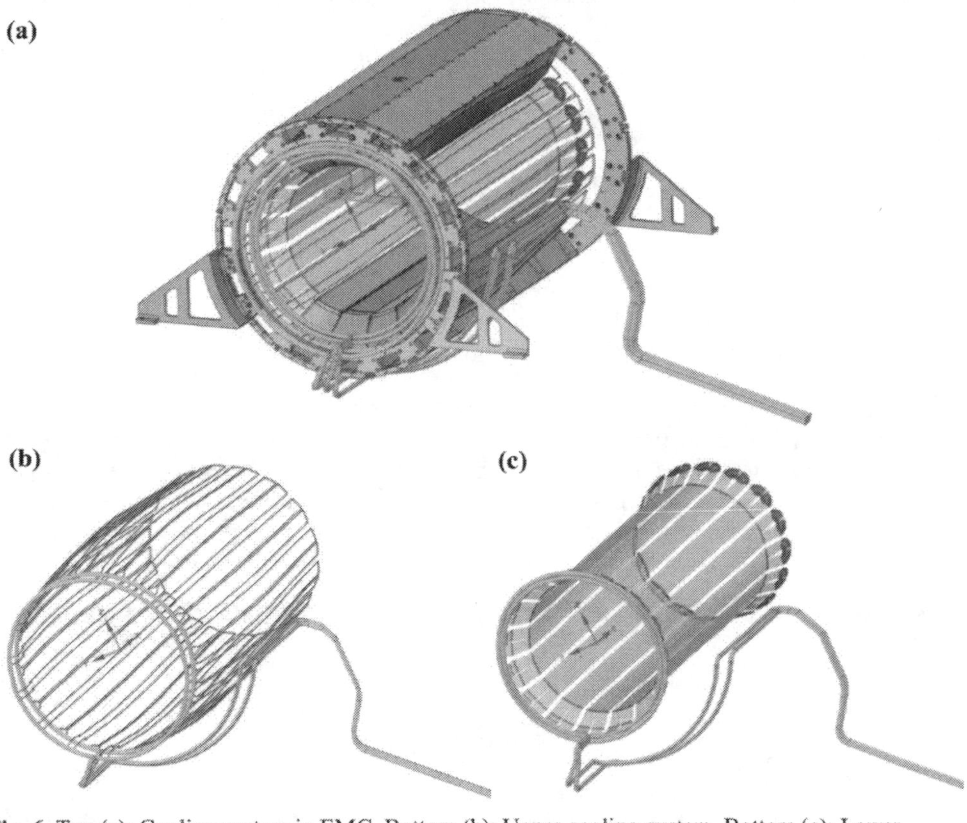

Fig. 6. Top (a): Cooling system in EMC. Bottom (b): Upper cooling system. Bottom (c): Lower cooling system

This setup was further simplified for the CFD simulation by neglecting all the structural components with no influence on the numerical simulation of the pressure drop. It should be noted that the structural components may be neglected because heat transfer is not simulated. Thus we created the upper and lower cooling systems.

As the heat transfer will not be evaluated in the numerical simulation of the flow field, the geometry of the upper and lower systems is further simplified by creating an inverse volume that is occupied by the cooling medium, i.e., neglecting the material of the tube structure.

5.1 Computational mesh

In order to assess the independence of the CFD simulation on the computational network, the network needs to be verified first. A domain consisting of one simplified upper circuit tube and a simplified part of the distribution pipe was prepared. Simplifying the domain reduces the computational demands, reducing the simulation preparation time and thus provides us with the possibility of testing different computational networks and their influence on our results. The verification domain consisting of the components mentioned above can be seen in Fig. 7.

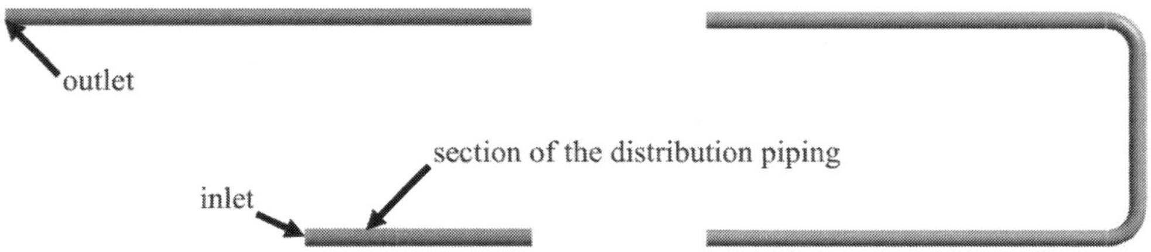

Fig. 7. Verification domain

Although the verification domain contains some geometrical and topological simplifications, it does correspond to the proposed pipeline. The subsequent simulation setup was applied to the verification domain prepared according to the specifications mentioned above. Tetrahedral elements of 3 mm with a first boundary layer thickness of 0.1 mm were applied to the part of the domain representing the distribution pipeline.

However, tetrahedral elements are not suitable in terms of stability of the computation and size of the computational mesh, so hexahedral elements with a size of 5 mm and with an identical first boundary layer thickness were used for the part of the domain representing the pipe. The conformal computational mesh set up according to the specifications can be seen in Fig. 8.

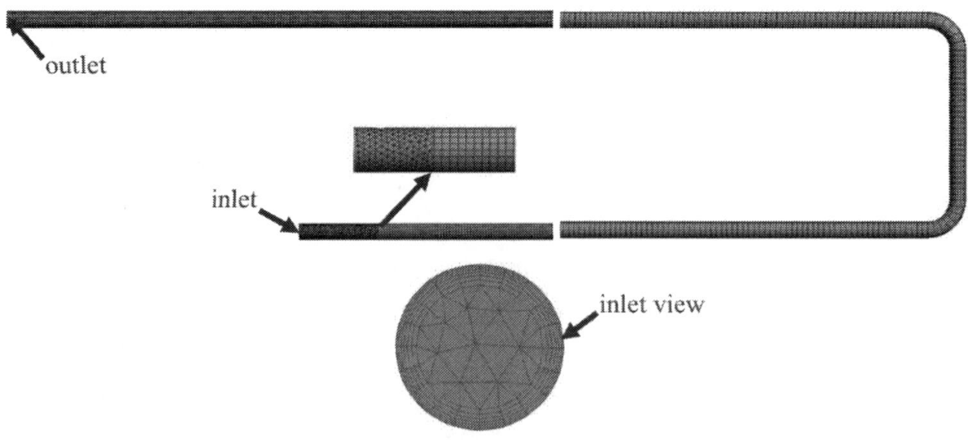

Fig. 8. Mathematical model of the verification domain

5.2 Simulation setup (pressure loss)

When setting up the numerical simulation, the inputs for the refrigerant entered into the calculation were similar to those used in the temperature field simulation. With the difference that the refrigerant mass flow rate was set to 3.84 m/s, the mass flow rate at the inlet to the cooling system thus set corresponds to a flow rate of 0.12 kg/s in one tube of the upper circuit. Another difference in the boundary condition setting is the refrigerant temperature, which has been set to -25 °C.

5.3 Results (pressure loss)

The results from the stationary CFD simulations of the flow field show that the customer's requirements for the total pressure drop of the cooling system were not met. As can be seen from Fig. 9a, the total pressure drop of the upper system is 3.9 bar. For the lower system, in Fig. 9b, the total pressure drop is 5.1 bar.

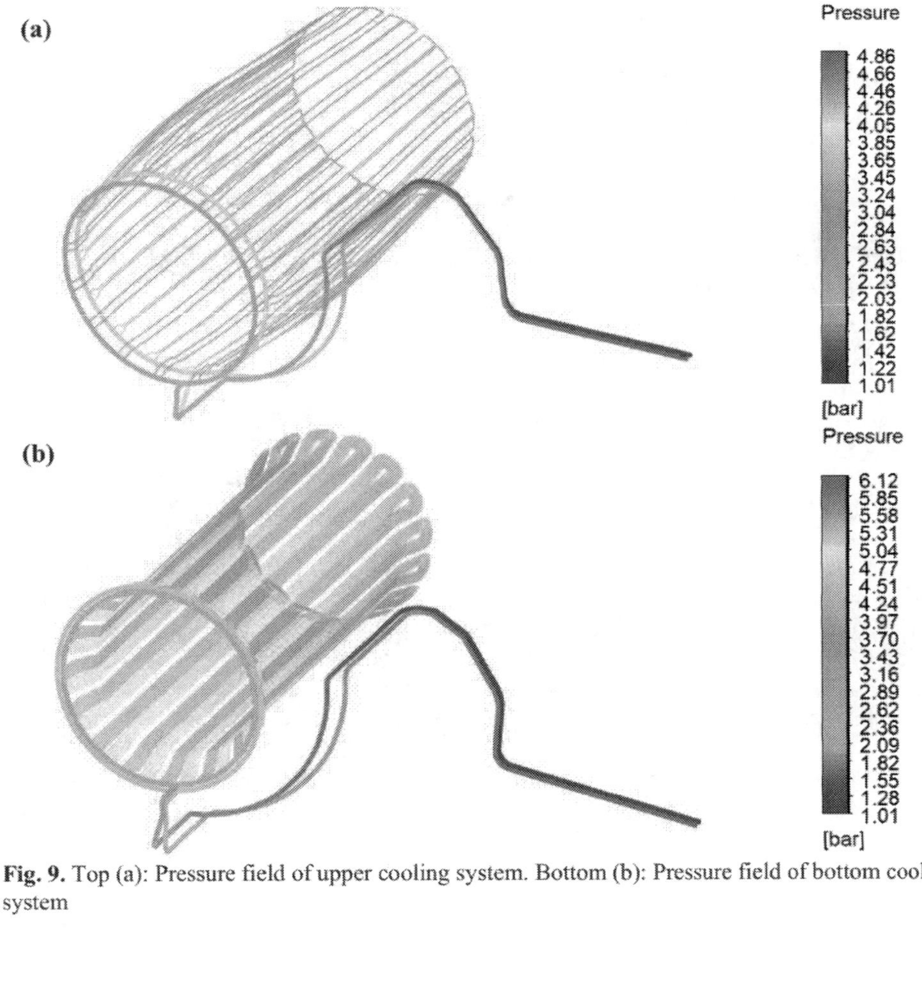

Fig. 9. Top (a): Pressure field of upper cooling system. Bottom (b): Pressure field of bottom cooling system

In the context of these results, it can be further stated that the customer's required pressure drop of the entire cooling system is an initial conceptual requirement. This consideration is also related to the finding that even the pressure drop of a single tube of the upper cooling circuit alone is not able to meet the requirements for a total pressure drop value of $\Delta p \leq 1$ bar.

This is because the pressure drop is mostly influenced by the mean velocity in the tubes. This velocity cannot, however, be reduced below the value of 2.5 m/s for a single tube, associated with the mass flow rate 0.12 kg/s, as lowering this velocity would lead to laminar flow in the tubes of the upper circuits, which is unacceptable from the point of view of the required heat transfer.

A compromise must therefore be made between the required heat output of the cooling system and the required pressure drop. This ultimately means that a significantly more powerful pump will be required than was assumed in the initial conceptual design.

6 Analytical calculation of pressure drop

The design also includes the creation of a tool to cover the design of the cooling system in terms of pressure losses with sufficient accuracy. For this purpose, we developed a 1D balance calculation based on known analytical formulas for local and frictional pressure losses. A major advantage of this calculation is a significant reduction in computation time, which is considerable for such a large task.

However, the balance calculation involves some simplifications that must be taken into account. Firstly, we neglect the elevation level at which the pipe is located, then we neglect the elevation difference between the inlet and the outlet of the medium. Furthermore, the possibility of different flow patterns in the pipe is not taken into account in the 1D calculation as we only consider one type of flow along the entire length of the pipe.

6.1 Cooling system variation analysis

The analysis was carried out in order to achieve the required (lowest possible) pressure drop of the whole system. It was done so that a reduction in pressure drop could be achieved either by reducing the length of the pipe or by increasing its diameter. This reasoning relates to the assumption that the dominant loss in the system is frictional loss, specifically, the local losses are approximately 10 times smaller for both cooling systems. Therefore, no further consideration will be given to changing the number or angle of the elbows in the distribution piping.

A graph of the dependence of the pressure drop on the diameter and the length of the pipe in the upper system can be seen in Fig. 10. It should be added that since we found that distribution pipe diameters ≤ 0.028 m have a very high pressure drop value, only diameters larger than this are included in the graph. It should also be added that the shortest distribution pipe length in the graph is 5 m, as the distribution pipe rings themselves are 4.7 m long. Thus, plotting a graph for lengths of 5-25 m corresponds to a significant number of distribution pipe design variations. The working point of the current design is plotted on the graph.

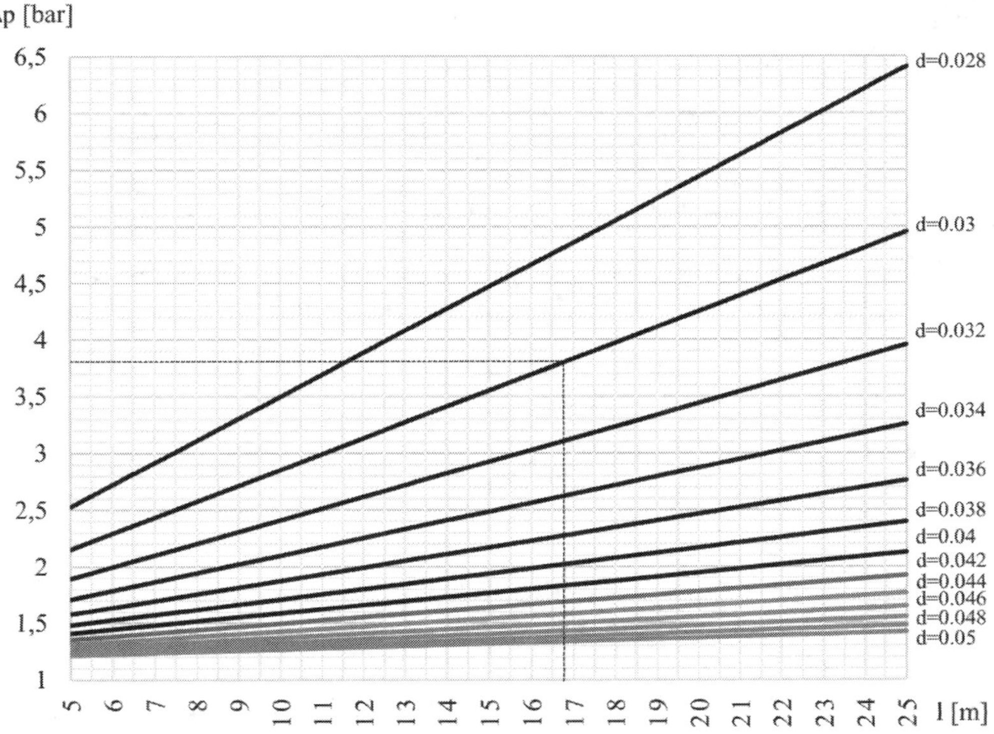

Fig. 10. Dependence of pressure drop on upper cooling system variation

7 Conclusion

The main task of this work was to design a cooling system that could keep scintillator crystals at a temperature between -30 °C and -20 °C during the operation of the calorimeter, so that the difference between the maximum and minimum temperature of all the crystals does not exceed 1 °C. From this point of view, this work was successful, as by adding a bottom cooling system to the initial computational domain, this condition was met, and the maximum temperature difference across all the crystals is now 0.3 °C.

In terms of pressure drop, the customer's requirements have not yet been met, but the requirement that the total pressure drop in the cooling system should not exceed 1 bar appears to be unrealistic, as even the pressure drop of a single tube in the system is approximately 1.13 bar. In view of this fact, the initial requirements have to be reconsidered and modified to take into account the priorities in the system design, which is temperature stability.

Lastly, it should be mentioned that this work also included the creation of a 1D computational tool which, as it turned out, is quite accurately able to cover the design of the cooling system in terms of pressure losses. Although this tool contains some simplifications, it can be used for additional studies dealing with further optimization of the cooling system.

The presented work was financially supported by the student project SGS-2019-021 (Improving the efficiency, reliability and service life of power machines and equipment 5).

Computational resources were supplied by the project "e-Infrastruktura CZ" (e-INFRA CZ ID:90140) supported by the Ministry of Education, Youth and Sports of the Czech Republic.

References

1. MESSCHENDORP, Johan. JPS Conference Proceedings. The PANDA Experiment at FAIR-Subatomic Physics with Antiprotons. [Online] 2016. [Citated: 25. 3. 2021.] Available from: https://doi.org/10.7566/JPSCP.13.010016.
2. BELIAS, Tassos a PETERS, Klaus.. PANDA Annual Report. [Online] 2021. [Citated: 5. 1. 2021.] Available from: https://panda.gsi.de/system/files/user_uploads/k.peters/RE-MGM-2020-002.pdf.
3. VOLF, Michal. AIP Conference Proceedings. CFD Simulation of The Cooling System of a Calorimeter Detector. [Online] 2021. [Citated: 25. 3. 2021.] Available from: https://doi.org/10.1063/5.0041387.
4. Chlazení Elktromagnetického Kalorimetru PANDA s Českou Stopou. Plzeň : ZČU, 2019.
5. ROSIER, P. FOSWIKI. Definition of The Chiller of The Cooling Plant. [Online] 2014. [Citated: 27. 3. 2021.] Available from: https://panda-wiki.gsi.de/foswiki/pub/EMC/Cooling/EMC-COOLING-Plant-Chiller-140113.pdf.

MATEC Web of Conferences **345**, 00016 (2021) https://doi.org/10.1051/matecconf/202134500016
Power System Engineering 2021

Software for hydraulic design of axial-flow pump impellers and its manufacturing documentations

László Kalmár[1*], *György* Hegedűs[2], *Árpád* Fáy[1], and *Norbert* Szaszák[1]

[1]Institute of Energy Engineering and Chemical Machinery, Department of Fluid and Heat Engineering, University of Miskolc, H-3515 Miskolc, Hungary
[2]Institute of Machine Tools and Mechatronics, Department of Machine Tools, University of Miskolc, H-3515 Miskolc, Hungary

Abstract. This article presents a hydraulic design procedure for axial-flow pump impellers, followed by their manufacturing documentations, all in one easy-to-use software named AXPHD V2.0 (AXial Pump Hydraulic Design) developed by one of the authors (Kalmár). After the user determined pump duty, the software offers input data which may be changed interactively. The hydrodynamic singularity method is used to compute the blade profiles on cylindrical surfaces. If the velocity and pressure distributions are accepted, then the body model of the impeller is produced by AUTODESK INVENTOR PROFESSIONAL 2019. Full manufacturing documentation is prepared including shop-drawings for traditional production, numeric modules for CAM, and files for 3D printing. A photo of an impeller made by 3D printing closes the paper.

1 Introduction

This paper has been written in an attempt to combine the hydraulic design procedure of axial-flow pump impellers with the preparation of their laborious production documentation within one easy-to-use software package.

There are many pump makers who use their well-proven techniques for the design and manufacturing of their pumps. However, new problems may arise, new techniques may appear, and so even the oldest firms face changes. It may happen, for example, that for a new duty, pumps are required not with top efficiency, but with extra good cavitation performance, and with very safe operation for about 30 years. In coping with such extra cases, the designer may wish to have a new impeller design within hours. The software named AXPHD V2.0 developed by one of the authors may fulfil this task.

The package is also applicable for jobs in common industrial use e.g., for low head irrigation or drainage pumps.

The software is not suitable for pumps of extra high efficiencies since these require special research. However, the software may provide help even in such cases as an auxiliary tool reducing the time.

* Corresponding author: aramka@uni-miskolc.hu

Special application of the software emerges in teaching. Getting familiar with the software the student meets all important variables of pump design, and producing a design is a concise job involving both hydraulic and manufacturing aspects.

2 Computation model

The main data of an axial-flow impeller are shown on Fig. 1. Outer diameter of the blades is D_K, the inner (hub) diameter D_B. Impeller speed is constant n, and the volume flow rate through the pump is Q.

A flow channel F is shown in Fig. 2 with middle surface of radius r. The blade profile (to be designed) is also shown with absolute velocities $c_{F\sigma}$ (σ, φ) and $c_{F\varphi}(\sigma, \varphi)$, σ is axial coordinate, and φ is angle measured from the inlet edge of one blade. Incompressible, frictionless flow, with zero radial velocity is assumed in the flow channel. The blade cascade (at radius r) evolved onto a plane is seen in Fig. 3 with the velocity triangles (r, σ_L, L, T in metres, **u**, **c**, **w** in m/s). The computations are executed for the flow through these blade cascades using the singularity method [1, 4 - 7].

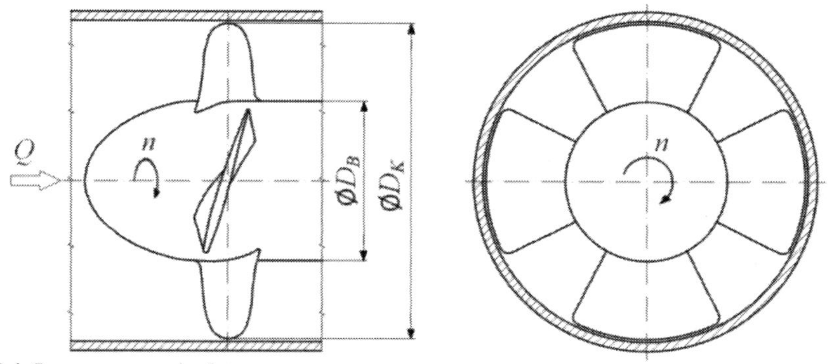

Fig. 1. Axial-flow pump, main data

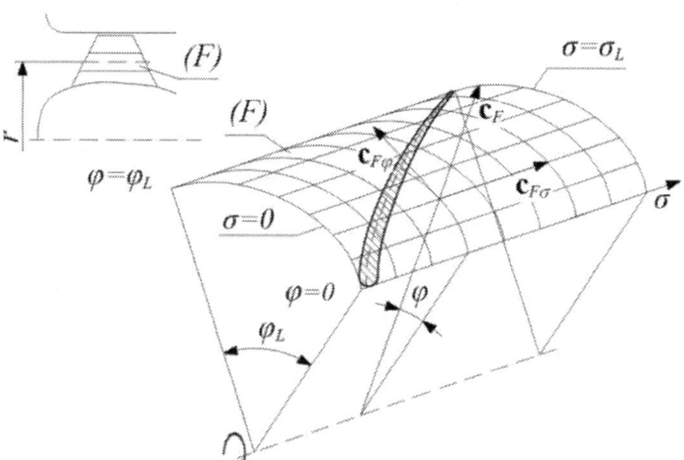

Fig. 2. Blade profile on the middle surface of a cylindrical channel [1,7]

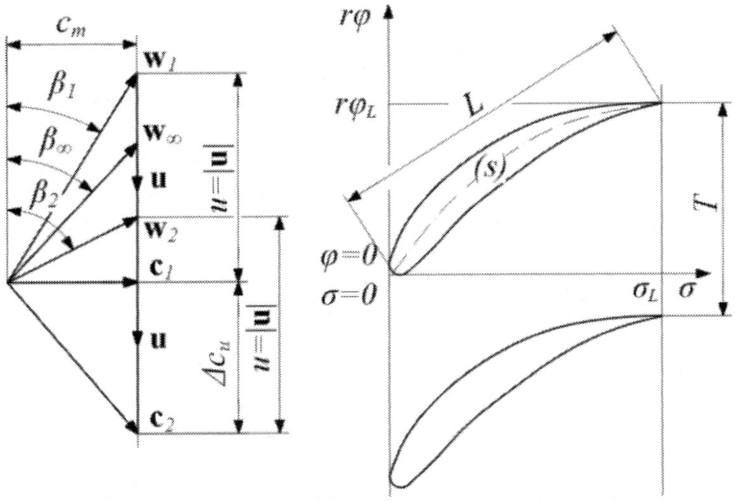

Fig. 3. Blade cascade evolved onto a plane, and the velocity triangles (**u** is peripheral velocity of the blade cascade at radius *r*, **c** and **w** are the absolute and relative velocities of the fluid, *1* at inlet and *2* at outlet respectively).

3 Main data

The main operating parameters of the axial-flow pump to be designed are:

- H: head of the pump [m],
- Q: volumetric flow of the pump [m^3/s],
- n: speed of the pump [rpm],
- ρ: density of the delivered fluid [kg/m^3],
- and the specific speed calculated therefrom is: $n_q = n\frac{\sqrt{Q}}{\sqrt[4]{H^3}}$.

Many axial-flow pumps work in practice with wide operating ranges. However, in order to simplify the design, the pump is associated with one single set of the above parameters called nominal or design parameters. The success of a new design depends largely on suitable selection of these parameters.

The usual head and discharge ranges of axial-flow pumps are:

$$2.5\text{ m} \le H \le 10\text{ m, and } 0.01\frac{\text{m}^3}{\text{s}} \le Q \le 10\frac{\text{m}^3}{\text{s}}.$$

Of course, there are known units operating at H = 20 m, but these are exceptional. The fluid is clean or raw water: ρ = 1000 kg/m^3. Some tender documents limit speed n because fear from excessive cavitation, and in selecting nominal data this must be respected. If n has not been fixed, using specific speed n_q is advised to be selected not too far from current industrial practice (Table 1):

Table 1. Suggestion for specific speed

H, meter	3	4	5	6	7	8
n_q, metric	330	270	230	200	170	140

Other design parameters are suggested by the software in Figs 4, 5, 6, based on industrial practice.

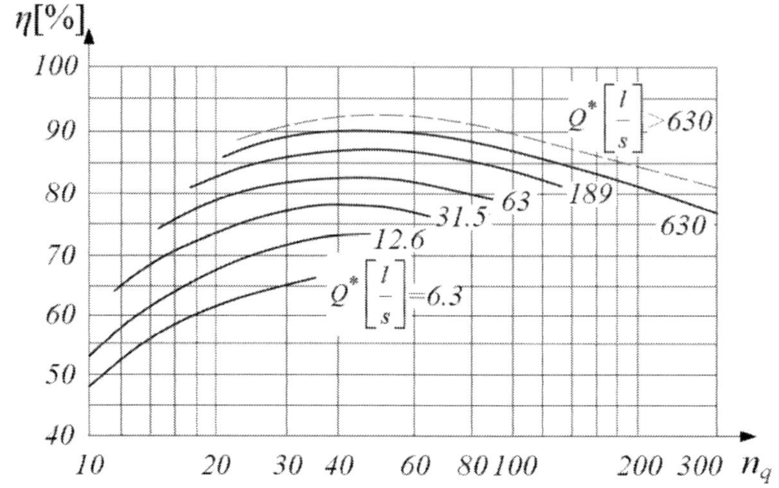

Fig. 4. Efficiency of the pump bowl [3]

The pump bowl includes impeller and guide vane row (η is pump efficiency, but without the loss of the discharge bend). The inlet flow coefficient ε is defined by: $c_m = \varepsilon\sqrt{2gH}$, where c_m is meridional flow velocity upstream of the blades. The outer diameter of the impeller D_K is calculated from $Q = \frac{(D_K^2 - D_B^2)\pi}{4} c_m$, while the hub diameter D_B accords with Fig. 6.

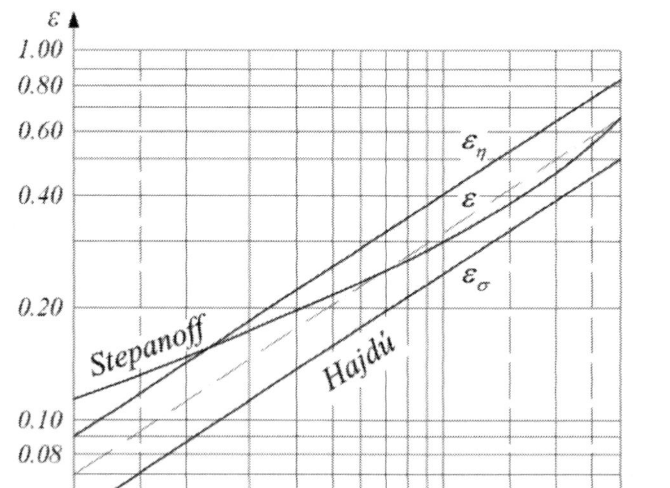

Fig. 5. Inlet flow coefficient [3], Stepanoff [8], Hajdú [9]

Figure 5 is used for suitable selection of ε. Using ε_η is recommended to achieve good efficiency, while ε_σ to reach good cavitation conditions, but the user may choose naturally another ε value.

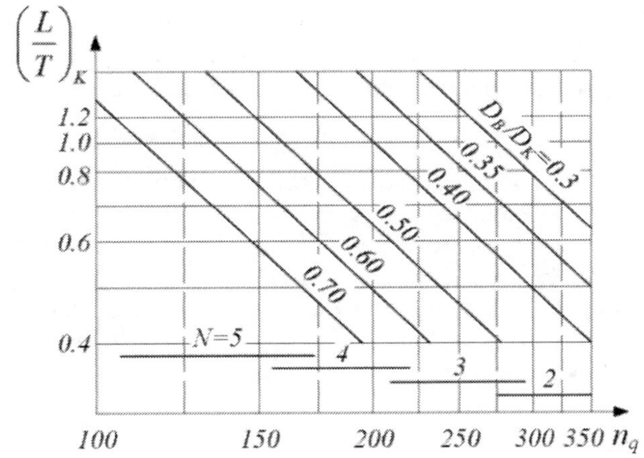

Fig. 6. Main geometric data of the impeller (*N* is number of blades) [3]

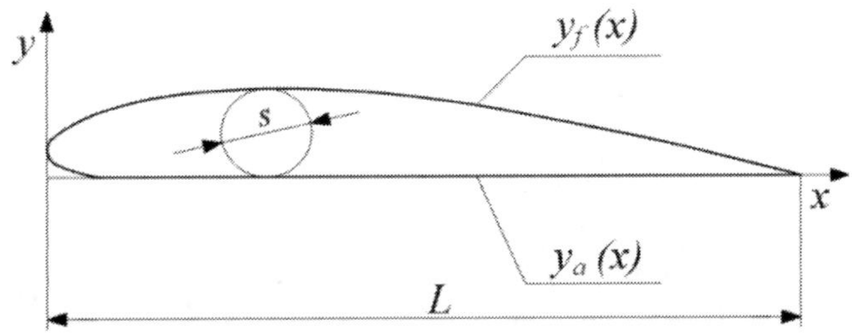

Fig 7. Profile section with maximum thickness

The software offers also values for $\left(\frac{L}{T}\right)_B$, $\left(\frac{s}{L}\right)_K$, $\left(\frac{s}{L}\right)_B$ (Fig. 7). The *peripheral velocity* is: $u = r\omega$, and Δc_u (Fig. 3) is calculated from: $u\Delta c_u = \frac{gH}{\sqrt{\eta}}$.

4 Computation of flow

For the flow (Fig. 2) conservation of mass and circulation-free conditions are required:

$$\frac{\partial c_{F\sigma}}{\partial \sigma} + \frac{1}{r}\frac{\partial c_{F\varphi}}{\partial \varphi} = 0 \qquad (1)$$

$$\frac{\partial c_{F\varphi}}{\partial \sigma} + \frac{1}{r}\frac{\partial c_{F\sigma}}{\partial \varphi} = 0. \qquad (2)$$

The partial differential equation system (Equs. (1) - (2)) together with the

$$\sigma = 0: c_{F\sigma} = c_{F1\sigma} = c_m; c_{F\varphi} = c_{F1\varphi} = 0$$

and

$$\sigma = \sigma_L: c_{F\sigma} = c_{F2\sigma} = c_m; c_{F\varphi} = c_{F2\varphi} = \Delta c_u(r)$$

boundary conditions is available to determine the unknown functions of $c_{F\sigma}$ (σ,φ) and $c_{F\varphi}$ (σ,φ) used to describe the velocity distributions along the middle surface (*F*) (see Fig. 2).

The middle surface of the flow channel (Fig. 2) evolved onto a plane, results in a straight blade cascade (Fig. 3) where the velocities have the same sizes as in Fig. 2. The camber lines of the profiles are computed (rendering ideal incidence flow to the impeller in pump nominal operating point). The singularities (vortex and source-sink distributions) are then placed on the camber line. The code offers a few distributions which the user may select. Thus, all parameters needed for the design are known, and the computations using the singularity method are carried out. The contour points obtained for the blades are then available, and non-dimensional velocity and pressure distributions are shown for approval. If accepted, the blade coordinates are used to build up the solid model of the impeller as shown below.

The code assumes frictionless flow. It may happen that the user intends to make a loss analysis (not included in the code, but any user's program can easily be adjoined). For such analysis the computed velocity field is easily accessible.

5 Software Modules

The AXPHD V2.0 design software was developed in the frame of Microsoft Visual Studio Community Edition using the C# language. The Code allows quick input, performs numerical algorithms, stores data, communicates to user written programs (Fortran, C/C, C++, VB, Java), and prepares many things for manufacturing the impeller. The Main menu consist of the following modules:

- Project management,
- Specification and definition of hydraulic and geometric characteristics,
- Implement the hydraulic design of the impeller,
- Development of the impeller's 3D body model,
- View and document the results.

Project management means selecting a new name or using one from the list of old names. Several impeller variants may be included in one project. Among others the following *functions* are used: *Input*, *Save*, *Save as*, *Archive*, *Delete*. The modules are left by pressing *Back to Main menu*.

Specification and definition of hydraulic and geometric characteristics. The module opens input Panels enabling easy identification of data. For impeller G162 (Fig. 11) designed for a model pump subject to hydraulic test, the following values were fed into Panel P1:

- Manometric head: H=4.4 [m],
- Volumetric flow: Q=0.437 [m^3/s],
- Speed of rotation: n=1150 [rpm],
- Density of fluid: ρ=1000 [kg/m^3].

The specific speed calculated therefrom becomes n_q=250 [rpm, m^3/s, m]. Other input data are fed using panels P2 – P7. After inputs are completed, output Panels may show the data. For G162 the main output parameters are: Number of blades: N=3, outer diameter: D_K=0.35 [m], inner diameter: D_B=0.154 [m]. Using 7 different radius r for the cylindrical sections has also been approved.

Implement the hydraulic design of the impeller. The singularity method computes the blade contours on the cylindrical sections only, and their final positions in the 3D space are determined as prescribed in the former module. Figure 8 shows the drawing of the hydraulic surfaces of the blade.

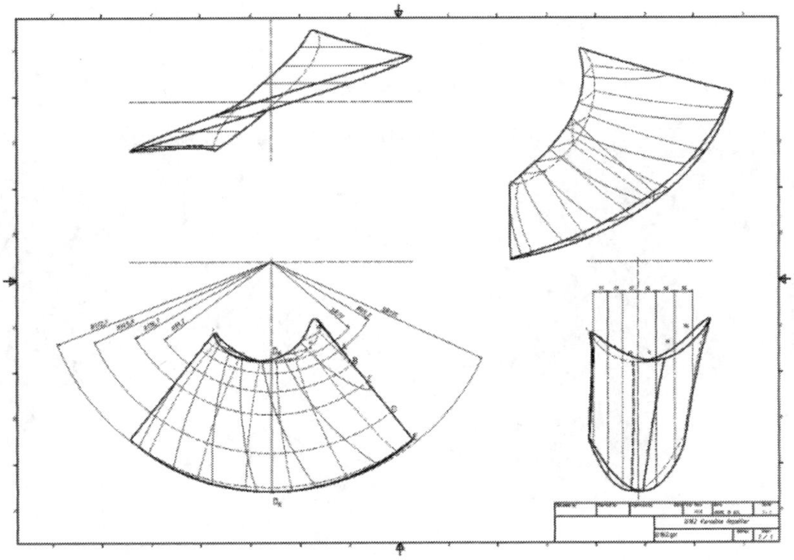

Fig. 8. Drawing of the blade hydraulic surfaces

Development of the impeller's 3D body model. This module offers first a list of possible documentations for selection. The blade is completed with parts joining to the hub (Fig. 9 a), which is also treated as a body ready for making (Fig. 9 b).

(a) (b)

Fig. 9. View of the body model of the blade

View and document the results. This module offers saving for any activities defined in the modules. Various libraries may be opened, and various formats (docx, 2Dpdf, 3Dpdf, stp) may be used.

6 Impeller produced by 3D printing

Files of format stp are suitable for producing 3D printing. The impeller G162, designed for a model pump (Fig. 10) by AXPHD V2.0, has been produced using 3D printing with the help of the specialists from KVINT-R Ltd. The accuracy proved to be excellent, even the threads produced by 3D printing, which fix the blades in the hub, worked appropriately with commercial nuts. This model pump impeller has been prepared for testing in a cavitation test rig of 350 mm impeller diameter.

The work outlined in this paper has been devoted to help pump design. The software is flexible, offers various options for general use, and moreover to these, major changes of the procedure may easily be introduced by user defined programs.

Fig. 10. Photo of axial-flow impeller designed by AXPHD V2.0 produced by 3D printing for model tests

This research was carried out as part of the GINOP–2.1.7-15-2016-00429 project with support by the European Union, co-financed by the European Social Fund.

References

1. Czibere, T.: The solution of the two main tasks of the hydrodynamic blade cascade computations by using the potential-flow theory (in Hungarian), Miskolc, (1965)., p. 154.
2. Czibere, T. – Kalmár L.: Softwares for turbomachines: solution of the first main task of cascade computations operating with compressible media (in Hungarian), Research report, Miskolc, (1975)., p. 54.
3. Czibere, T.: Turbomachines, lecture notes, (J14-500), Budapest, (1977).
4. Fáy, Á.: Simple performance model for pumps, ASME Symposium "Pumping Machinery 1993", Fluids Engineering Conference, Washington D.C. June 20-24. (1993).
5. Lewis, R.I.: Vortex element methods for fluid dynamic analysis of engineering systems, (2005), books.google.com.
6. Kalmár, L.: Computer aided design of axial pump impellers, Research report, Miskolc, (2014).
7. Kalmár, L. Hegedűs, G. Czibere, T.: Computer Aided Designing of Blading of the Axial Flow Pump Impeller, Pumps, Compressors, Vacuum Pumps, Hungary, vol. XXIII, no. 1, pp. 53–68, (2016).
8. Stepanoff, A.J.: Centrifugal and axial flow pumps, (1957), John Wiley and Sons.
9. Hajdú, S. Szalay, Gy.: Parameters for the qualifications of irrigation and drainage pumps, Papers on Water Management (1967), No. 4

Modelling fluid flow and heat transfer tasks for testing PIR motion Sensor

Jakub Kazda[1], *Richard* Matas[1*], and *Jan* Sedláček[1]

[1]New Technology – Research Centre (NTC), University of West Bohemia, Univerzitni 2732/8, 301 00 Plzen, Czech Republic

Abstract. The aim of this paper is to describe the methodology of modelling fluid flow and heat transfer tasks for testing electronic motion sensors, especially PIR (passive infra-red) motion sensors. The motion sensor measures changes in electromagnetic radiation in the infrared range caused by the movement of warmer objects compared to its surroundings. Distortion of the perception of the environment can lead to false alarms or insensitivity of the detector. PIR motion sensors should be designed and located to minimize heating of the lens by ambient air. The sensor may be exposed to conditions not specified in the basic standards. The practical part of this article deals with a non-standard test. A single-phase transient simulation in a specific test cabinet for two specific motion sensor models was performed using the computational fluid dynamics (CFD) method. The aim of these simulations was to evaluate the effect of hot flowing air on the lens and to compare the designs of both models. Comparison and analysis of the results showed that the test configuration does not guarantee a defined flow at the outlet of the pipeline, so a new configuration was designed.

1 Introduction

Nowadays, there are many ways to protect and secure property against fire or other damage (e.g. theft). One way is to use a PIR (passive infrared) motion sensor. A PIR sensor is a device working on the pyroelectric principle. The sensor measures different values of electromagnetic radiation in the infrared range, which enter through an optical lens to the pyroelement. Objects with a temperature higher than absolute zero (-273.15 °C) radiate their heat in the form of infrared radiation, which is invisible to the human eye but detectable to electromagnetic sensors. The size of the space monitored by the sensor is determined by its viewing angle. The pyroelement evaluates the change of infrared radiation, which leads to a change to the output voltage and, if necessary, an alarm is triggered [1, 2, 3].

PIR sensors are tested according to EN 50131-X [4]. In normal operation, the sensor may be exposed to conditions other than those specified in the standard, such as placement near windows, various heat sources, fans, air conditioners, or above heating elements. This topic is not yet reflected in the available information sources.

This article deals with a non-standard test whose configuration was inspired by specific measurements in a laboratory. A single phase transient simulation in a test cabinet for two

* Corresponding author: mata@ntc.zcu.cz

different models of motion sensors was carried out using computational fluid dynamics (CFD). The aim of these simulations was to evaluate the effect of hot flowing air on the lens and to compare the designs of the models. The second aim was the assessment of the quality of the test rig design in terms of stability of the velocity and temperature field. The project was therefore designed as a pilot to explore the potential of engineering simulations to identify the impact of off-design conditions on PIR sensors.

2 Geometry and computational domain

Two models of sensors marked Type 1 and Type 2 were chosen for simulations as shown in Figure 1. The computational domains around the sensors are determined by the dimensions of the test cabinet – see Figure 2. The arrangement of the cabinet does not directly comply with the EN standards; however, it provides stricter conditions for testing the sensors than the standards.

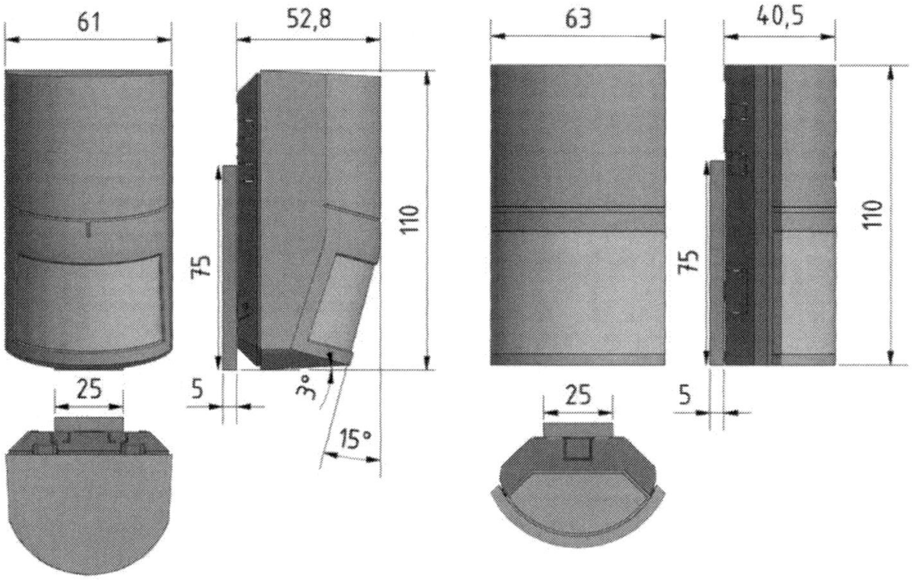

Fig. 1. PIR motion sensor models

The computational domain contains an air outlet into which the air flow enters through a 90° pipe and a washer for mounting the sensors. The aim of this task is not to monitor the heat transfer through the walls of the cabinet, so only the inner surfaces of the cabinet were considered in the simulations. It is appropriate to include the volume of the PIR body in the transient simulation, due to heat accumulation, as the thermal properties of the volume are important for the surface temperature to monitor the warming of the body surface. Therefore the volume of the lens and the entire detector was also modelled to monitor the warming of the lens surface (Figure 2).

The output surfaces (ambient atmosphere) are marked in green and the input area in red – the orange colour indicates the location of the BOI body (body of influence) for setting up better local discretization.

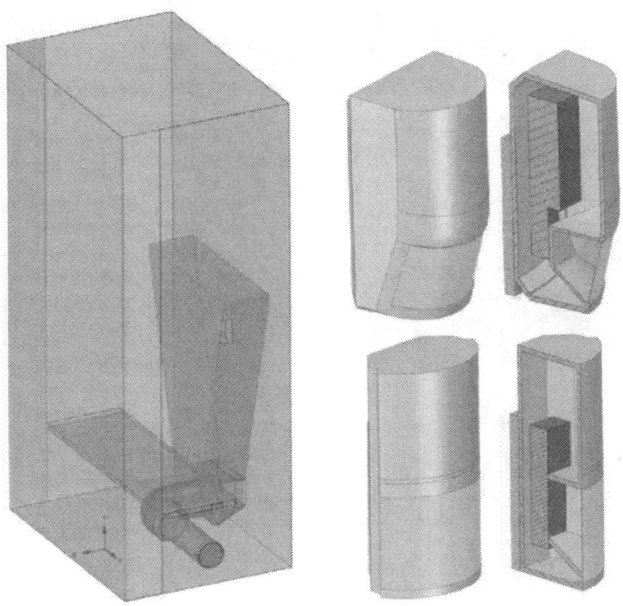

Fig. 2. Computational domain (left) and simplified models of sensor bodies (right)

3 Solver setup and boundary conditions

The CFD system ANSYS FLUENT 2020 R2 [5] was used for the simulations. The SST k-ω [6, 7] was chosen as a turbulent model, the energy equation was included as well. For the mesh independence study, a stationary type of calculation with averaging of values was chosen. A transient solver was chosen for the final simulations, because it is important to capture the vortices and fluctuations around the sensors.

The gravitational acceleration was set to 9.81 ms^{-2}. The material properties of all the elements were set according to the manufacturer's data. The flowing air was set as an incompressible ideal gas – the density depends only on the temperature.

A turbulence intensity of 2 % and a (Kolmogorov) turbulence scale of $4.8 \cdot 10^{-4}$ m were set. The inlet air temperature was set at 35 °C with a flow velocity of 0.7 ms^{-1}. The temperature was set to 20 °C and the pressure to 101.325 kPa at the outlet. A no-slip condition was set on the walls and the heat transfer coefficient was set to 10 W/(m^2K).

4 Computational mesh

The detector interior, solid bodies and whole test cabinet domain were meshed with conform polyhedral meshes, because these sections are geometrically suitable for these types of elements. The polyhedral elements are able to surround complex geometric shapes and have a high number of adjacent cells, so it achieves a generally higher accuracy and stability then tetrahedral elements.

In order to ensure the accuracy of the CFD results from the simulation, a mesh independence study was made. Nine mesh versions of the Type 1 model were created for this study. All the meshes satisfy the recommended quality limits for polyhedral meshes. The mean values and RMSE fluctuations of velocity magnitude and total temperature were monitored (in stationary simulations) at specified surfaces, see Figure 3.

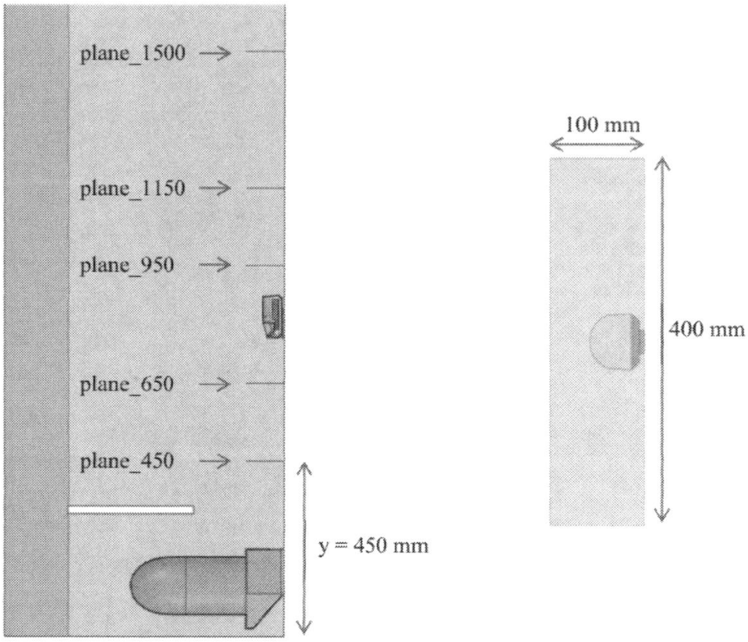

Fig. 3. Surfaces for evaluating the mesh independence study

The results of the mesh independence study are shown in Figures 4 and 5. It can be seen that the values obtained are quite strongly mesh dependent for the first five versions of the mesh and only weakly dependent for the next four versions. For the purpose of this work, the sixth version of the computational mesh with about 8.7 million polyhedral cells was chosen as the applicable discretization level.

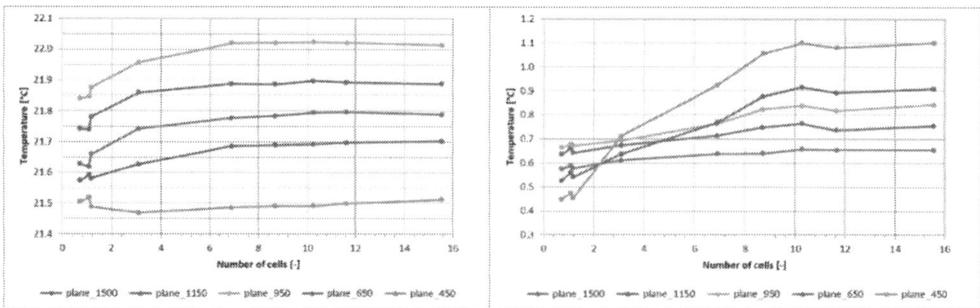

Fig. 4. Mean temperature (left) and RMSE temperature (right)

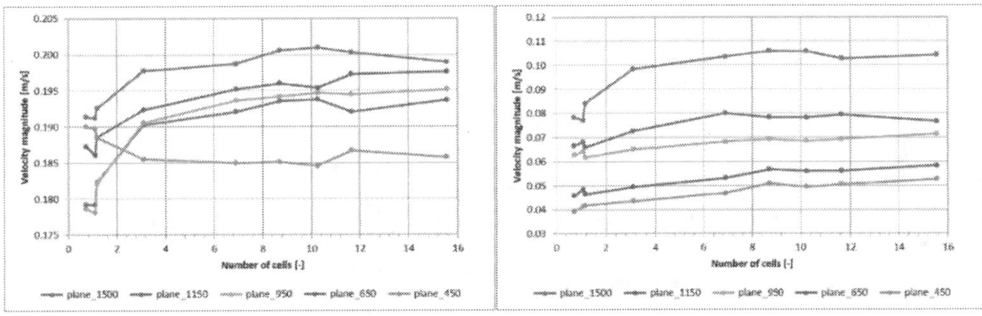

Fig. 5. Mean velocity magnitude (left) and RMSE velocity magnitude (right)

Below in Figure 6 is a preview of the selected mesh. After evaluating the optimal mesh, the same mesh generation setting was also implemented on Type 2.

Fig. 6. Computational mesh – detail and general views

5 Methodology of evaluating the results

Figure 7 shows the planes and points and their locations where important quantities were monitored. The layout is the same for both sensor models. The xy plane is at a distance of 685 mm from the front (in the middle of the pipe opening). The planes for the mesh study are also used for evaluation.

A steady-state simulation was used for the first run together with an additional 30 time steps of non-steady-state simulation. The evaluation was carried out during an additional 300 time steps of $5 \cdot 10^{-4}$ s.

All the results were evaluated relatively, thus the values are normalized by the maximum value from the stationary simulation of Type 1.

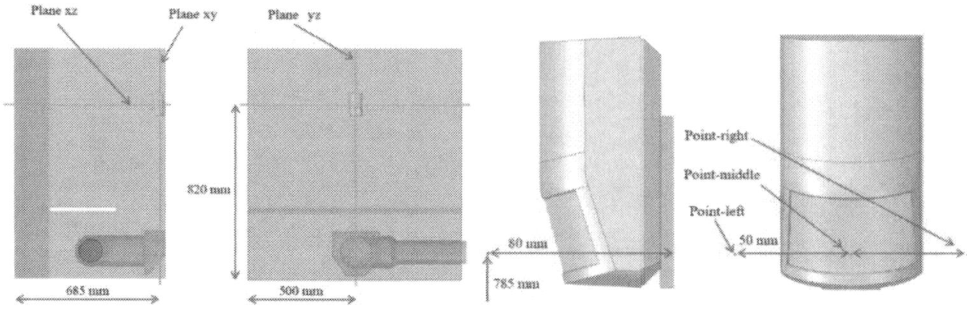

Fig. 7. Methodology of evaluation

6 Results of the CFD simulations

The normalized temperature contours in Figure 8 show a slightly non-stationary stream of warmer air. This heat propagation caused by flow along the wall also indicates the importance of proper placement of the detector in the space, as heating of the lens can adversely affect the function of the sensors. By comparing the generated contour time sequences over time, the pulsating nature of the outgoing warm air stream was observed

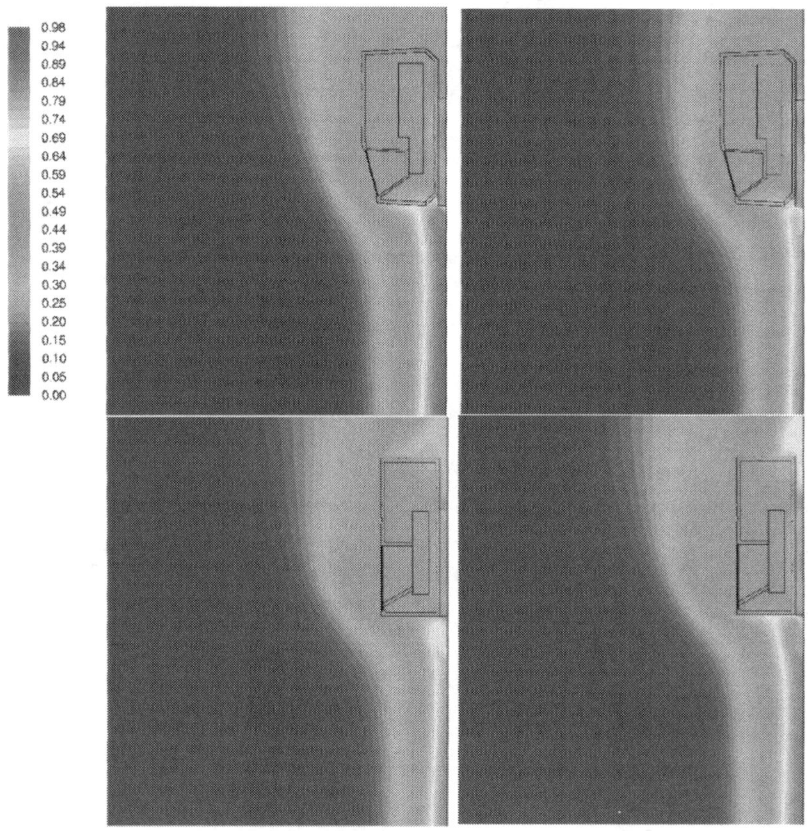

Fig. 8. Contours of the normalized temperature at 7 s and 12 s, plane yz – Type 1 (up) and Type 2 (down)

It was expected (and confirmed by calculations) that after the impact of the flowing medium on the lower edge of the detector, the flow splits. This situation is shown in Figure 9 and it also shows the inhomogeneous flow in the test cabinet.

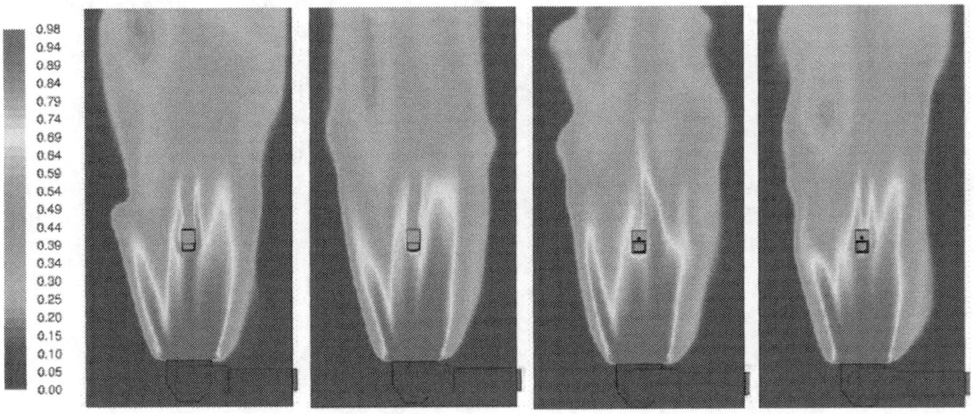

Fig. 9. Contours of the normalized temperature at 7 s and 12 s, rear wall – Type 1 (left) and Type 2 (right)

Figure 10 shows the flow field around the sensor and the separation of the flow in front of the lens. The flow separation is more extensive for the Type 2 model. This effect is caused by the geometry of Type 2 which is more sensitive to the flow separation than Type 1.

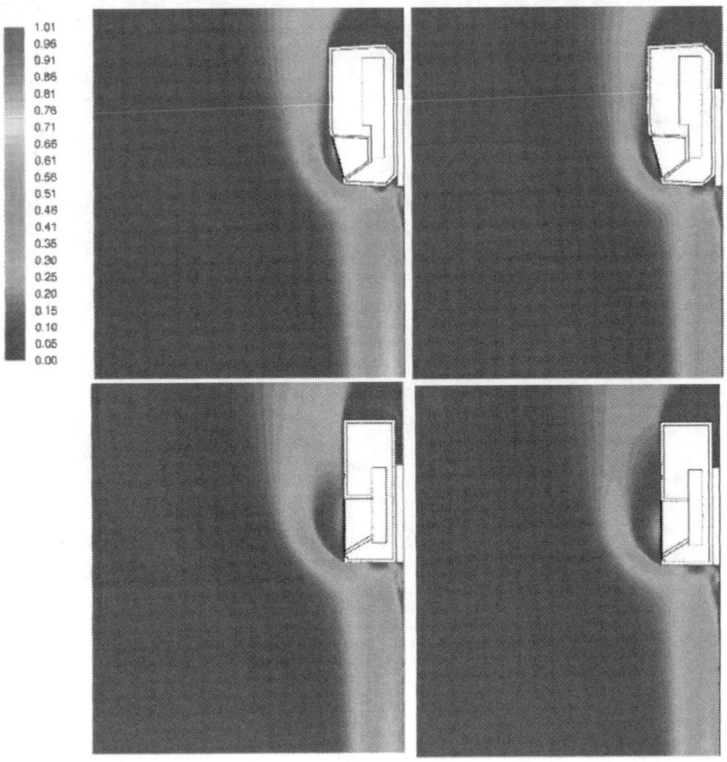

Fig. 10. Contours of the normalized velocity mag. at 7 s and 12 s, plane yz – Type 1 (up) and Type 2 (down)

The results in Figure 11 show that the flow for Type 2 is more unstable than Type 1. The results also show, similar to the results in Figure 9 for temperatures, that the flow from the inlet pipe is not symmetrical or stable.

This shows that the design of the test cabinet and especially the air outflow from the supply pipe is not satisfactory. The geometry of the circular duct (especially the elbows and the outlet) does not allow a homogeneous flow of the warm air in the cabinet; the flow is unstable and undefined.

The shape of the Type 2 sensor also causes higher fluctuation compared to Type 1, but the differences are not large. This finding is supported by the monitoring of temperatures and velocities during the unsteady simulation – see the normalized values from various points in the computational domain of both types in Figure 12. More detailed information about the modelling and the results can be found in the work [8].

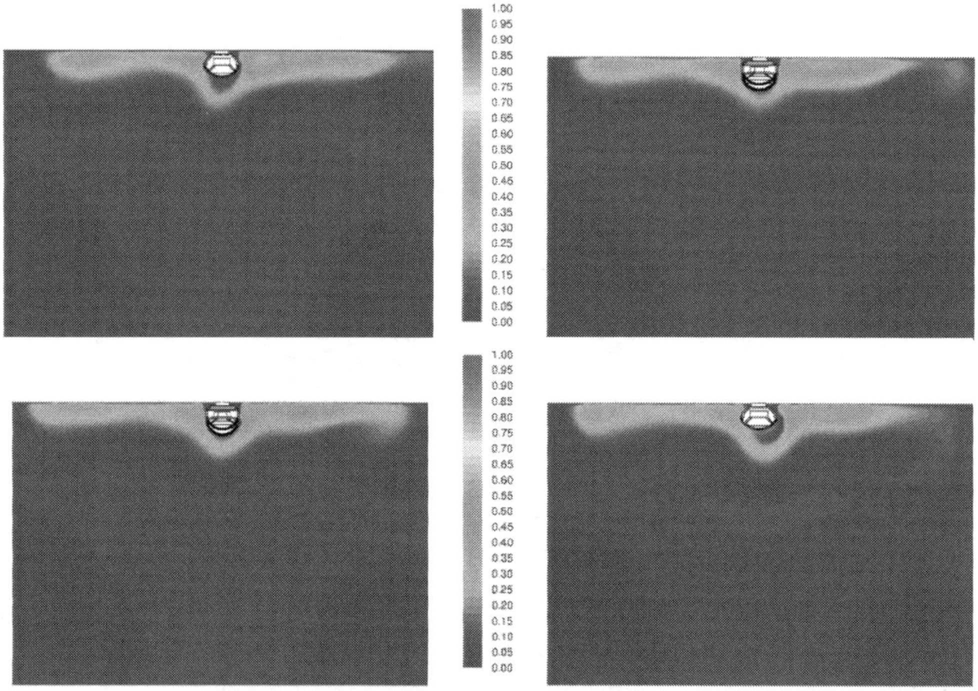

Fig. 11. Contours of the normalized velocity mag. at 7 s and 12 s, plane xz, Type 1 (top) and Type 2 (bottom)

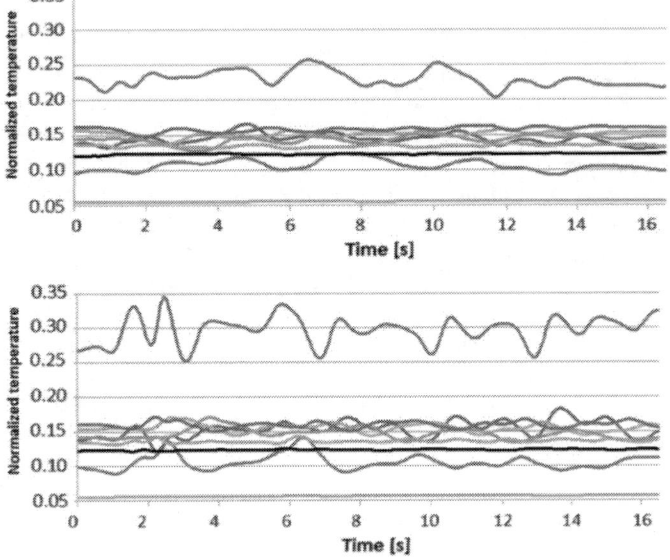

Fig. 12. Temperatures monitored during unsteady simulation – Type 1(up) and Type 2 (down)

7 New design of the test cabinet

Based on the findings described above, a new test cabinet for non-standard PIR testing was designed. It has a more suitable design of the inlet pipe, which guarantees a homogeneous flow at the outlet of the pipe. A fan is located behind the inlet, followed by a heater, an elbow with internal inserts and a flow conditioner. To change the flow pattern, a properly designed element can be placed at the outlet of the supply pipe. The PIR sensors can be mounted in various ways. A preview of the new configuration can be seen in Figure 13.

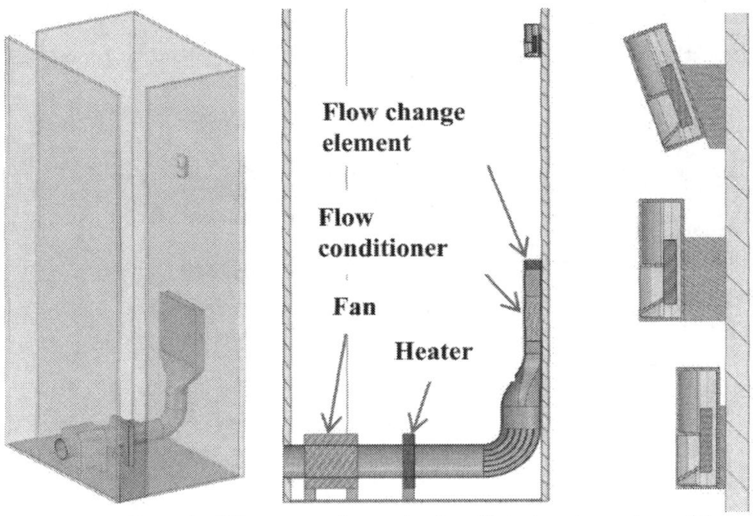

Fig. 13. The new test cabinet for PIR sensors for testing in off-design thermal conditions

8 Conclusions

PIR motion sensors should be constructed and, above all, positioned so that the lens is warmed as little as possible by the ambient air. Different mounting conditions of PIR sensors that may occur in practice are as yet not well researched. The knowledge gained can help manufacturers and users to address the placement of sensors outside the standard conditions. Basic engineering principles regarding the enveloping of PIR sensors by warm flowing air were tested using numerical simulations. The results of the simulations can help in selecting a suitable location for measuring velocity or temperature in experimental measurements. They also show the importance of the shape of the sensor with respect to the interaction with the flowing warmed air. Finally, the test cabinet used for non-standard testing of sensors was analysed and based on the results, a new design was proposed for testing sensors in non-standard conditions, eliminating the shortcomings of the original design.

The results were developed within specific research (UWB SGS-2019-021) and contract research.

References

1. G. Cirino, L. G. Neto, "Design of cubic-phase distribution lenses for passive infrared motion sensors", Sao Paulo: The International Society for Optical Engineering, 2013. 10.1117/12.487786.
2. M. Bori, "Možnosti ověření vlastností PIR detektorů", Zlín: Bakalářská práce, Univerzita Tomáše Bati ve Zlíně, 2010 (in Czech).
3. L. Michalec, "PIR detektor: skvělý sluha, ale zlý pán", *Vyvoj.hw.cz.* [Online] 19. 3 2013. [Citation: 3. 12. 2020] https://vyvoj.hw.cz/automatizace/pir-cidlo-skvely-sluha-ale-zly-pan.html (in Czech).
4. EN 50131-X, Alarm systems - Intrusion and hold-up systems.
5. ANSYS Fluent User's Guide 2020 R2, ANSYS, Inc., 2020.
6. F. R. Menter, "Zonal Two Equation k-ω Turbulence Models for Aerodynamic Flows", AIAA Paper 93-2906.
7. F. R. Menter, "Zonal Two Equation k-ω Turbulence Models for Aerodynamic Flows", AIAA Paper 93-2906.
8. J. Kazda, "Modelování úloh s prouděním a sdílením tepla určených pro testování elektronických čidel", Plzeň: Diplomová práce, Západočeská univerzita v Plzni, 2021 (in Czech).

Sizing the thermal energy storage (TES) device for organic Rankine cycle (ORC) power systems

Piotr Kolasiński[1*], and *Sindu* Daniarta[1, 2]

[1]Department of Thermodynamics and Renewable Energy Sources, Wrocław University of Science and Technology, Wybrzeże Wyspiańskiego 27, 50-370 Wrocław, Poland.
[2]Department of Energy Engineering, Faculty of Mechanical Engineering, Budapest University of Technology and Economics, Műegyetem rkp. 3, H-1111 Budapest, Hungary

Abstract. Thermal energy storage (TES) became one of the main research topics in modern power engineering. The design of TES devices and systems depend on their application. Different thermal energy storage materials (e.g., solids, liquids, or phase change materials) can be applied in TES devices. The selection of the thermal energy storage material depends mainly on the thermal power and operating temperature range of the TES device. These devices and systems are applied in different energy conversion systems, including solar power plants or combined heat and power (CHP) stations. The application of TES devices is also considered in the case of other industries, such as metallurgy. The possible application of TES devices is particularly promising in the case of organic Rankine cycle (ORC) systems. These systems are often utilizing floating heat sources such as solar energy, waste heat, etc. TES device can be therefore applied as the evaporator of the ORC system in order to stabilize these fluctuations. In this paper, the possible thermal energy storage materials used in TES devices applied in ORCs are discussed. Moreover, the modelling results are reported related to assessment parameters which can be applied to size the TES device for ORC system utilizing different low-boiling working fluids. The thermal properties of working fluids are taken from CoolProp. The function of heat capacity of different TES materials is also provided and the calculation is computed by employing MATLAB. The result shows that based on the simulation, the gradient of the natural characteristic of TES with working fluids ($\zeta(T_b)$) tends to decrease. The presented result in this paper gives a new point of view which can be used by scientists and engineers during the design and implementation of TES evaporators dedicated to ORC power systems.

1 Introduction

Intensive research aimed at methods and technologies which can be utilized by the modern power industry for environmental protection and efficiency improvement is currently proceeded. One of the most important challenges is still reducing fossil fuels consumption. Complete phase-out of fossil fuels is not possible in the case of many industrial processes (e.g., chemical factories, metallurgy, etc.), however, their negative impact can be limited in

* Corresponding author: piotr.kolasinski@pwr.edu.pl

some processes by using renewable and waste energy sources, rationalization of energy management and increasing the energy conversion efficiency. Various forms of energy can be obtained from alternative sources, while thermal energy sources are particularly interesting in terms of their diversity and possible applications. Thermal energy can be easily converted into different energy products, such as electricity, cold or process heat thanks to the application of different power systems and technologies, for example, organic Rankine cycle (ORC) systems. The principle of ORC system operation is the same as in the case of standard steam power plant operating according to Clausius-Rankine cycle [1]. However, a low-boiling working fluid is used instead of water. Thanks to this change it is possible to extract the thermal energy from heat sources featuring lower thermal potential than in the case of standard steam power plants. ORCs currently became one of the most important research topics among energy conversion systems that can convert thermal energy from alternative energy sources. Practically, thermal energy is transferred by carriers that are characterized by different thermal potentials and physical forms. These include high temperature (500-1500 °C), medium temperature (250-500 °C) and low temperature (40-250 °C) renewable and waste heat carriers. The temperature level of the heat carrier and its output are defining the exergy level and power of the heat source, and thus its quality and suitability for an application. The high- and medium-temperature carriers are featuring the highest quality, while the quality of low-temperature carriers is lower. High- and medium-temperature waste heat carriers are released in large industrial plants and factories [2], while renewable carriers featuring such thermal potential are limited to seismically active areas [3]. Concentrated solar power can also feature this level of temperature [4]. In the industry, various technologies such as waste heat recovery boilers and steam power plants are applied to harvest the heat contained in high- and medium-temperature waste heat carriers [5]. The recovered energy can be then used to cover the own needs of technological processes, (e.g., process steam production in recovery boilers, heating of combustion substrates, drying processes) or can be applied to generate electricity, heat or cold. Unlike high- and medium-temperature carriers, low-temperature renewable and waste carriers are common and spatially dispersed and are rarely used, mainly due to the limited availability of efficient recovery technologies. What is more, low-temperature renewable and waste carriers are often featuring fluctuating thermal and output characteristics which can negatively influence the operating conditions of the recovery system.

Thermal energy storage (TES) systems and devices can potentially be applied to balance the fluctuating heat source. TES devices can differ in terms of the design and applied storage substance. Different solids and liquids featuring high thermal capacity, as well as phase change substances (i.e., PCM – phase change materials) can be applied in TES devices. Applied storage substances should also be chemically stable and resistant to cyclical processes of charging and discharging. The applied solids are, for example, rocks (granite or soapstone), reinforced concrete, cast iron, cast steel, as well as silica and magnesite refractory bricks. The liquids used are water, organic liquids and various types of oils (including hardening oils). Applied PCM's are organic, inorganic and eutectic substances (such as hydrated salts, paraffin, etc.). Thanks to the beneficial effect of latent heat release during a phase transition use of PCM's as storage substances results in higher efficiency and performance of the TES device. Basic technical parameters (i.e., operating temperature range, heat capacity, power, efficiency, storage period and investment cost) of the TES device depend on its type and storage substance used. The operating temperature of TES devices ranges from 200 to 1200 °C (solid accumulating substances), 20–260 °C (liquid accumulating substances) and 20–120 °C (PCM's) [6]. Single TES devices can be combined into a larger TES system which can be charged by various heat sources (i.e., conventional as well as renewable and waste). Depending on the type of TES, heat can be stored for a different period of time, starting from several hours (short-term storage) up to several months (long-term storage). The most important properties of TES devices are heat capacity, power, performance, charging and discharging times, and investment costs. The design and principle of operation of the TES device may be similar to the design of regenerators or classic shell and tube heat exchangers. The difference is that the heat is transferred from the heat carrier to the storage substance instead of the second working fluid. The comparison of the temperature range of heat sources utilized by ORC systems with the operating temperatures of different types of TES devices shows that TES devices using solids and liquids as heat accumulating substances are the most promising to be applied in ORC systems. The application of PCM-based TES devices may be limited in ORC due to the low operating temperature range (up to ca. 120 °C). The TES device investment costs depend on its design, operating parameters, storage material used and the place of application. These costs vary between 10-100 EUR/kWh [7].

The most important technical data related to different types of TES devices [7, 8] are reported in Table 1. By the type of storage substance used, TES devices can be divided into types listed in Table 2 [7–11].

One of the possible applications of TES devices in ORCs can be the use of a TES device instead of an evaporator (or as an additional heat exchanging device). Such an application of the TES device gives the opportunity for balancing the fluctuating heat source (such as solar radiation, heat obtained from biomass combustion, geysers, etc.). In the literature studies, some articles have reported the TES device instead of an evaporator for ORC power systems [7, 12]. Many configurations of TES in ORC presented in the articles, however, there is still few data concerning the modelling. Therefore, in the following part of the paper, the results of modelling of the size of the TES device which is considered to be applied as the main heat source for the ORC system are presented for different low-boiling working fluids, storage materials and operating temperature ranges. This paper also shows the gradient of sizing as a function of outlet temperature before the expansion process.

Table 1. Classification of TES devices by the physical phenomenon being the principle of TES device operation [7, 8].

Phenomena	Heat type	Storage medium
Thermal phenomena	Sensible heat	Liquids Solids
	Latent heat	Solid-liquid Liquid-gas Solid-solid
Chemical phenomena	Thermal chemical pipeline	-
	The heat of chemical reaction	-
	Chemical heat pumps	-

Table 2. Typical parameters of TES devices [7–11].

Type of device	Thermal capacity (kWh/t)	Temperature range (°C)	Power (MW)	Efficiency (%)	Storage period	Cost (EUR/kW)
Sensible (hot water)	10–50	40–100	0.001–10.0	50–90	Days/months	0.1–10
Phase-change material (PCM)	50–150	40–220	0.001–1.0	75–90	Hours/months	10–50
Chemical reactions	120–250	400–1200	0.01–1.0	75–100	Hours/days	8–100

2 Systems, models and mathematical description

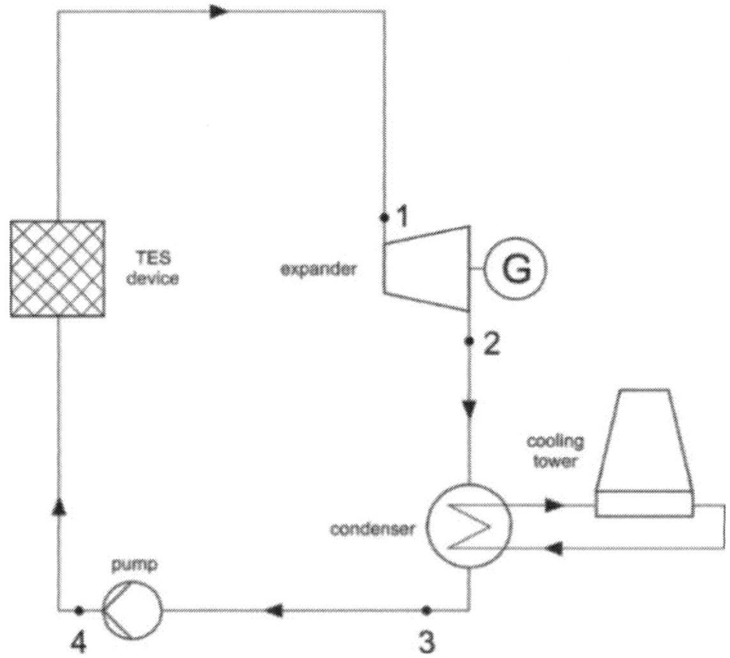

Fig. 1. Schematic layout of the ORC system with TES device [12].

Figure 1 shows the scheme of the modelled system. TES device is applied in the system as the heat supply. In the real ORC system, thanks to such an application of TES device it would be possible to supply the heat to the system continuously in case of powering the system via heat sources featuring floating thermal and output characteristics (e.g., solar thermal energy). The ORC system is therefore composed of a TES device, expander, condenser and pump. A low-boiling working fluid flow is forced by a pump to the inlet of the TES device where it evaporates. Obtained vapour flows via pipelines to the inlet of the expander (turbine or volumetric machine) where during expansion its thermal parameters are reduced and thermal energy is converted into mechanical energy of the expander's rotating shaft. After expansion, working fluid is flowing through the pipelines to the inlet of the condenser, where it rejects heat to the cooling fluid and is liquified. The obtained liquid is then directed again to the pump and the working cycle therefore completes. Figure 2 shows the *T-s* diagram of the ORC cycle (red lines). Shaded regions in Figure 2 visualize the heat needed for working fluid preheating, evaporation and superheating.

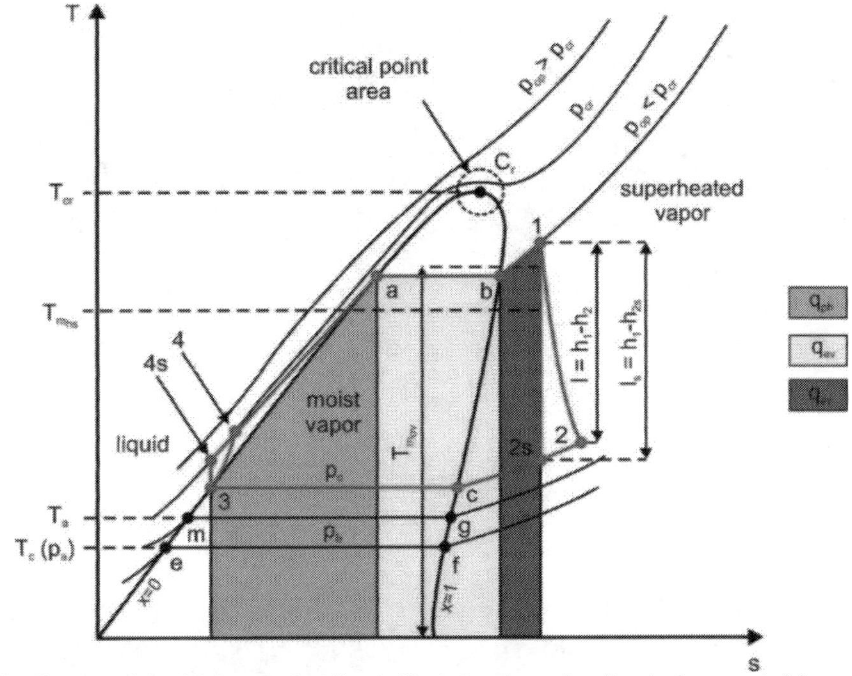

Fig. 2. *T-s* diagram of the ORC cycle (red lines). Shaded regions visualize the heat needed for working fluid preheating (q_{ph}), evaporation (q_{ev}) and superheating (q_{ov}) [12].

Different low-boiling working fluids and different thermal storage substances can be applied in the considered system. The working fluid selection can be based on different selection criteria. The most important are the environmental impact of the substance (i.e., ODP and GWP), safety (e.g., toxicity, flammability and explosivity) and thermal properties. It is of great importance that working fluid should be safe both for the environment and humans. The list of working fluids that can be applied in ORCs was significantly limited by international regulations and many of the working fluids which were successfully applied in ORCs (e.g., R11 and R123) are currently banned or being withdrawn from the application. Therefore, in the present study, we decided to consider these working fluids which negative impact on the environment is significantly limited. The list of considered working fluids is reported in Table 3. Taking into account the thermal properties of the selected fluids we decided to limit the treatment to the subcritical cycles, therefore the modelling is proceeded for the TES device temperature ranging from 373.15 $\leq T_{\text{upper}} \leq T_{\text{crit}}$-20 K (see Table 3). For these conditions of the heat source the heat amount needed to preheat and evaporate the working fluid (per 1 kg of substance) can vary for different working fluids and could be calculated using Equations (1) and (2).

$$q_{\text{pre}} = h_{\text{a}} - h_4 \quad (1)$$

$$q_{\text{ev}} = h_{\text{a}} - h_{\text{b}} \quad (2)$$

The thermal energy of the TES device can be calculated using the following Equation (3).

$$Q_{\mathrm{TES}} = m_{\mathrm{sm}} \cdot c_{\mathrm{psm}} \cdot T_{\mathrm{sm}} \quad (3)$$

Where m_{sm} is the mass of the heat storage material, c_{psm} is the specific heat capacity of storage material, T_{sm} is the temperature of storage material.

Table 3. The thermal properties of the considered low-boiling working fluids taken from CoolProp [13], the boundary condition and the steps of calculation.

No	The selected working fluid	CAS no.	T_{crit} (K)	T_{triple} (K)	T_{boiling} (K)	T_{lower} (K)	T_{upper} (K)	Steps (K)
1	Butane	106-97-8	425.12	134.89	272.66	308.00	$373.15 \leq T_{\mathrm{upper}} \leq T_{\mathrm{crit}} - 20$	0.01
2	Diethyl ether	60-29-7	466.70	270.00	307.60	308.00		0.01
3	Isohexane	107-83-5	497.70	119.60	333.36	T_{boiling}+10		0.01
4	Neopentane	463-82-1	433.74	256.60	282.65	308.00		0.01
5	Novec 649	756-13-8	441.81	165.00	322.20	T_{boiling}+10		0.01
6	Pentane	109-66-0	469.70	143.47	319.20*	T_{boiling}+10		0.01
7	R1233zd(E)	102687-65-0	439.60	195.15	291.41	T_{boiling}+10		0.01
8	R245ca	679-86-7	447.57	196.00	298.41	T_{boiling}+10		0.01
9	R365mfc	406-58-6	460.00	239.00	313.34	T_{boiling}+10		0.01

*The boiling temperature at the pressure of 140,805 Pa.

The heat that can be extracted from the TES device during its cooling from its initial temperature of T_A to the final temperature T_B can be calculated using Equation (4).

$$\Delta Q_{\mathrm{TES}} = m_{\mathrm{sm}} \cdot c_{\mathrm{psm}} \cdot (T_{\mathrm{A}} - T_{\mathrm{B}}) \quad (4)$$

In this study, we decided to omit the analyses related to superheating of the working fluid as this process is not often applied in practice and for many working fluids its application leads to lowered cycle efficiency [14, 15]. On the basis of the selected temperature of the TES device the range of TES operating temperature was selected as reported in Table 4. As it was mentioned earlier TES devices can use different solids, liquids and PCMs as storage materials. In this study, we focus on the TES device temperatures higher than 100 °C. Therefore, PCMs were not included in the present study, and the focus was paid to selected solids and liquids which were reported by different authors as promising materials for thermal energy storage. The list of these materials is reported in Table 4.

To size a thermal storage device, it is possible to define an assessment parameter using which it is possible to calculate the mass of thermal storage material which is needed to store a certain quantity of heat and on this basis compare different thermal energy storage materials and ORC working fluids. The other, below defined, assessment parameters can be useful for making the comparison of the different TES devices which are applied in ORC power systems that utilize different working fluids.

The specific heat capacity of thermal energy storage materials is not constant and is more or less dependent on temperature. High variability of this parameter is especially visible during phase transitions. Specific heat capacity of material can be determined experimentally (for example differential scanning calorimetry/DSC can be applied for this purpose). The obtained experimental results are then approximated by polynomial equations. The general polynomial equation for the specific heat capacity of a material can be given as Equation (5), the selected TES material is given in Table 4.

Table 4. The list of thermal energy storage materials which were selected for the modelling and the polynomial equations for the specific heat capacity of these materials.

No	TES material	T (K)	C_0 (J/kg·K)	C_1 (J/kg·K²)	C_2 (J/kg·K³)	C_3 (J/kg·K⁴)	C_4 (J/kg·K⁵)	Refs.
1	Monel metal	273.15 – 573.15	422.8668	0.1842	-	-	-	16
2	Stainless steel 446	273.15 – 773.15	451.3370	0.4497	-	-	-	16
3	Sandstone	308 – 550	$c_p(T) = 1135.532\left(\frac{T}{239.01163}\right)^2\left(\frac{\frac{T}{239.01163}}{1+\left(\frac{T}{239.01163}\right)^3}\right)$					17
4	Granite	300 – 1200	$c_p(T) = -228.24 + 3380.1\cdot 10^5\cdot T^{-2} - 3474.8\cdot 10^7\cdot T^{-3} + 110880\cdot T^{-0.5} - 2378100\cdot T^{-1}$					18
5	Chrome brick	293.15 – 1973.15	800	0.3	-	-	-	19
6	Concrete	300 – 600	957.55	0.3403	-	-	-	20
7	Chamotte brick	293.15 – 1623.15	880	0.23	-	-	-	19
8	Silicon carbide	293.15 – 1973.15	960	0.146	-	-	-	19
9	Magnesite brick	293.15 – 1923.15	1050	0.3	-	-	-	19
10	Rapeseed oil	298.15 – 513.15	2098.5	-5.976	$14.933\cdot 10^{-2}$	$-8.753\cdot 10^{-4}$	$1.621\cdot 10^{-6}$	21
11	Soybean oil	298.15 – 513.15	1966.4	-0.409	$5.491\cdot 10^{-2}$	$-3.705\cdot 10^{-4}$	$0.807\cdot 10^{-6}$	21
12	Sunflower oil	298.15 – 513.15	1950.6	1.20	$4.174\cdot 10^{-2}$	$-3.31\cdot 10^{-4}$	$0.76\cdot 10^{-6}$	21
13	Cotton oil	298.15 – 513.15	2118.6	-5.443	$14.06\cdot 10^{-2}$	$-7.924\cdot 10^{-4}$	$1.373\cdot 10^{-6}$	21
14	Jatropha oil	298.15 – 513.15	1960.8	0.441	$12.947\cdot 10^{-2}$	$-10.423\cdot 10^{-4}$	$2.262\cdot 10^{-6}$	21

$$c_p(T) = C_0 + C_1T + C_2T^2 + C_3T^3 + \cdots + C_nT^n \quad (5)$$

where $C_0 \ldots C_n$ are polynomial approximation constants and T is the temperature.

Table 4 reports the polynomial equations for the specific heat capacity of thermal energy storage materials which were selected for the modelling. The heat transfer process can be treated as isobaric. Therefore, the mass and enthalpy change of thermal energy storage material needed to proceed with the preheating, evaporation and superheating of the 1 kg of low-boiling working fluid can be obtained from the following energy balance of TES device, Equation (6).

$$m_{WF} \cdot (h_1 - h_4) = m_{sm} \cdot c_p(T_{sm}) \cdot (T_{TES1} - T_{TES2}) \quad (6)$$

where T_{TES1} and T_{TES2} are the initial and the final temperature of the TES device.

Therefore, the dimensionless storage mass parameter as an assessment that can be applied for sizing the thermal energy storage device used to preheat, evaporate and superheat 1 kg of low-boiling working fluid can be calculated by Equation (7).

$$\zeta_{sm} = \frac{m_{sm}}{m_{WF}} = \frac{h_1 - h_4}{c_p(T_{sm})\Big|_{T_{TES1}}^{T_{TES2}} \cdot (T_{TES1} - T_{TES2})} \quad (7)$$

where m_{WF} is the mass of the low-boiling working fluid (in the present study it is assumed as m_{WF}=1 kg). Furthermore, the value of m_{sm} (which is the mass of the heat storage material) can be determined from the result of dimensionless storage mass parameter from the ratio of enthalpy changes.

As mentioned above, in the present study we treat the ORC cycle as evaporation (superheating is not considered). Therefore, in this modelling the dimensionless storage mass parameter is calculated by Equation (8).

$$\zeta_{sm} = \frac{m_{sm}}{m_{WF}} = \frac{h_b - h_4}{c_p(T_{sm})\Big|_{T_{TES1}}^{T_{TES2}} \cdot (T_{TES1} - T_{TES2})} \quad (8)$$

3 Results and discussion

The obtained modelling results related to the variation of the dimensionless storage mass parameter ζ vs. temperature of the working fluid at the outlet of the TES device (T_b) are summarized in Figure 3 for different working fluids and different solids used as the thermal energy storage materials. The analysis of the modelling results presented in Figure 3 gives the possibility to compare and estimate the size of the TES device. The value of $\zeta(T_b)$ parameter gives the information about the mass of thermal energy storage material which is needed to preheat and evaporate 1 kg of low-boiling working fluid. Therefore, the lower the value of this parameter is, the smaller mass of thermal energy storage material will be needed to evaporate the working fluid. The obtained values of $\zeta(T_b)$ parameter ranges between 1.09 and 22.19. The highest value of this parameter is obtained for Monel metal and neopentane for the TES device temperature of T_b=360 K. Obtained modelling results can be discussed on the examples enlarged in Figure 3. For example, if butane is selected as the working fluid, the obtained values of the $\zeta(T_b)$ parameter ranges between 2.98 and 12.16 for different thermal energy storage materials. The lowest values of this parameter are obtained for granite, while the highest values are obtained for Monel metal. The variation of the $\zeta(T_b)$ parameter calculated for selected thermal energy storage material and different working

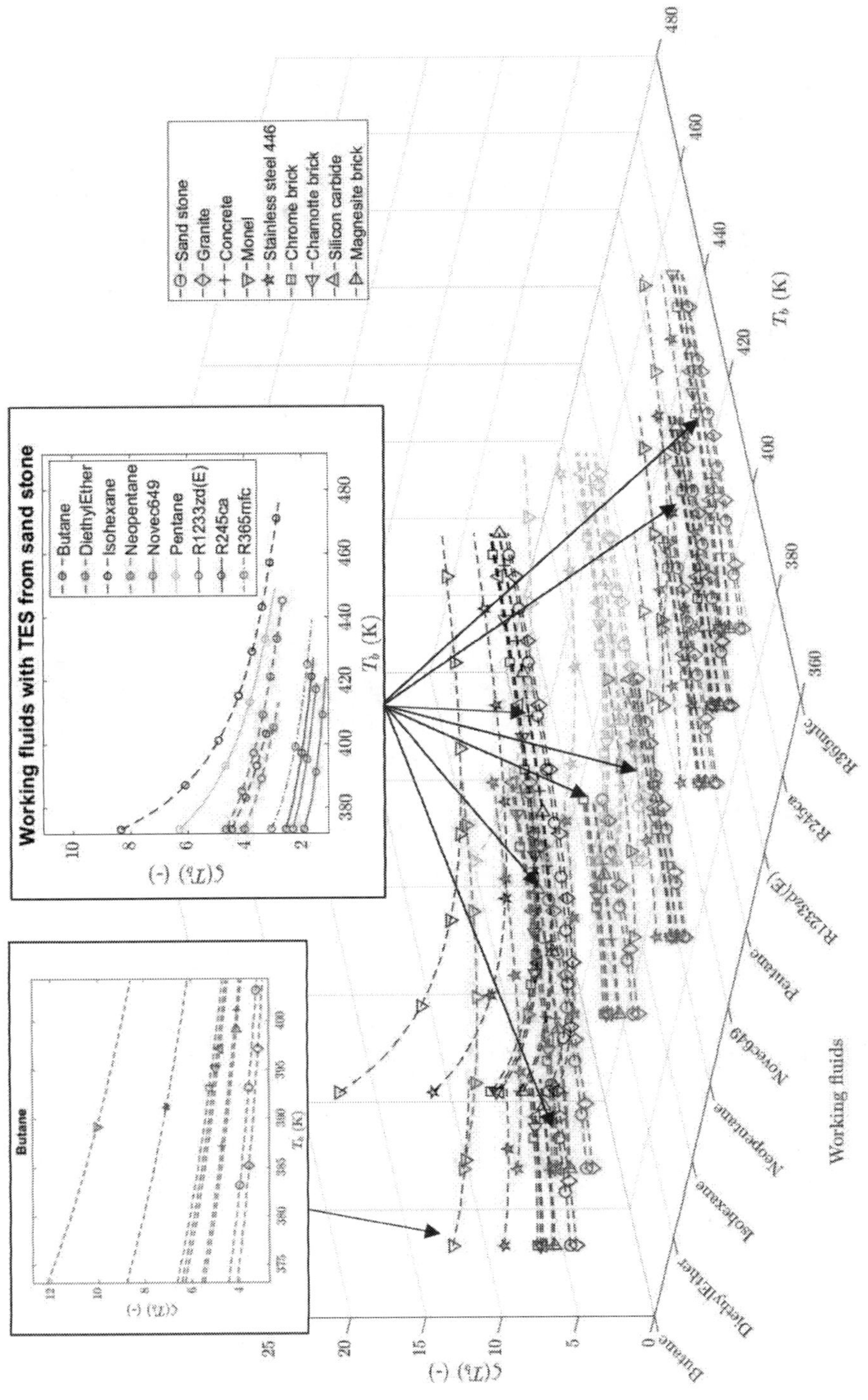

Fig. 3. A 3D diagram of the variation of $\zeta(T_b)$ parameter vs. the temperature of the working fluid at the outlet of TES evaporator (T_b) for different working fluids and different solid thermal energy storage materials.

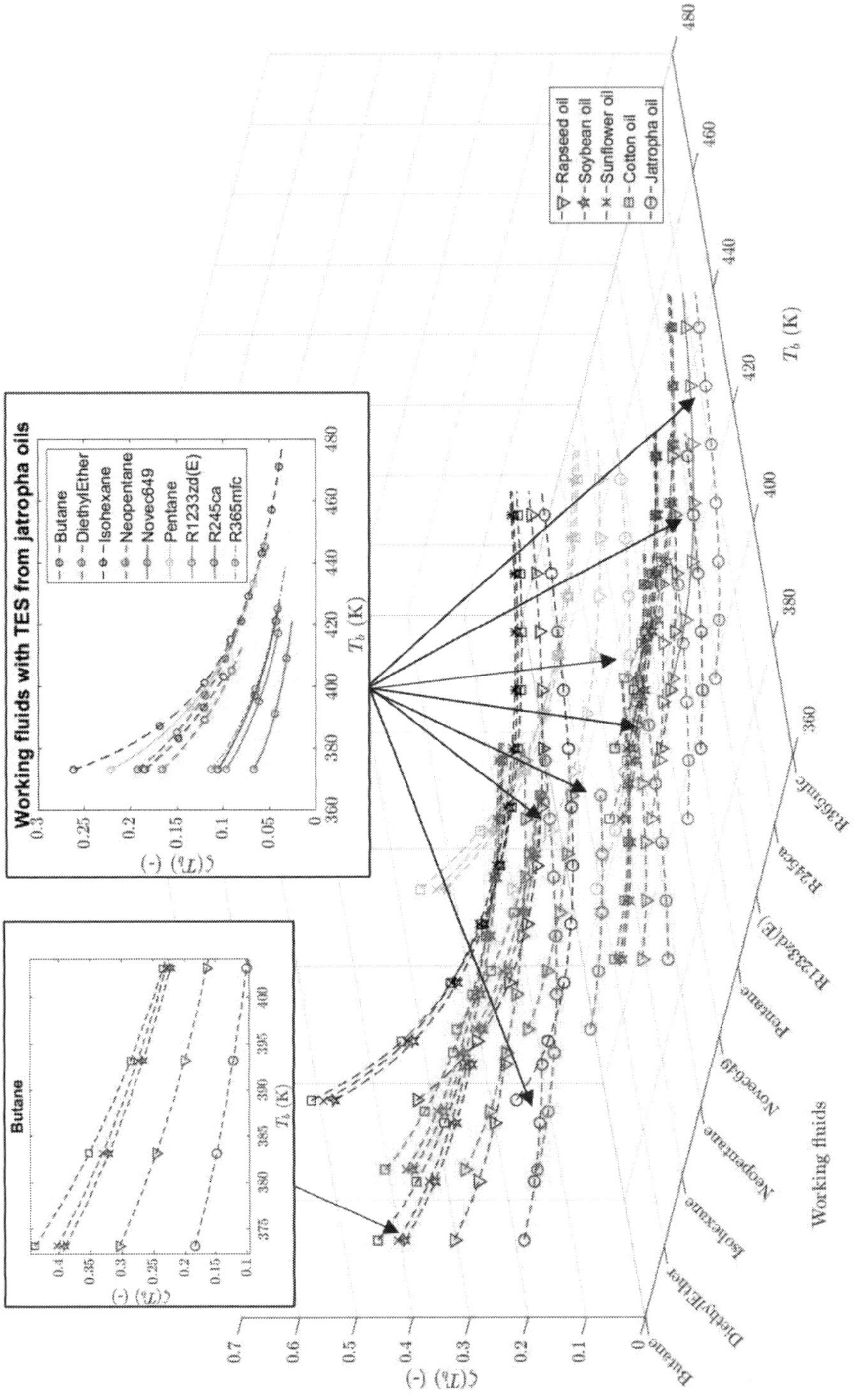

Fig 4. A 3D diagram of the variation of $\zeta(T_b)$ parameter vs. the temperature of the working fluid at the outlet of TES evaporator (T_b) for different working fluids and different vegetable oils thermal energy storage materials

fluids can be discussed based on the sandstone as it is enlarged in Figure 3. The obtained values of the $\zeta(T_b)$ parameter ranges between 1.17 and 8.34 for different working fluids and TES device temperature range. The lowest values of this parameter are found for Novec649, while the highest values are obtained for isohexane. Therefore, for the TES device utilizing sandstone Novec639 seems to be the working fluid which application gives the possibility of limiting the size and mass of the TES device if the maximum temperature of the working fluid at the outlet of the TES device is T_b=420 K. If the higher temperature of the working fluid at the outlet of TES device is considered R245ca, R365mfc, diethyl ether, pentane and isohexane are the working fluids for which the $\zeta(T_b)$ parameter ranges between 2 and 5.

The TES material using vegetables are separately drawn which can be seen in Figure 4. Generally, since the heat capacity of these vegetable liquids is high, the sizing using these vegetable liquids are relatively smaller than the selected rocks and metals.

Figure 4 visualizes the variation of the dimensionless storage mass parameter ζ vs. temperature of the working fluid at the outlet of the TES device (T_b) for different working fluids and different vegetable liquids (i.e., oils) used as the thermal energy storage materials. For these thermal energy storage materials, the obtained values of $\zeta(T_b)$ parameter ranges between 0.01 and 0.62. The lowest values of this parameter are obtained for Novec639 and jatropha oil, while the highest values are obtained for isohexane and cotton oil. Therefore, for the TES device temperature ranging between 370 K and 420 K selection of Novec639 as ORC working fluid and jatropha oil as thermal energy storage material can give the opportunity of limiting the size and mass of the TES device. If the higher temperature of the working fluid at the outlet of the TES device is considered R365mfc with jatropha oil, R365mfc with rapeseed oil, pentane with jatropha oil, and isohexane with jatropha oil are the pairs of ORC working fluids and thermal energy storage materials for which the $\zeta(T_b)$ parameter ranges between 0.02 and 0.07.

In this study, if the simulation is calculated separately by increasing the evaporation temperature, the enthalpy change of evaporator gradually drops while the enthalpy change of preheater is gradually rising (this phenomenon occurs as an effect of the boundary condition based on the saturated curve). It is found that the enthalpy changes of the combined evaporator, preheater and TES slightly rise. Furthermore, the specific heat capacity of the selected TES materials significantly increases. By knowing these conditions, the result shows that the dimensionless storage mass parameters tend to decline in Figure 3 and 4.

4 Conclusions

This paper has presented the results of the analysis on sizing the TES evaporator for ORC systems and its gradient as a function of temperature before the expansion process. Thermal energy storage devices are especially promising for powering the ORC systems utilizing the heat sources with floating output and temperature characteristics, such as waste or solar heat. The dimensionless storage mass parameter which can be used for sizing the TES evaporator for the ORC system was defined and introduced. This parameter can be applied for assessing the applicability of different thermal energy storage materials for different organic working fluids and sizing TES devices. The modelling results show that depending on the temperature of the working fluid at the outlet of the TES device and applied working fluid the value of the dimensionless storage mass parameter may vary in the range between 1 and 20 and differs for different working fluids and thermal energy storage materials. The obtained modelling results show that Novec396 seems to be a promising working fluid for applications in which the temperature of the working fluid at the outlet of the TES evaporator is limited to T_b=420 K, while R245ca, R365mfc, pentane and isohexane are the working fluids which application can be promising in case of higher temperature of the working fluid at the outlet of TES evaporator. Among the considered thermal energy storage materials, results of modelling related to granite and liquid oils give promising, low values of dimensionless storage mass parameters.

The analysis presented in this paper gives the results which can be used by scientists and engineers during the design and implementation of TES evaporators dedicated to ORC systems.

References

1. T. Zygmunt Kaczmarczyk, Energy Convers. Manag. **244**, 114437 (2021).
2. H. Jouhara, N. Khordehgah, S. Almahmoud, B. Delpech, A. Chauhan, and S.A. Tassou, Therm. Sci. Eng. Prog. (2018).
3. M.H. Dickson and M. Fanelli, *Geothermal Energy: Utilization and Technology* (Routledge, 2013).
4. H.L. Zhang, J. Baeyens, J. Degrève, and G. Cacères, Renew. Sustain. Energy Rev. **22**, 466 (2013).
5. P. Kolasiński and B. Rogosz, in *32nd Int. Conf. Effic. Cost, Optim. Simul. Environ. Impact Energy Syst. ECOS 2019*, edited by S. W., G. P., W. S., and A. W. (Institute of Thermal Technology, Department of Thermodynamics, Theory of Machines and Thermal Systems, Wrocław University of Science and Technology, Wrocław, Poland, 2019), pp. 4607–4617.
6. A. Sharma, V. V Tyagi, C.R. Chen, and D. Buddhi, Renew. Sustain. Energy Rev. **13**, 318 (2009).
7. P. Kolasiński, Therm. Sci. Eng. Prog. **19**, 100586 (2020).
8. I. Sarbu and C. Sebarchievici, Sustain. **10**, (2018).
9. S. Ushak, A.G. Fernández, and M. Grageda, in *Woodhead Publ. Ser. Energy*, edited by L.F.B.T.-A. in T.E.S.S. Cabeza (Woodhead Publishing, 2015), pp. 49–63.
10. S.C. Johnson, F. Todd Davidson, J.D. Rhodes, J.L. Coleman, S.M. Bragg-Sitton, E.J. Dufek, and M.E. Webber, in edited by H. Bindra and S.B.T.-S. and H. of N.E. Revankar (Academic Press, 2019), pp. 119–175.
11. L. André, S. Abanades, and G. Flamant, Renew. Sustain. Energy Rev. **64**, 703 (2016).
12. P. Kolasiński, Układy Organic Rankine Cycle (ORC) Małej i Średniej Mocy Wykorzystujące Rozprężarki Wyporowe (Oficyna Wydawnicza Politechniki Wroclawskiej, Wrocław, 2021).
13. I.H. Bell, J. Wronski, S. Quoilin, and V. Lemort, Ind. Eng. Chem. Res. **53**, 2498 (2014).
14. G. Qiu, Renew. Energy **48**, 565 (2012).
15. T. Yamamoto, T. Furuhata, N. Arai, and K. Mori, Energy **26**, 239 (2001).
16. T.B. Douglas and J.L. Dever, J. Res. Natl. Bur. Stand. (1934). **54**, 15 (1955).
17. I.M. Abdulagaov, Z.Z. Abdulagatova, S.N. Kallaev, and Z.M. Omarov, Geomech. Geophys. Geo-Energy Geo-Resources **5**, 65 (2019).
18. S.Q. Miao, H.P. Li, and G. Chen, J. Therm. Anal. Calorim. **115**, 1057 (2014).
19. E. Kostowski, *Przepływ Ciepła* (Wydawnictwo Politechniki Śląskiej, Gliwice, 2010).
20. J. Pan, R. Zou, and F. Jin, Energies **10**, (2017).
21. J.-F. Hoffmann, G. Vaitilingom, J.-F. Henry, M. Chirtoc, R. Olives, V. Goetz, and X. Py, Sol. Energy Mater. Sol. Cells **178**, 129 (2018).

On the rational form of rotary control diaphragms for steam turbines with industrial and cogeneration steam extraction

Arkadiy Zaryankin[1], *Alexander* Akatov[1], *Ivan* Lavyrev[1*], and *Mikhail* Cherkasov[1]

[1]National Research University "Moscow Power Engineering Institute, Institute of power engineering and mechanic, Department of steam and gas turbines, Krasnokazarmennaya 14, 111250 Moscow, Russian Federation

Abstract. The Russian park of power steam turbines contains a large number of turbines with adjustable steam extraction from the flow path where rotary control diaphragms are used as flow controllers, which structurally easily fit into the flow path of these turbines. This method of regulating the steam flow rate is accompanied by a decrease in the efficiency of the subsequent stages of the turbine and causes the appearance, at reduced loads, of additional disturbing forces acting on the rotor blades. In the presented materials, a variant of a post-sampling stage with a radial rotary control diaphragm is considered. The performed mathematical modeling of the working fluid flows in such a stage showed that in this case, at all turbine loads, a relatively uniform velocity field is provided when steam enters the nozzle apparatus, which naturally entails the elimination of the noted drawbacks.

1 Introduction

In the Russian thermal power industry, along with condensing steam turbines, a large place is occupied by turbines with cogeneration and industrial steam extraction from the flow paths of these turbines.

This circumstance gives rise to a number of new requirements, both for the formation of their flow paths and for the modes of their operation.

In particular, the central requirement is the condition of the turbine operation independence both in terms of the heating or industrial steam extraction schedule, and in terms of the electrical load schedule.

This condition entails the need to vary the steam flow rate through the turbine cylinders in a very wide range.

If the pre-selection stages of the considered turbines operate mainly in about a fifty percent range of variation in steam flow rates, then in the post-selection stages this range reaches 90-95%. Accordingly, the regulating body providing such a wide range of operating modes of the post-extraction stages inevitably generates a very large unevenness of the velocity field in front of the inlet section of the nozzle apparatus of the subsequent stage,

* Corresponding author: lavirev@mail.ru

thereby greatly reducing the vibration reliability of the working blades of this stage. The specified working body for regulating the costs is in a very difficult situation from the point of view of ensuring the reliability of post-selection stages.

Here, in the final analysis, the central problem is the problem of providing a uniform velocity field in the mode of reduced loads when steam enters the nozzle apparatus of the post-extraction stage.

The existing bodies for regulating steam extraction from the flow path of the indicated turbines do not solve this problem. In this regard, it seems appropriate to consider the existing design solutions and consider possible ways to maintain uniform velocity fields in front of the working blades of the post-extraction stage with reduced steam consumption through such stages.

2 Typical rotary diaphragms of steam turbines with adjustable steam extraction and their disadvantages

In steam turbines with industrial and cogeneration steam extraction from their flow paths, axial rotary diaphragms are widely used as a regulator of these extractions. A typical design of such diaphragms is shown in Fig. 1 [1].

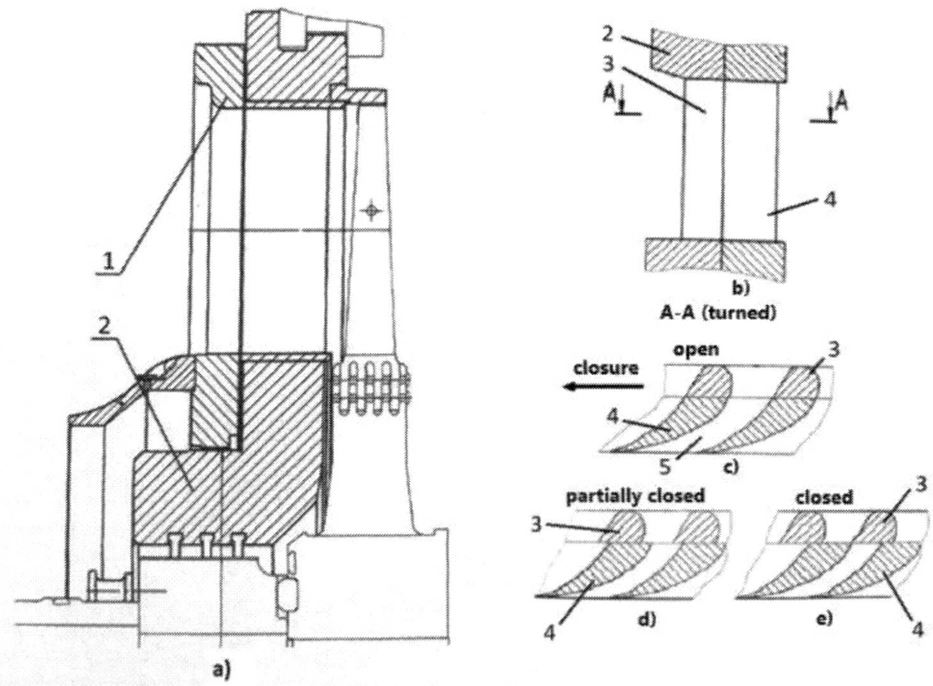

Fig. 1. The design of the PND rotary-regulating diaphragm. a, b - radial section, c, d, e - rotary-regulating diaphragm.

Structurally, the regulating diaphragm 1 is part of the nozzle apparatus 2 of the post-sampling stage, since in the open position, the profile baffles 3 of the diaphragm 1 are the inlet parts of the profiles of the nozzle 4. This solution provides an extremely low influence of the regulating body itself on the efficiency of the post-selection stage. The shape of the interscapular channel of the nozzle apparatus in this position of the control diaphragm is shown in Fig. 1c.

In aerodynamic terms, the reduced shape of the interscapular channel differs little from the shape of the interscapular channel in the lattice of profiles of conventional nozzles. According to the data of [2], the coefficient of profile losses of the considered lattice at a dimensionless velocity $M_{1t} < 0,9$ does not exceed 4.5%.

The situation changes dramatically when the diaphragm is turned and the inlet sections of the profiles are displaced to the position shown in Fig. 1d. The structural diagram of the flows in the channel formed in this way, when the diaphragm is opened by 25% [2] and the supercritical pressure drop across the nozzle apparatus is shown in Fig. 2.

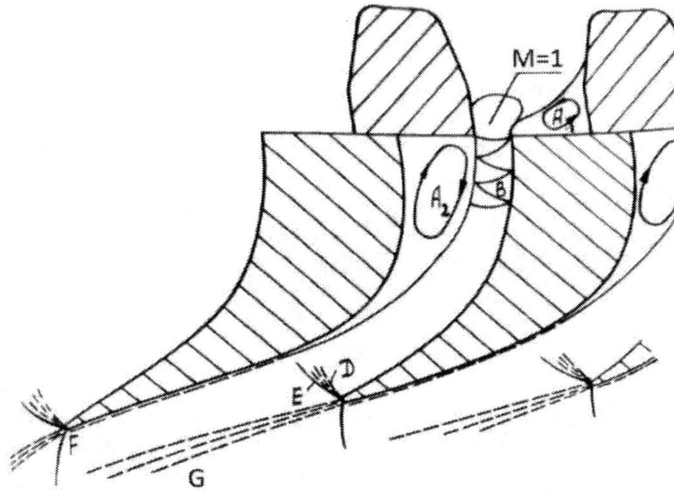

Fig. 2. Scheme of vortex zones. A1, A2 - vortex zones, B - system of shock waves and rarefaction waves behind the adjustable section, D - rarefaction waves, E - internal edge shock, F - external edge shock, G - edge vortex wake

Such an overlap of the inlet section in the channel of the nozzle array entails an inevitable transition from stationary to unsteady flow with the formation of closed separation zones with intense vortex flow.

As a result, the rotor blades of the post-sampling stages perceive an unsteady flow with a highly irregular velocity field in the circumferential direction and low-frequency pulsations of velocities and pressures.

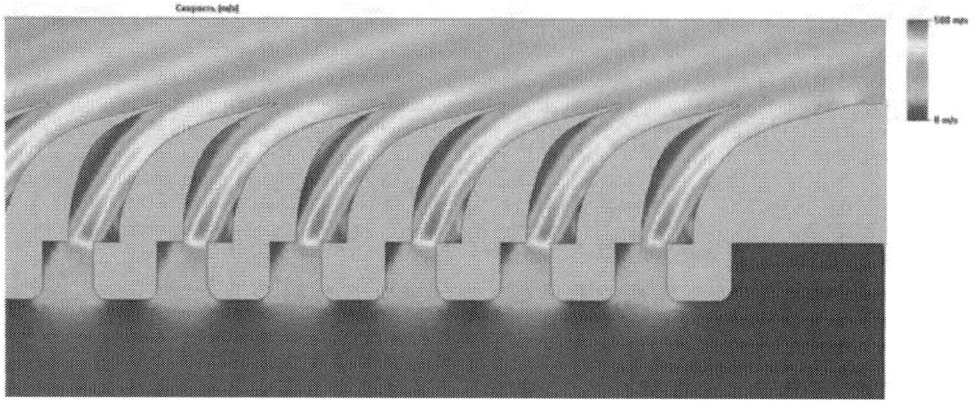

Fig. 3. Velocity field in the rotary diaphragm

The velocity field behind the nozzle apparatus with partial opening of the rotary diaphragm, shown in Fig. 3, obtained by mathematical modeling of flows, clearly confirms what has been said and agrees very well with the flow diagram shown in Fig. 2. If, to assess the degree of unevenness of the velocity field behind the nozzle apparatus of the post-selection stage, we use as the coefficient of unevenness $\gamma_{н}$ the difference between the maximum velocity behind the nozzle apparatus C_{max} and the minimum velocity C_{min}, referred to the average velocity $C_{ср}$ ($\gamma_{н} = \frac{C_{max}-C_{min}}{C_{ср}}$), then when the rotary diaphragm is closed by 50%, the specified coefficient $\gamma_{н}$ turns out to be equal to $\gamma_{н} = 0{,}72$. At the same time, according to experimental data [2], the profile losses in the annular lattice increase from 5% at full opening of the rotary diaphragm to 55%.

Naturally, with such a velocity field in the inlet section, the impeller blade unit experiences very large dynamic loads, which sharply reduce the vibration reliability of the post-sampling stages.

A very original way of improving the aerodynamic qualities of a nozzle apparatus with a rotary diaphragm is considered in [2], where not the inlet part of the nozzle lattice profiles was connected to the rotary diaphragm, but its entire part, which forms the convex surface of the profiles.

In this case, when the diaphragm is turned, the shape of the entire interscapular canal changes as shown in Fig. 4.

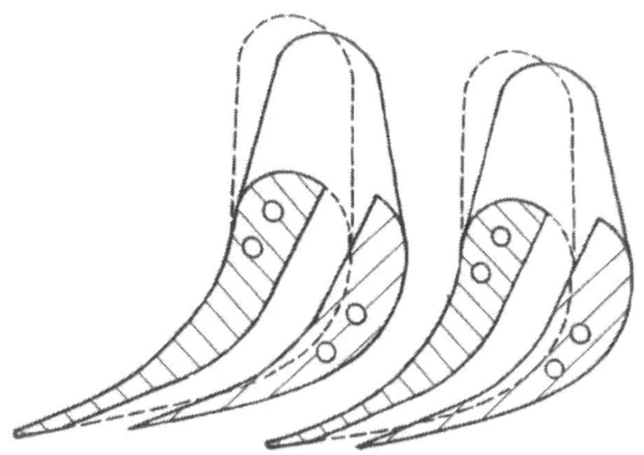

Fig. 4. Rotary diaphragm with variable channel shape

If in the original rotary diaphragm, when it is rotated, the steam inlet into the subsequent interscapular channel is carried out with a sudden expansion (Fig. 1), then in this case, when the passage area of the interscapular channel changes, a smooth bypass is maintained and, accordingly, the profile losses (compared to the original version) decreased by 15% (when closing by 50%) with a simultaneous decrease in the coefficient of unevenness of the velocity field.

However, due to the constructive complexity, such a solution turned out to be not rational.

The solution considered by A.A. Zhinov turns out to be simpler and, at first glance, quite reasonable. in the Kaluga branch of the Bauman Moscow State Technical University, where the rotary diaphragm is separated from the nozzle apparatus of the post-selection stage. At the same time, in order to reduce losses from throttling, the new rotary diaphragm provided the nozzle principle of throttling steam in a partial turbine operation mode. Here, with limited axial dimensions, the principle of nozzle steam distribution was implemented, which is characteristic of high-pressure cylinders of most powerful steam turbines. Accordingly, all the disadvantages inherent in this type of steam distribution [3] have remained, the main of which is the partial (in the circumferential direction) of the steam supply to the chamber between the nozzles of the regulating bodies (rotary diaphragm) and the steam inlet into the nozzle array of the post-extraction stages.

As shown in [3], even with relatively large axial dimensions of the said chamber, the subsequent stage perceives the flow with practically the same circumferential unevenness of the velocity field that is generated by the control stage. In this case, such a generator of circumferential irregularity is an axial rotary diaphragm.

In confirmation of the above, Fig. 5 shows the results of mathematical modeling of the velocity fields in the flow path of the entire complex under consideration, consisting of a rotary diaphragm, an equalization chamber and a nozzle apparatus, performed at the Kaluga branch of the M.V. Bauman. It is clearly seen here that, indeed, the unevenness of the velocities [2], which is generated by the axial rotary diaphragm, passes through the axial space of the alignment chamber without a noticeable decrease in the initial unevenness of the velocity field.

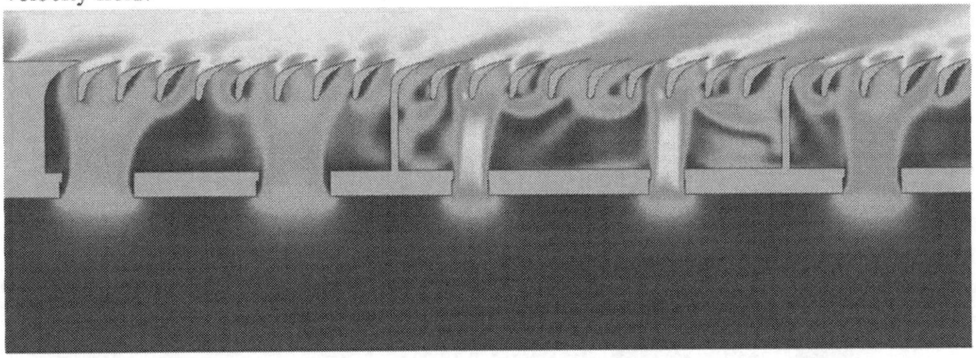

Fig. 5. Velocity field when steam enters the post-extraction stage (full opening of the control diaphragm)

Thus, the considered constructive solutions do not provide an increase in the efficiency and reliability of the blade apparatus of the post-selection stages, while maintaining the axial rotary diaphragms as a regulating body.

In aerodynamic terms, the problem under consideration can be solved by switching from axial to radial diaphragms, which were considered earlier in [4, 5, 6].

3 Constructive version of the radial rotary diaphragm

As an object for aerodynamic research by the method of mathematical modeling of the efficiency of using radial rotary diaphragms, a control radial rotary diaphragm was developed, inscribed in the flow path of the T-56/73-7,8/0,04 KTZ turbine, instead of the existing rotary axial diaphragm.

The structurally investigated rotary diaphragm connected with the nozzle apparatus of the post-sampling stage is shown in Fig. 6.

In this case, the steam supply to the nozzle apparatus 1 with a blade apparatus 2 comes from a special annular chamber 3, consisting of a lower annular profiled bypass 4 connected to the body of diaphragm 2, an upper annular bypass 5 with windows 6 for steam passage on the annular side surface of the bypass 5 , and end cap 7. The minimum axial size of the chamber is determined from the conditions of equality of the total area F_{01}, windows 6.

A rotary radial diaphragm 8 with windows 9, counter windows 6 on the outer cylindrical surface of the chamber 3, were installed on this surface and fixed with annular grooves on the end surface of the cover 7 and in the upper bypass 5.

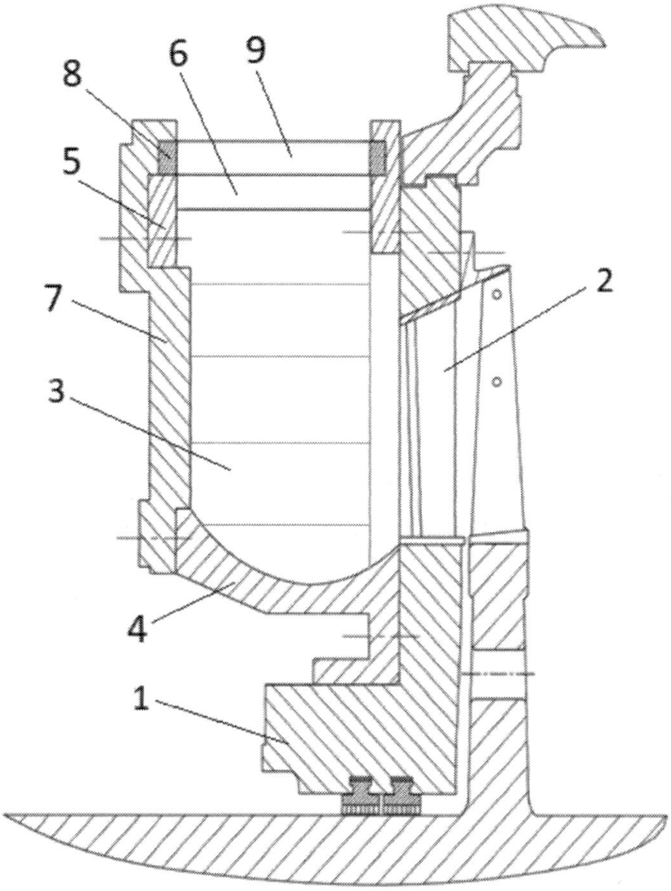

Fig. 6. Rotary diaphragm design

When the diaphragm 8 is turned, the windows 6 on the stationary cylinder 5 are overlapped by the walls of the diaphragm 8, thereby reducing the flow rate of steam supplied to the post-extraction stages. With the complete overlap of windows 6, a certain leakage of steam remains through the inevitable gap between the cylindrical surface of the bypass of the chamber 3 and the cylindrical inner surface of the diaphragm 8. Since in a purely heating mode it is necessary to supply a certain amount of steam to the post-extraction compartment to cool its blade apparatus, in this case the leak steam through the inlet diaphragm is not a disadvantage.

The proposed modernization of the executive bodies for regulating the steam flow rate in the cogeneration extraction of a particular turbine does not require changing the existing rotary diaphragm drive system.

In addition, when using a cylindrical rotary diaphragm, there is no need to unload it from axial forces, which is very important for turbines with industrial steam extraction.

The extent to which the proposed solution solves the problem of ensuring a uniform velocity field when steam enters the nozzle apparatus of the post-sampling stages can be judged from the results of the mathematical modeling carried out below for the steam flow in the flow part of the chamber 3 connected to the nozzle apparatus of the post-sampling stage.

4 Methodology for the study of steam in the nozzle apparatus of the post-extraction stage using a radial rotary diaphragm and its results

To carry out numerical simulation of the rotary diaphragm together with the post-sampling stage, a 3D model of their vapor volume and a computational grid were built (Fig. 7a). To improve the accuracy of the calculation, the mesh was locally refined in the area of the nozzle apparatus, the passage windows of the diaphragm and the mixing chamber.

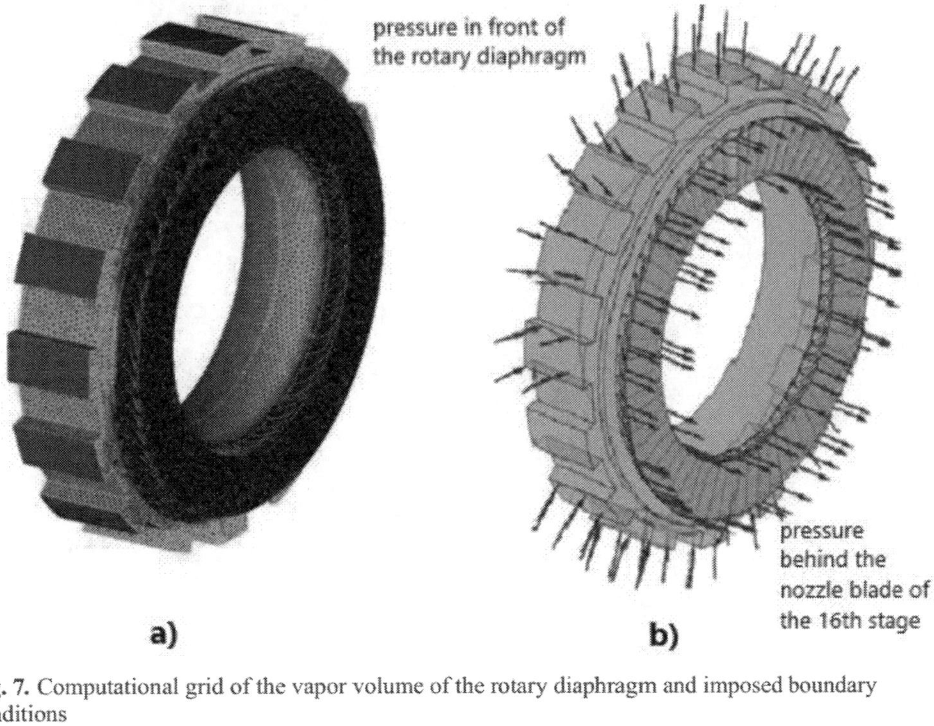

Fig. 7. Computational grid of the vapor volume of the rotary diaphragm and imposed boundary conditions

The boundary conditions imposed on the model (Fig. 7b) were the thermodynamic parameters of the steam in front of the rotary diaphragm and behind the nozzle apparatus of the first post-extraction stage of the turbine. The classical equations of conservation of mass, energy and momentum were taken as the calculation equations, and the k-ε model was used as the turbulence model.

The results of mathematical modeling of the velocity fields in the meridional section of the rotary diaphragm in the condensation mode are shown in Fig. eight.

It is clearly seen here that with the full opening of the diaphragm, a practically uniform field of steam velocities at the entrance to the nozzle apparatus is actually provided, similar to the velocity field when using the original axial diaphragm.

However, the most interesting is the mode with partial overlap of supply windows 6 (Fig. 6)

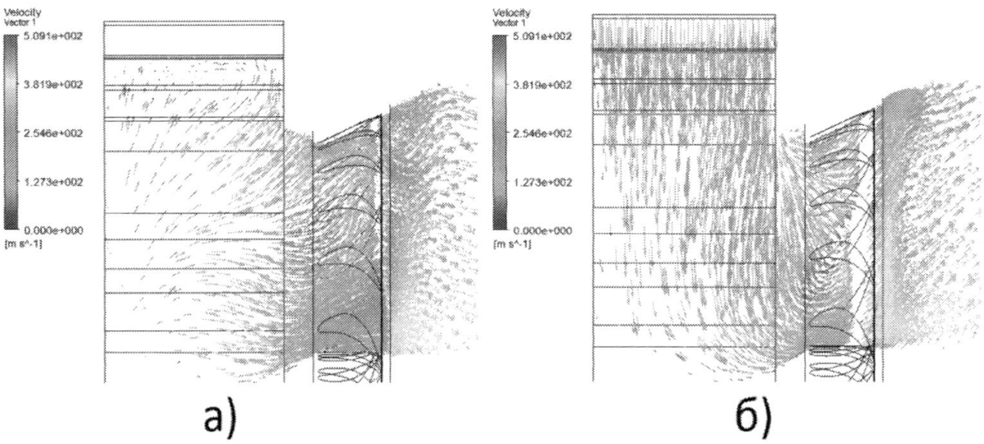

Fig. 8. Velocity vector field a) in the meridian section passing between the entrance windows b) in the meridional section passing through the entrance window

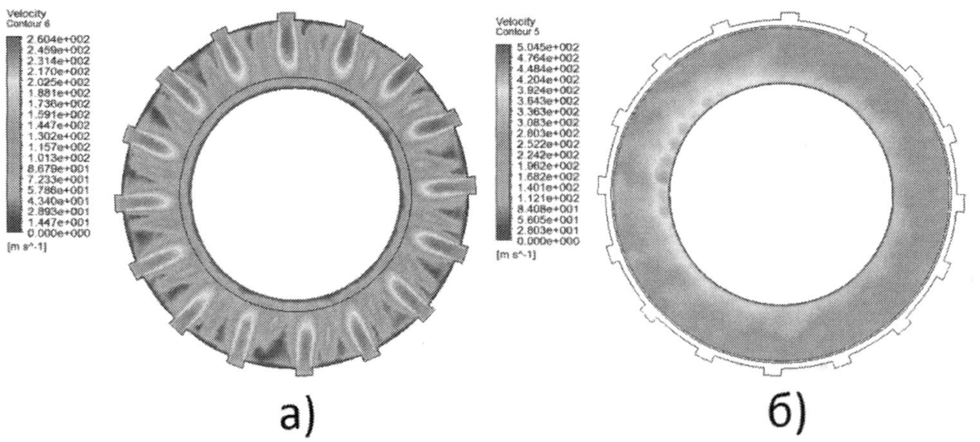

Fig. 9. Velocity field in the inlet and outlet sections when the diaphragm is opened by 50%

For this particular case, Fig. 9 shows the flow pattern in the meridional section of the post-sampling stage with an attached control diaphragm and the end velocity field when entering the steam directly into the nozzle apparatus.

It is clearly seen that in chamber 3 there is indeed an almost uniform velocity field (Fig. 9a). The velocity field in the circumferential direction at the outlet of the nozzle apparatus also changes very little (Fig. 9b), and in the radial direction, due to an increase in the step between the profiles in the nozzle lattice, the velocities naturally decrease.

It is important to note that in the transition from a flat rotary diaphragm to the considered annular rotary diaphragm, the additional dynamic load on the impeller blades of the post-sampling stage practically disappears, due to the influence of the uneven flow of the impeller profiles of the stage impeller entering the array.

5 Conclusions

1. The analysis of the known methods of reducing the dynamic loads on the rotor blades of the post-extraction stages caused by the disturbances introduced into the steam flow by the regulating rotary diaphragms showed that these methods, while maintaining the axial rotary diaphragms, cannot solve this problem.
2. The carried out mathematical modeling of the flow in a complex consisting of a radial rotary diaphragm, a chamber for leveling the velocity field and a subsequent nozzle apparatus showed that with such a combination of a regulating body with a subsequent turbine stage, a uniform velocity field is provided in all turbine operation modes as when steam enters the nozzle apparatus of the post-extraction stage, and when the steam leaves it.
3. The flows obtained as a result of the conducted studies of the flows in the nozzle apparatus of the post-extraction stage with an upstream radial rotary diaphragm clearly show that when using these diaphragms, not only the problems of increasing the efficiency of the post-extraction compartments of turbines with industrial and cogeneration steam extraction are solved, at reduced costs, but also increased vibration reliability of the rotor blades of the entire post-sampling section of the turbine.

References

1. Barinberg. G. D. *Steam turbines and turbine plants of the Ural Turbine Works*; edited by prof., doctor of technical sciences. Yu.M. Brodov and Ph.D. A.Yu. Kultyshev; 3rd ed., *Revised and enlarged*. Yekaterinburg, 2017.
2. Haveman Y. Improvement of steam distribution bodies for steam turbines with steam extraction. Abstract for cann. Diss. M. 1987.
3. Zaryankin A.E., Zroychikov N.A., Rogalev N.D., Rogalev A.N., Mitrokhova O.M. *Influence of the type of steam distribution on the efficiency of high-pressure cylinders of power turbines*. MPEI Bulletin 2015 No. 5 p.5-10.
4. Weller V.N. *Automatic regulation of steam turbines*. Publishing house "Energia" Moscow 1977. pp. 178-180.
5. *Steam turbines of low power* M. Publishing house "Energia" Moscow, 1987. pp. 129-131
6. Haritsky G.F. *Power engineering* 1979 №3 p.37-39.

Verification tests of a new blade flutter research facility

Jan Lepicovsky[1*], *David* Šimurda[1], and *Petr* Šidlof[2]

[1]Institute of Thermomechanics of the Czech Academy of Sciences, v.v.i., Dolejškova 1402/5, 182 00 Praha 8, Czech Republic
[2]Technical University of Liberec Studentská 1402/2, 460 01 Liberec 1, Czech Republic.

Abstract. Long term strategic changes in power generation approaches will require more flexibility for large power generating turbines as an unavoidable consequence of the increasing share of power generated by alternative energy sources. Demanded flexibility for the power turbine output will augment undesired flow phenomena in the low-pressure turbine module, which will consequently enhance blade flutter problems of long slender blades in turbine last stages. In order to advance the understanding of blade flutter onset conditions, the Institute of Thermomechanics of the Czech Academy of Sciences instigated an advanced research program on blade flutter research in high-speed turbomachines. A new innovative test facility for Blade Forced Flutter research was designed and built in the High-Speed Laboratory of the Institute of Thermomechanics. The concept of the new test facility is based on extensive experience with an older Transonic Flutter Cascade facility operated at the NASA Glenn Research Center in Cleveland, Ohio. At present, the first phase of verification tests of the new facility is in progress. The ongoing steady-state tests are intended for exploration of a newly proposed quasi-stationary method to investigate instigating flow conditions leading to an onset of intense blade flutter. Results of some opening tests under steady flow conditions are presented in the paper. The blade drive mechanism for unsteady tests with oscillating blades has not yet been installed in the facility. The presented paper is a work-in-progress report on the ongoing research of complex blade flutter problems.

1 Introduction

A new advanced test facility for Blade Forced Flutter research (BFF) was designed and built in the High-Speed Laboratory (HSL) of the Institute of Thermomechanics (IT) at the Czech Academy of Sciences (CAS) [Lepicovsky et al., 2020]. The concept of the new test module is based on extensive experience with an older Transonic Flutter Cascade (TFC) facility operated at the NASA Glenn Research Center in Cleveland, Ohio [Boldman & Buggele, 1978; Shaw et al., 1986]. Both, the BFF as well as TFC facilities are dedicated to flutter research on linear blade cascades. The new BFF facility features several advanced

* Corresponding author: jandrsc@gmail.com

modifications, namely an independent mechanical drive for each blade and improved instrumentation of tested blades using less expensive miniature commercial pressure transducers. Further, the new facility has provision for blade torque measurements and it will enable installation of optical windows on both test-section side walls for interferometer investigations of the cascade flow field. The BFF blade chord is 120 mm as opposed to 90 mm of the TFC blades, which permits more detailed blade instrumentation. On the other hand, a disadvantage of the BFF cascade is that it consists of 5 blades only versus the 9-blade TFC cascade [Lepicovsky et al. 2002]. The smaller number of blades is dictated by the size and the flow-rate capacity of the IT CAS wind tunnel. Consequently, the BFF cascade tuning for acceptably uniform flow periodicity will be more demanding due to a smaller number of blades

At present, two sets of blades have been fabricated and are being prepared for testing. The first set of blades consists of simple planar blades (FLA). This blade row is intended for exploration of a newly proposed quasi-stationary method modeling the distribution of cascade blade loadings caused by incidence of a single blade at steady-state flow. Data will be used as an initial computational input to predict evolution of flutter oscillations of the remaining blades in a cascade. The purpose of these tests is to generate a reliable experimental dynamic data base needed for better understanding of the onset conditions leading to dangerous blade flutter oscillations and for further refinement of the prediction computation codes.

2 Test facility status

The aim of this paper is to report an ongoing progression of the BFF facility build-up as well as to present some initial experimental data acquired up-to-date. The BFF research facility is in the first phase of its development and built-up. Only steady-state flow tests with no blade oscillations are being performed at present. The test section case with the investigated cascade has been completed and is installed in the modular suck-down wind tunnel [Luxa, 2001; Luxa et al., 2002] in the IT laboratory in Novy Knin (Fig. 1). The tunnel consists of an inlet nozzle (A), located behind a silica gel drying bed, and followed by a transition segment (B) from a circular to a rectangular cross-section flow path. The new cascade test section (C) is located past the transitional segment. Tailboards placed in the exit module (D) are provided for fine tuning of the cascade operation regime. A downstream transition piece (E) connects the entire test section to an exit duct leading to an underground vacuum chamber. The main flow regulator and a shut-off valve, which are not shown in the figure, are located in the main exit duct (F). An enlarged view of the test section is in Fig. 2; the side walls of the exit module (D) are removed to allow installation of the tailboards. A suitable setting angle of the tailboards can to a degree improve flow uniformity in the tested cascade [Lepicovsky et al., 2000; Lepicovsky et al., 2002].

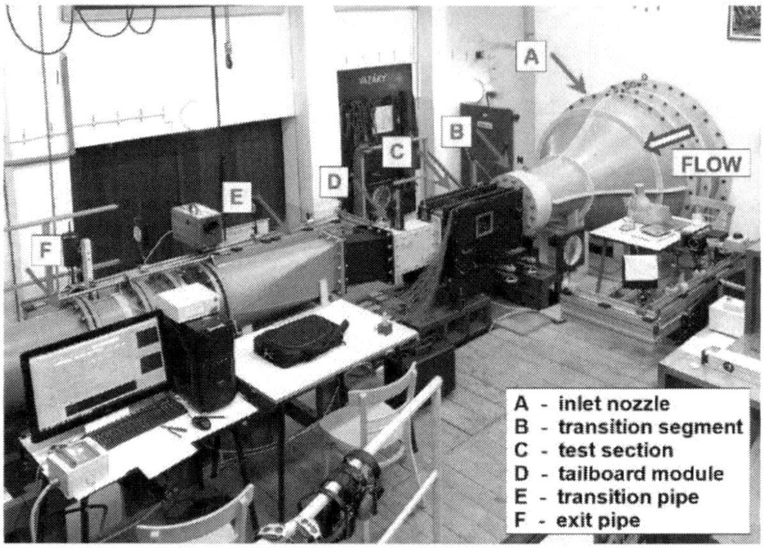

Fig. 1. Blade forced flutter research test facility in the high-speed laboratory of IT CAS.

Fig. 2. Enlarged view of the BFF test module.

A detail sketch showing the cascade test section (CTS) inner arrangement is in Fig. 3. The test section is provided with two blade-supporting bridges located outside of the section flow path walls. The blades, provided with pinions on both ends, are inserted into the slide bearings in both bridges. This arrangement allows setting the blades to a desired incidence angle and around the pinion axis to simulate the blade torsional flutter motion. The right side of the test section, in the flow direction, is labeled the driver side, whereas the left side is labeled the instrumentation side. From an operational perspective it is preferable to separate the blade driven side from the blade instrumented side in particular for easier mutual exchange of test blade installation in the cascade. In the NASA TFC facility the driver and the blade instrumentation was constructed from the same blade side which proved to be cumbersome and very labor intensive for any test cascade reconfiguration. A view of the CTS driver side is in Fig. 4. Blade drive pinions protrude through the CTS flow path inner wall. The driving mechanism has been completed, however has not yet been mounted on the CTS side wall. The driver crankshaft will be affixed to the hexagon tip of the blade pinion. Ends of flat blades can also be seen protruding through the cut-outs in the flow path wall. View of the section instrumentation side with the inner wall removed is in Fig. 5. Five flat blades are visible here. The adjustable flow floor can be also seen in this figure.

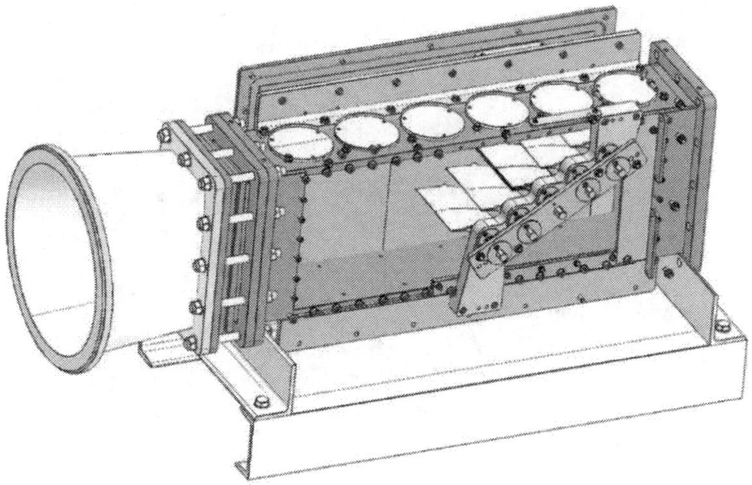

Fig. 3. View of the cascade arrangement and the bridge supporting blades in the BFF test module.

Fig. 4. View of the test module driver side with blade pinions protruding through the cascade inner wall.

The middle blade is set at an incidence angle offset of 3 deg. This blade is instrumented with 10 surface pressure static ports; the connection tubes stick out from the blade body. The second blade above the floor is affixed with six channels ready for miniature high-frequency pressure transducers to be inserted in the blade for the upcoming dynamic testing.

The flow-path inner side wall on the CTS left side is instrumented with a row of wall static taps upstream of the cascade and a similar row of the wall taps downstream of the blade cascade. A bundle of pressure tubing connected to the taps in the section inner wall is shown in Fig. 6. The distribution of the wall static taps and blade suction side taps for the FLA cascade are shown in Fig. 7. Some of the main dimension are also included in the same figure for a better idea about the cascade size. The red tubes connect wall static taps upstream and downstream of the cascade, whereas the white tubes are connected to the middle blade surface pressure taps. The bulkhead passage seal in the lower left corner secures the tubing connection out of the test module to the data acquisition equipment. Not all the tube connections are active for a given test run due to a limited number of reading pressure ports.

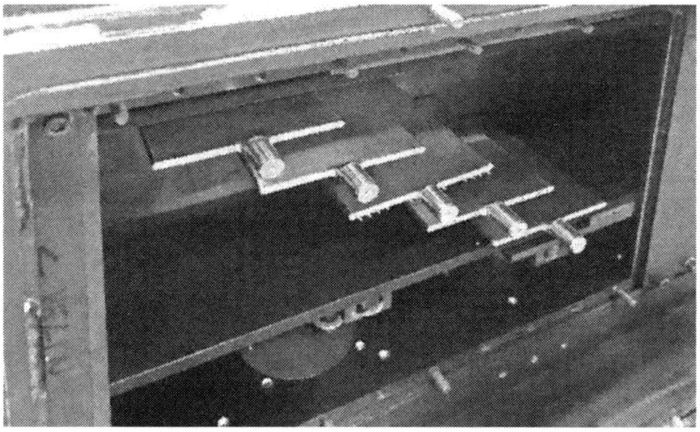

Fig. 5. View of blade cascade instrumentation side with removed section inner wall.

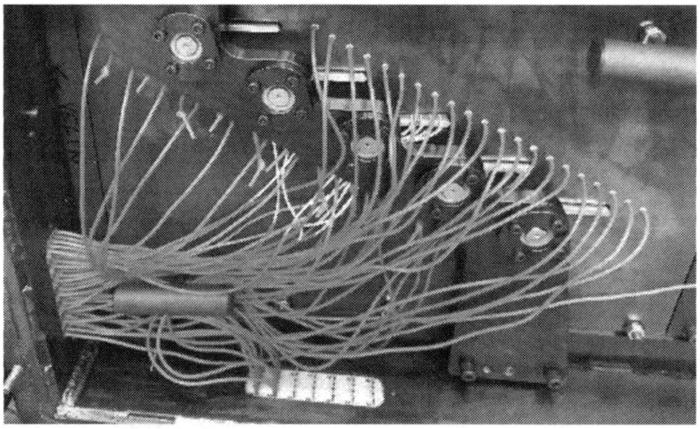

Fig. 6. Routing of connecting tubes from wall pressure taps to a pressure scanner.

3 Experimental results

Cascade inlet and exit Mach numbers as well as the overall cascade pressure ratio were determined during the initial sets of experiments. The average values of the inlet total pressure, measured by a single probe past the inlet nozzle in the channel center, and averages of wall static pressures measured upstream and downstream of the cascade (see Figs. 6 and 7) were used to determine cascade operation conditions. The cascade inlet and exit Mach numbers are isentropic values calculated from the above-mentioned data. A linear cascade (FLA) comprised of five flat blades was used initially. The FLA blade cascade exhibits features of a turbine blade row. A quick glance over data in Figs. 8 and 9 confirms this remark. The flow Mach number increases and the pressure decreases while passing through the cascade.

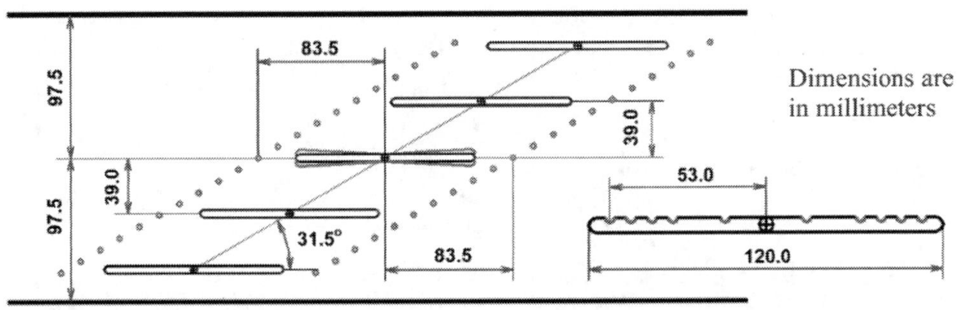

Fig. 7. Locations of side wall pressure taps upstream and downstream of the cascade.

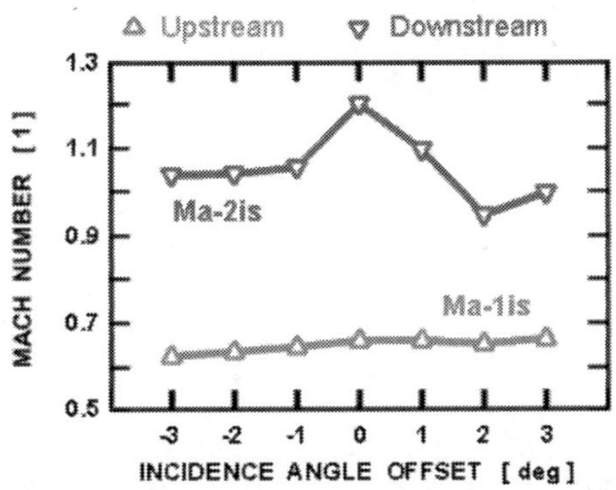

Fig. 8. Effects of middle blade incidence angle offset on FLA cascade peak performance.

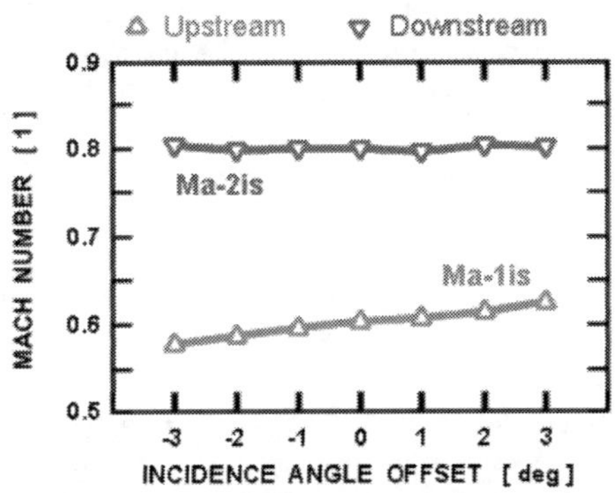

Fig. 9. Effects of middle blade incidence offset on FLA cascade inlet flow at exit Mach number of 0.8.

The effects of changing the incidence angle offset (ICO) of the cascade middle blade are depicted in the above-mentioned Figs. 8 and 9. The incidence angle offset of the middle blade was gradually adjusted from -3 deg to +3 deg in steps of 1 deg while the the tunnel control valve was opened to the maximum through-flow rate. As seen in Fig. 8 the inlet Mach number (Ma-1) is practically not affected by the incidence angle change, rising very mildly from the level of 0.62 up to 0.66, whereas the average exit Mach number (Ma-2) first increases from a level of 1.04 at the incidence angle offset of -3 deg, then peaks up to 1.2 at the zero-incidence offset and eventually drops down to 0.99 for the incidence angle offset of +3 deg. Data presented in Fig. 9 shows results of similar investigations when the exit Mach number was kept constant at a level of 0.8. The inlet Mach number was steadily rising for this operation regime from a level of 0.58 to 0.63 while the incidence offset angle varied from an angle of -3 deg up to the level of + 3 deg.

The effects of the middle blade variations of the incidence offset (BIO) on the static wall pressure levels upstream of the FLA cascade appear to differ noticeably between the lower and the upper halves of the cascade as can be seen in Figs. 10 and 11. The ratios of local pressures and the total inlet pressure are plotted here for the cascade upstream and downstream positions. The position of the blade pin axes are marked by small black circles in the figures.

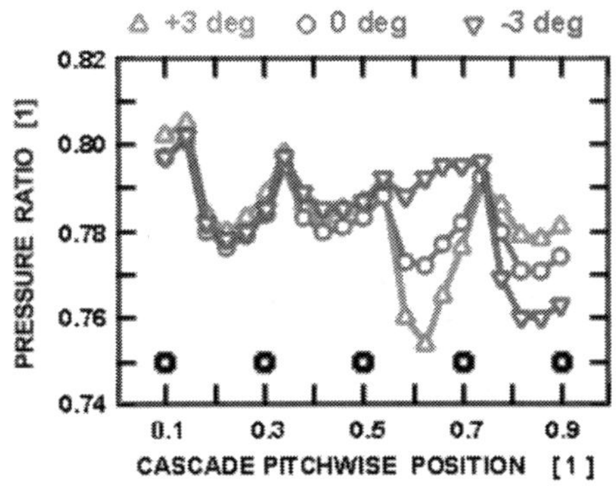

Fig. 10. Effects of middle blade incidence angle offset on wall pressure distribution upstream of FLA blade cascade (Ma-in = 0.6).

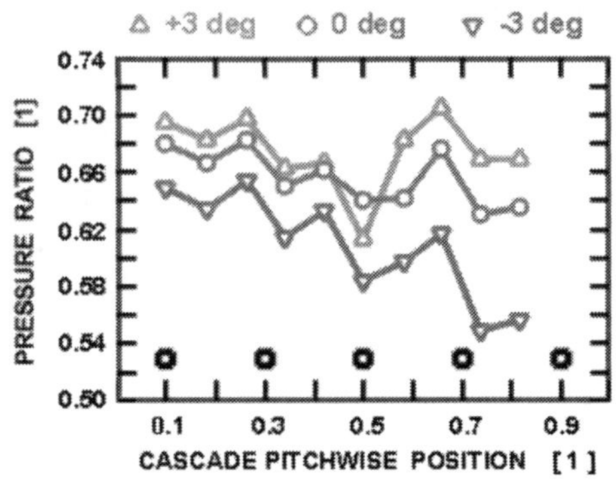

Fig. 11. Effects of middle blade incidence angle offset on wall pressure distribution downstream of FLA blade cascade. (Ma-in = 0.6)

As can be seen in Fig. 10, the incidence offset of blade 3 in the cascade (middle blade) has no effect on the wall pressure distribution for the lower cascade half (blades 1 through 3). On the contrary the upper half of the cascade span between blades 4 and 5 is overwhelmingly affected by the blade 3 incidence offset. In the blade channel 3-4 the pressure level rapidly plunges for the incidence angle offset upturn and it rapidly increases for the incidence angle offset decline below the nominal value. The incidence angle offset effects on the wall static pressure distribution in the channel 4-5 are just reversed to the changes in the blade channel 3-4. The effects of varying offset angle setting for blade 3 are noticed downstream of the cascade along the entire span of the blade row. These changes are stronger over the entire span of the cascade, albeit the changes are again more pronounced for the upper cascade half (blades 3 through 5) than for the lower cascade half (blades 1 through 3).

At present, the test blades are being instrumented with static taps as well as with miniature high-frequency pressure transducers [Lepicovsky, 2005]. The blade instrumentation is not yet completed. To date, only one FLA blade is instrumented with static taps. The blade is instrumented on one side only. However, thanks to the blade symmetry, the blade can be turned by 180 deg in the cascade and consequently the surface pressure distribution can be eventually obtained on both sides of the blade. The blade loading distributions are presented in terms of pressure coefficients as usually defined for compressible flow:

$$C_p = \frac{p_i - p_1}{0.5 \cdot k \cdot p_1 \cdot M_1^2}$$

where p_1 is the average static pressure as measure upstream of the cascade, p_i is a local pressure measured at a given location on the blade surface, k is the ratio of specific heats, and M_1 is the inlet Mach number calculated from the measured upstream data.

The measured pressure coefficient distributions on both surfaces of the middle FLA blade #3 are shown in Figs. 12 through 14. Data shown in these figures are for the blade incidence angle offsets of -3 deg, 0 deg, and +3 deg, respectively. Data reveals huge changes in the blade pressure coefficient values even for such relatively small changes of the blade incidence angle offsets. Acquisition of unsteady data for the blade pressure side as well as the blade loading of the adjacent blades was not competed at the time of writing this paper. Therefore no substantiated judgment of the entire blade loading can be stated at present. The same set of data is replotted in Figs. 15 and 16 where the pressure coefficient distributions are plotted separately for the upper and lower side of the cascade middle blade. The effects of the blade incidence offset angle variations are clearly demonstrated here. Nevertheless, partial data in Figs. 12 through 16 strongly indicate that the changes in the blade loading during the flutter oscillations will be significant and strong enough to modify the instant flow pattern not only in the blade vicinity but even in an extended range of the entire cascade. Consequently, fast flow pattern changes will affect the loading on the adjacent blades in the row and will excite the flutter oscillations of additional blades in the cascade.

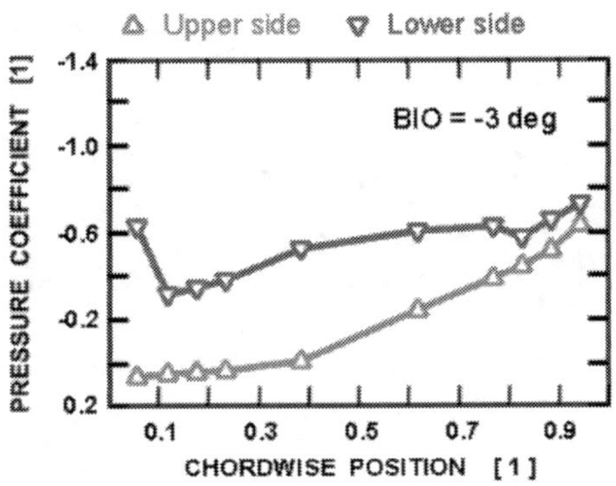

Fig. 12. Distribution of surface pressure coefficient on blade #3 for an incidence angle offset of -3 deg.

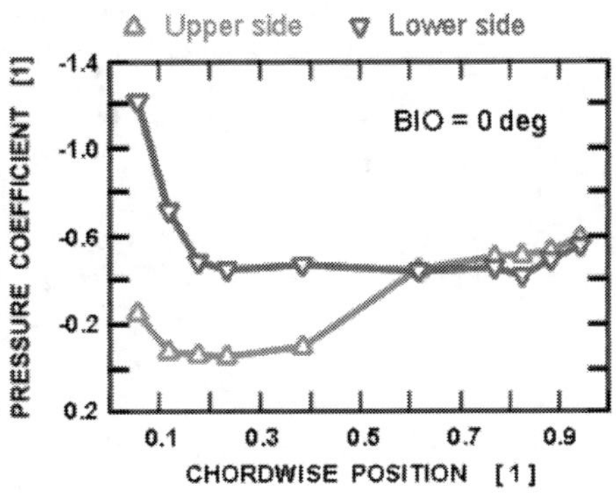

Fig. 13. Distribution of surface pressure coefficient on blade #3 for no incidence angle offset.

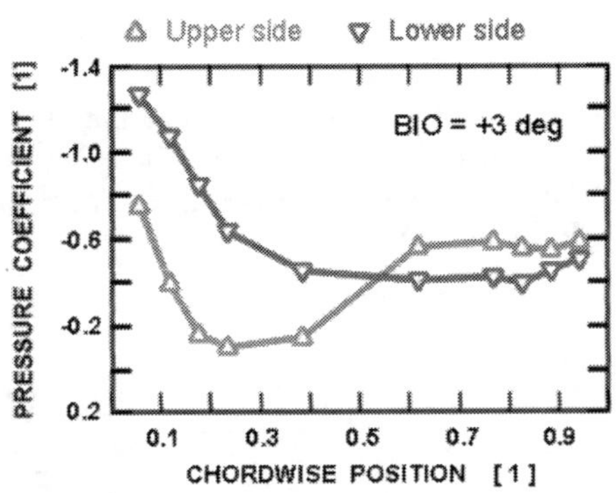

Fig. 14. Distribution of surface pressure coefficient on blade #3 for an incidence angle offset of +3 deg.

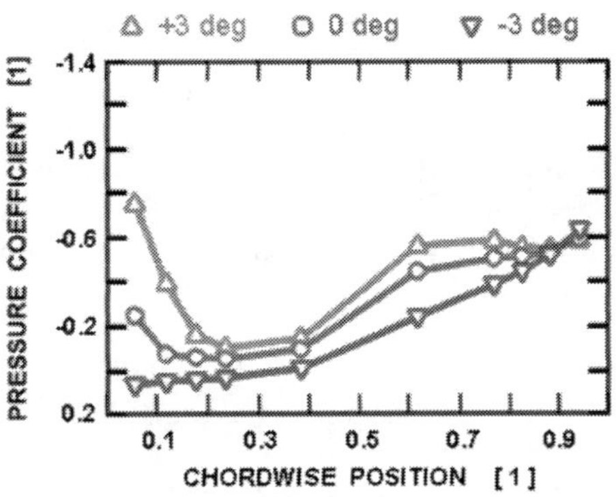

Fig. 15. Effects of middle blade incidence angle offset on pressure distribution on the upper side of the middle blade in cascade (Ma-in = 0.6).

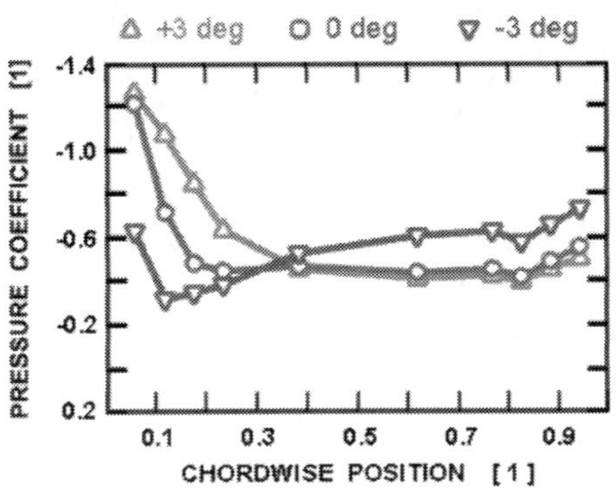

Fig. 16. Effects of middle blade incidence angle offset on pressure distribution on the lower side of the middle blade in cascade (Ma-in = 0.6).

A computational study of the effects of varying cascade middle blade incidence angle offset is being performed in parallel with the experimental effort. The aim of this complex project is to verify validity of the proposed alternative quasi-stationary approach to blade flutter onset research. This method is based on a phased-out stationary measurements of blade loadings at a range of blade incidence angle offsets, which can be determined more easily and efficiently than collecting data from a blade under flutter vibrations.

4 Conclusions

The presented paper is a work-in-progress report on the initial phase of the ongoing research on complex blade flutter problems. Nevertheless some conclusions may be already drawn based of the few preliminary results.

- The modular arrangement of the entire test section enables easy and fast reinstallation of this test module in the wind tunnel frame depending on the wind tunnel test section availability and operation schedule.
- The newly built test section proved to be well suited for advanced research of a single or multiple blade incidence angle "misalignment" effects on cascade performance characteristics under steady flow conditions. Setting the incidence angle offsets for individual blades can be arranged from the outside of the test section and is easy and fast.
- The test section is provided with adjustable floor and ceiling in front of the cascade and adjustable tailboards downstream of the cascade. The initial tests were performed for fixed positions of these adjustable articles. Data in Figs. 10 and 11 indicate that some adjustment may be needed to improve the uniformity of the flow periodicity in the cascade. Boundary layer suction upstream of the tested cascade is intended, albeit not available yet. Common experience is that boundary layer suction is the most effective way for flat cascade flow periodicity improvement [AGARD-AG-328, 1983, Song & NG, 2007].

The work was supported by the Grant Agency of the Czech Republic under grant no. 20-11537S. Institutional support of the Institute of Thermomechanics of the Czech Academy of Sciences RVO: 61388998 is also gratefully acknowledged.

References

1. AGARD-AG-328, "Advanced Methods for Cascade Testing", Linear Cascades, Chap. 2, pp. 13-73, Neuilly sur Seine, (1983).
2. Boldman, D.R. and Buggele, A.E.: "Wind Tunnel Tests of a Blade Subjected to Mid-chord Torsional Oscillations at High Subsonic Stall Flutter Conditions", NASA TM-78998,(1978).
3. Luxa, M.: "Budovani noveho aerodynamickeho tunelu v Novem Knine", Symposium o anemometrii, ISBN 80-238-7138-2, Praha (2001) (in Czech)
4. Luxa, M., Jonas, P., and Safarik, P.: "Eintrittentwurf des WindKanals", Fluid Mechanics and Thermodynamics Symposium, ISBN 80-902714-9-9, Praha (2002) (in German).
5. Lepicovsky, J., McFarland, E.R., Chima, R.V, and Wood, R.V.: "On Flow Periodicity in the NASA Transonic Flutter Cascade, Part-I, Experimental Study", ASME 2000-GT-0572, (2000).

6. Lepicovsky, J., McFarland, E.R., Capece, V.R., and Senyitko, R.G.: "Methodology of Blade Unsteady Pressure Measurement in the NASA Transonic Flutter Cascade", NASA/TM-2002-211894, (2002).
7. Lepicovsky, J.: "Effects of Transducer Installation on Unsteady Pressure Measurements on Oscillating Blades", ASME-2005, DOI:10.1115/GT2005-6900, (2005).
8. Lepicovsky, J., Sidlof, P., Simurda, D., Stepan, M., and Luxa, M.: "New Test Facility for Forced Blade Flutter Research", XIX International Conference Power System Engineering PSE 2020, p. 20, Pilsen (2020).
9. Shaw, L.M., Boldman, D.R. Buggele, A.E., and Buffum, D.H. "Unsteady Pressure Measurements on a Biconvex Airfoil in a Transonic Oscillating Cascade", J. of Eng. for Gas Turbines and Power, Vol. 108, pp. 53-59, (1986).
10. Song, B. and Ng, W.F.: "The Role of AVDR in linear cascade testing", Journal of Aerospace and Power, Vol. 22, No. 6, (2007).

MATEC Web of Conferences **345**, 00021 (2021) https://doi.org/10.1051/matecconf/202134500021
Power System Engineering 2021

Design of an experimental device for preheating combustion air

Lucia Martvoňová[1*], *Mária* Polačiková[1], *Juraj* Drga[1], and *Alexander* Backa[1]

[1]University of Zilina, Department of mechanical engineering, Department of power engineering, Zilina, Slovakia, Univerzitná 8215/1 01026, Zilina, Slovakia

Abstract. The principle of the proposed device is to use part of the waste heat, which otherwise leaves through the chimney unused into the air, to preheat the combustion air. reducing chimney losses and preheating the combustion air will increase the efficiency of the heat source. the device is actually a gravity loop heat pipe with an evaporating part located behind the furnace and a condensing part in the combustion air supply duct. Heat transfer is realized by means of phase transformation of the working medium in the proposed device.

1 Introduction

Fireplace inserts as a heat source produce heat output by burning fuel. In the form of flue gases, a large amount of waste heat is released through the chimney into the air, which also reduces the efficiency of the fireplace insert itself [1]. It is possible to remove part of the heat from the flue gas, which can be used for water heating [2]. When using these applications, it can be a problem to ensure the safe operation of the fireplace insert in the system and therefore it is necessary to include safety elements such as safety valves, expansion vessels, etc. Another disadvantage is the need for storage tanks, which not only take up space but also increase investment costs. The most advantageous solution therefore seems to be the use of waste heat directly in the combustion processes.

1.1 Loop Thermosyphon

Loop thermosyphons are passive heat transfer devices using phase transformation with almost negligible heat loss. The appropriate combination of the working medium used and the construction packaging defines the operating temperature range of the heat pipes [3].

Heat pipes have proven to be an effective solution especially in high temperature applications and insulation, where there is any combination of uneven heat load, limited air flow through the heat generating components as well as space or weight constraints. The loop thermosyphon is used to transfer heat over relatively long distances without additional mechanical devices. The advantage is also flexibility in a limited space and the ability to work in all working conditions [4].

* Corresponding author: lucia.martvonova@fstroj.uniza.sk

This type of heat pipe can only work in the position of the condenser above the evaporator. The movement of the working substance from the condenser to the evaporator occurs due to the weight of the working substance. The inner surface of the heat pipe is smooth and does not contain a capillary structure. The amount of working substance must be sufficient for perfect wetting of the evaporator walls, but not too large due to the formation of bubble boiling in the evaporator [5].

2 Equipment to increase the efficiency of the fireplace insert

Important factors for calculating the efficiency of a fireplace insert include the combustion air temperature and the flue gas temperature. By inserting a heat pipe with a closed loop into the fireplace insert system, we can use part of the heat escaping in the form of flue gases to preheat the combustion air. By removing part of the heat from the flue gas, the chimney losses are reduced and by transferring this heat through the heat pipe, the combustion air is heated, which also reduces the need for fuel.

When designing the device, we used the construction of the fireplace insert: FIG.1. The evaporating part of the heat pipe is located in front of the chimney mouth, where it removes heat from the flue gas, which is transferred in the form of saturated steam through a steam pipe to a condenser located in the combustion air supply channel. Due to the location of the condenser above the evaporator, a loop thermosyphon was used, where the working medium can move under its own weight.

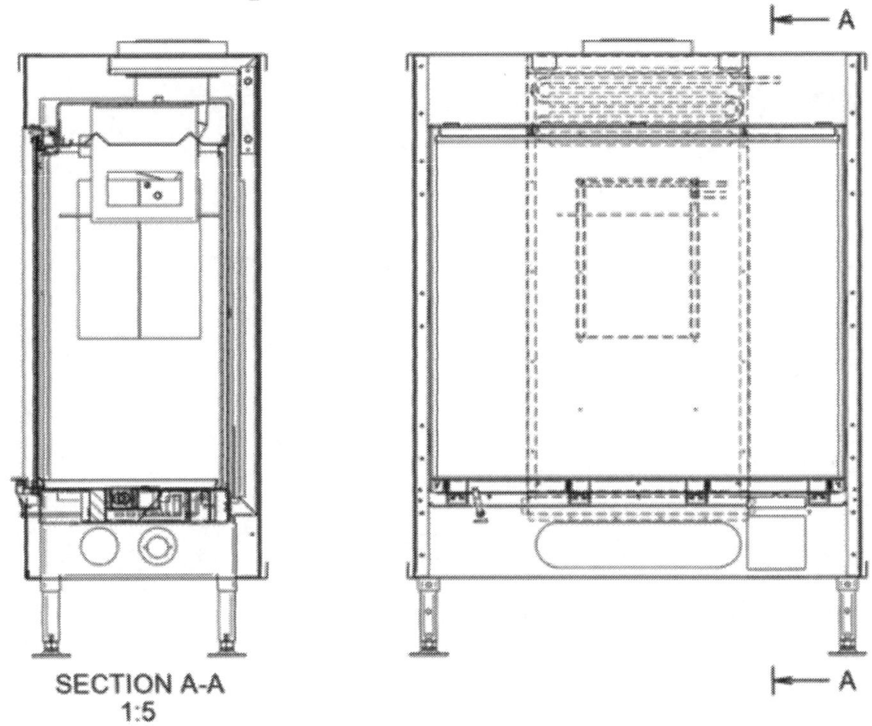

Fig. 1. Placement of a loop thermosyphon in the fireplace insert.

FIGURE 2. shows the dimensioned external dimensions of the evaporator together with the dimensions relating to the location of the evaporator relative to the condenser. Due to the atypical chimney of the used fireplace insert, we had to adjust the duct for flue gas discharge from the evaporator space.

In addition to the design, the correct choice of heat pipe material and working medium. The material of the heat pipe must be subject to several conditions and, in addition to the required high thermal conductivity, compatibility with the working medium must also be ensured [6]. At the same time, the device must not corrode and must withstand higher chimney temperatures. The ideal combination for this application is the use of copper as a wall material and water as a working medium. The results of chemical and metallographic tests show that copper and water are compatible. The release of inert gases occurred only with the use of surfactants to improve the wettability of the copper capillary systems and with the incorrect removal of the residues of these substances.

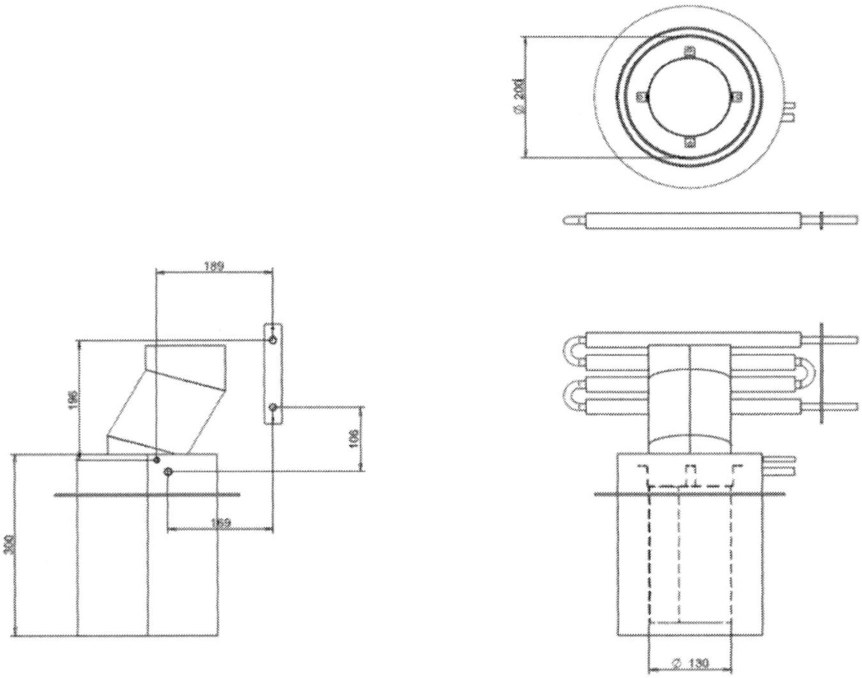

Fig. 2. Dimensions of the device to increase the efficiency of the fireplace insert.

2.1 Evaporator

By placing the evaporator at the chimney inlet, we can use the largest amount of heat from the flue gas. In order to prevent the flow of flue gases through the center of the evaporator, we placed a deflector in the evaporator: FIG. 3. It directs the flow of flue gases so that they bypass the inside of the evaporator. By reducing the cross-section, the flue gas flow rate will also increase, which will also give us a higher Reynolds number and a higher heat transfer between the flue gas and the evaporator.

Due to the direct contact of the evaporator with the flue gases, we chose stainless steel as a material, which is also compatible with the working medium, water.

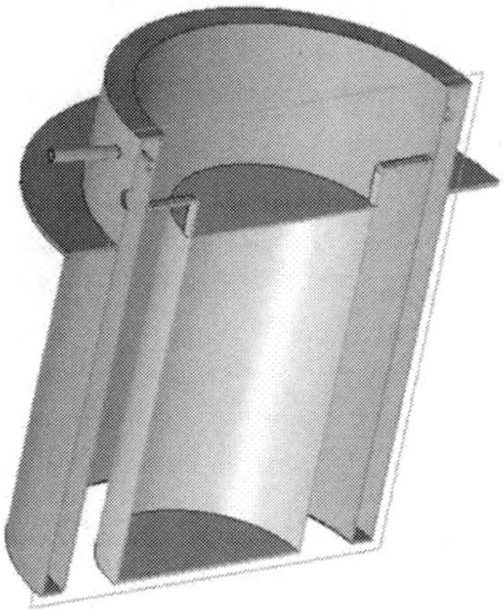

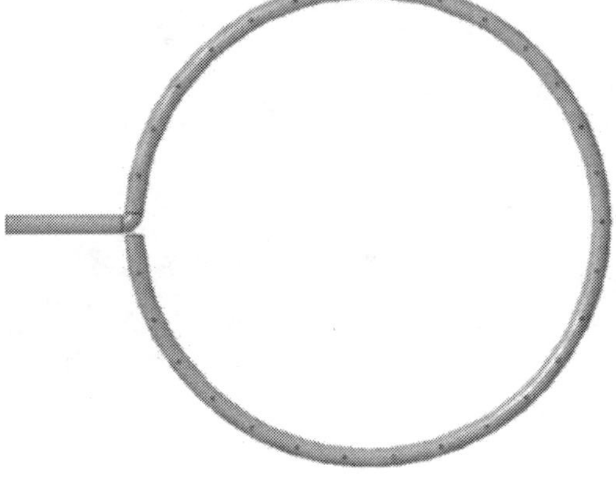

Fig. 3. Evaporator and deflector cut. **Fig. 4.** Distribution pipe.

In order to use the full potential of the evaporator through the distribution pipe: FIG. 4. with 30 holes with a diameter of 1.5 cm evenly distributed around its circumference, condensate is supplied. If condensate flows into the evaporator only directly from the condensing line, it could happen that the condensate is concentrated in only one part of the evaporator and the condensate does not evaporate evenly.

2.2 Condenser

The condenser has been designed according to the combustion air supply duct in which it will be located. The wall material of the condenser is the same as for the heat pipe, i.e. copper. In order to ensure a higher heat transfer between the condenser and the supply air, we used rolled aluminum fins: FIG. 5.

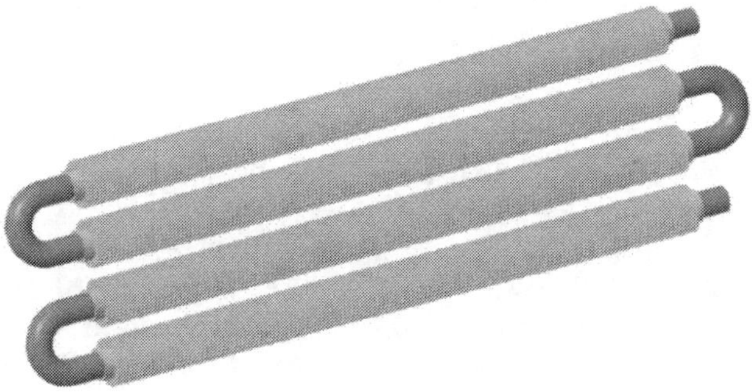

Fig. 5. Condenser.

2.3 Evaporator and condenser cooperation

The principle of the loop thermosyphon: Fig. 1 consists in the phase transformation of the working substance. In our case, the water in the evaporator is heated by heat from the flue gas until it reaches the boiling point and changes to steam. Due to the lower density of steam, it spontaneously flows further into the condenser located above the evaporator through a steam pipe with a diameter of 12 mm, where it transfers the heat obtained and turns it back into a liquid. Due to its own weight, the condensed steam flows through an 8 mm diameter condensing pipe, which is connected to a distribution pipe for even condensate flow into the evaporator and the cycle is repeated. Thanks to this natural flow, it is not necessary to include a pump or capillary structure in the system to get the condensate back into the evaporator.

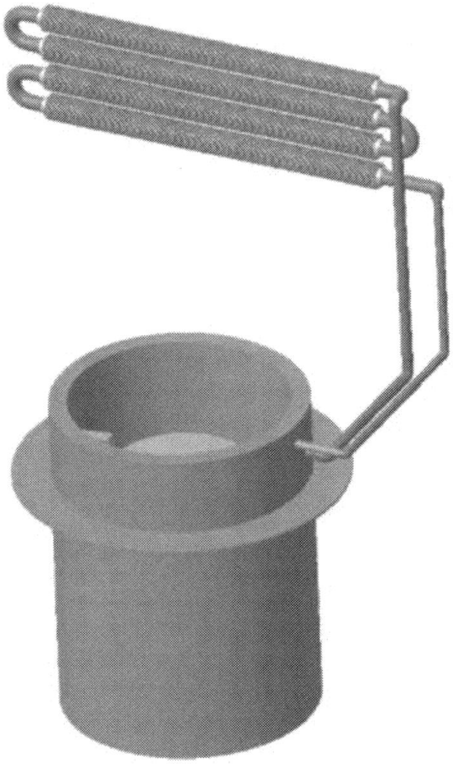

Fig. 6. Condenser and evaporator connection.

Despite the relatively simple construction of this loop thermosyphon, a higher demand for vacuum tightness can be a disadvantage in production. However, the device constructed in this way is low-maintenance, which is a great advantage today.

The device constructed in this way is automatic and in the case of shutting down the fireplace insert, the supply of heat for heating the working substance is interrupted, which gradually stops the phase transformation in the heat pipe. If, on the other hand, there was too much heat transfer between the flue gas and the evaporator and the condenser could not remove the heat obtained by the evaporator, the pressure in the heat pipe would gradually increase, which would also increase the boiling point. Thanks to this natural cycle, it is not necessary to include a large number of safety elements in the system, nor is there a need for an accumulation tank.

3 Conclusion

Increasing the efficiency of heat sources by improving the combustion process is slowly reaching its limits, so we must focus on other possible solutions. Only by using part of the heat from the flue gas to heat hot water can we reduce chimney losses, which have a great impact on the efficiency of the heat source. However, if we use this heat to preheat the combustion air, we can significantly increase the efficiency of the heat source.

In the case of fireplace inserts with solid fuel combustion, a more significant increase in efficiency is expected; when burning solid fuels, chimney temperatures reach higher values than when burning gaseous fuel.

The device with a heat pipe with a closed loop also opens up possibilities for increasing the efficiency of already standing fireplace inserts and small heat sources, where with a relatively small investment we can easily increase the efficiency of a heat source for heating.

This work was supported by projects KEGA 033ŽU-4/2018 Heat sources and environmental pollution and APVV-17-0311 "Research and development of waste-free technology for decomposition and selection of undesirable components from the process gas generated by the gasifier".

References

1. R Nosek, M Holubčík. Energy properties of air dry firewood. Acta Facultatis Xylologiae Zvolen, 58/1, 105-112 (2016)
2. R Nosek, M Holubčík, Š Papučík. Emission Controls Using Different Temperatures of Combustion Air, Scientific World Journal, 487549 (2014)
3. Nemec, P. – Caja, A. - Lenhard, R. (2009) Analysis of heat transfer limitation of wick heat pipe, Conference: International Conference on Experimental Fluid Mechanics Location: Liberec, CZECH REPUBLIC, EXPERIMENTAL FLUID MECHANICS 2009, PROCEEDINGS OF THE INTERNATIONAL CONFERENCE Pages: 230-235
4. Novomestský, M. – Kapjor, A. – Papučík, Š. – Siažik, J. (2016) Heat pipe thermosyphon heat performance calculation Conference: THE APPLICATION OF EXPERIMENTAL AND NUMERICAL METHODS IN FLUID MECHANICS AND ENERGY 2016: XX. Anniversary of International Scientific Conference
5. HLAVAČKA, V., POLÁŠEK, F. a ŠTULZ, P. a ZBOŘIL, V., 1990. Tepelné trubice v elektrotechnice. 1. Vyd. Praha: STNL – Nakladatelství technické literatúry. ISBN L25-B3-IV-4/52783.
6. Lenhard, R. – Kaduchová, K. – Papučík, Š.(2014) Analysis of the fill amount influence on the heat performance of heat pipe. [online], 1608(1), 146-152 [cit. 2020-09-07]. ISSN 0094-243X.

MATEC Web of Conferences **345**, 00022 (2021) https://doi.org/10.1051/matecconf/202134500022
Power System Engineering 2021

Adaptable test rig for two-stage axial turbine

Martin Němec[1*], and *Tomáš* Jelínek[1]

[1]VZLU – Czech Aerospace Research Centre, Beranových 130, 199 05, Prague – Letňany, Czech Republic

Abstract. This contribution describes a new test rig for a two-stage axial turbine built in the VZLÚ. The test rig has replaced an original facility used for a full stage aerodynamics investigation. The motivation for the design of the new test facility was the limitations of the original one. The design is briefly discussed, and then the first measurement results are presented. The first operation was performed with a turbine stage already measured in the original facility. This allows the comparison of the most important quantities.

1 Introduction

The investigation of flow fields in turbomachines is one of the basic fields of research that VZLÚ has been engaged in for a long time. It is partially due to long-term cooperation with the world's leading producers of turbomachines. Since 2010, a measuring device has been put into operation at the VZLÚ, which enabled flow research on a single-stage axial turbine stage: measurement of integral characteristics using a water dynamometer and measurement of flow fields by traversing behind the stator and rotor blades [1], [2]. Thanks to this device, it was possible to bring new knowledge to the research of flow in the turbine stages and a lot of data to refine the producer's design models [3], [4].

During its operation, the device was gradually improved. Its arrangement was expanded to a one-and-half stage by adding a subsequent stator behind the stage. The shape and positions of both shaft and rotor shroud seals were also improved over the past decade according to the requirements of individually designed projects [5], [6].

At present, in cooperation with industrial partners, there is an increasing demand for research into the mutual interaction of purge flows emerging from seals with the main flow in the turbine and mutual interaction in applications of multistage arrangements. Research of design-overloaded stages with non-axial inlet is also a topic of today.

The above-mentioned new demand in the field of research, together with the broad experience gained by operating the existing experimental turbine, has necessitated the creation of a new test rig concept that would allow research on a two-stage axial test stage.

* Corresponding author: nemec@vzlu.cz

2 Basic concept of the new test rig

The basic concept of the newly created device is based on the original device (section shown in Fig. 1), which was part of a closed-loop wind tunnel (depicted in Fig. 2) driven by a turbo-compressor with a power of 1.3 MW and providing the mass flow of 7 kg/s. To minimize costs, it was decided to use the existing Froude-Hoffman F249-RACE-GT water dynamometer for loading the turbine stage with a maximum loading power of 700 kW, and a maximum speed of 15,000 rpm.

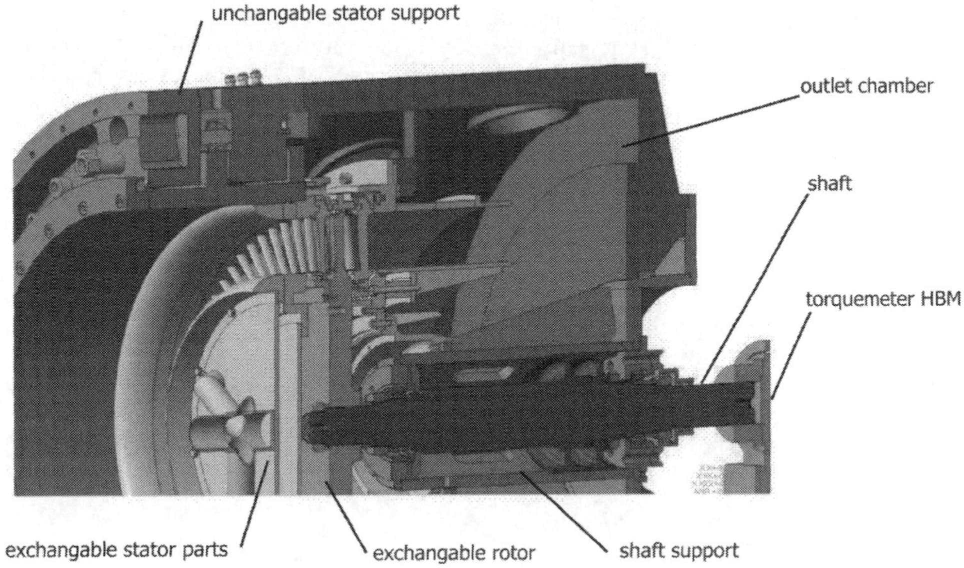

Fig. 1. The original turbine - the one-and-half-stage configuration

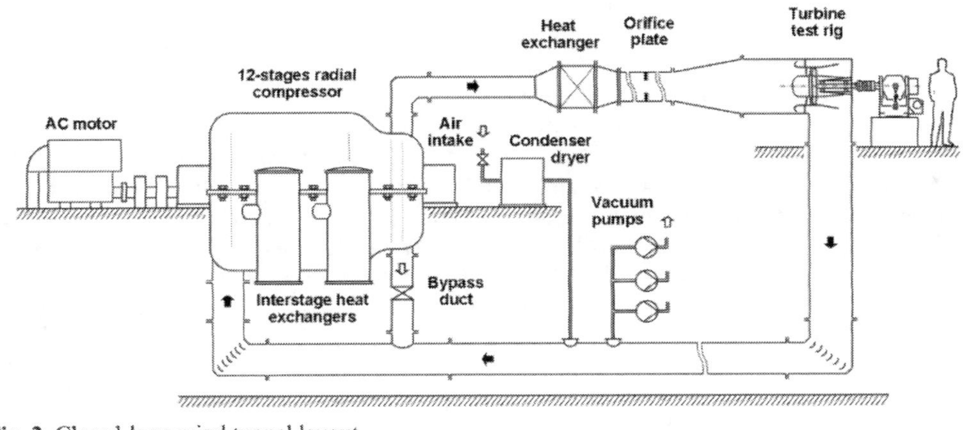

Fig. 2. Closed-loop wind tunnel layout

In addition to enabling the measurement of the characteristics of two axial turbine stages, the goal for the new plant was to eliminate the following disadvantages and limits of the original concept: i) Critical speed, ii) Refinement of mechanical bearing losses, iii) Removal of double skin arrangement.

The original device had a limitation especially in terms of critical speed, which was caused by insufficient rigidity of the facility structure – frame, outlet chamber, shaft support. Furthermore, it did not allow mechanical measurement of shaft bearing losses. The losses due to bearing friction were estimated by measuring the bearing temperature. Time-consuming stabilization was one of the disadvantages of such a loss determination.

Another limitation was the double-walled arrangement, which caused problems in setting up the probes and makes optical access difficult when using optical methods. On the other hand, the double-walled arrangement did not require thermal insulation of the stage from the surroundings, and in the case of measurements using thermocouple probes, the error caused by conducting heat from the outside to the sensor itself was reduced.

Two-stage turbine research is typically focused on the interactions of purge flows at the rotor tip (influence of clearance, blade tip shape in the case of shroud-less stages, shroud geometry, shroud output shaping), hub area geometry (shaft seal and its outlet), clocking of both rotors and stators, and the effects of axial clearances of the blade rows. The new facility should therefore allow these activities to be carried out.

To gain an overview and inspiration for the solution of individual key elements of the new equipment as well as for the solution of more detailed structural units, a search of operated two-stage turbines was performed in the literature and on the websites of the most important research institutes focused on the field of turbines.

In general, test equipment for two-stage arrangement is most often focused on the research of the two stages interaction of internal combustion turbines used for aircraft, mostly the transition module between the high-pressure and low-pressure stages. This structural unit is characterized by the fact that the high-pressure and low-pressure turbines have different speeds, which corresponds to the concept of test equipment - double-shaft turbines with each shaft on a separate dynamometer. The following is an overview of some two-stage test turbines.

- DLR - Institute of Propulsion Technology - NG Turb equipment [7]
- Onera, Modane Avriex Center (CMA) - TURMA equipment [8]
- Graz University of Technology - Transonic Test Turbine Facility [9]
- University of Notre Dame - Multistage Turbine Test [10]
- ETH Zurich - LISA equipment [11]

The initial considerations about the basic concept of the facility were mainly based on the analysis of previous activities performed on the original facility within the solved projects. The most common activity of the VZLÚ turbomachine laboratory is the testing of numerous variants of turbine stages for various industrial partners - turbine producers. Therefore, the basic requirement was that the device must be designed to allow a relatively simple exchange of tested stage geometries and to allow the installation of stages of different diameters, limited by a hub diameter of about 220 mm and a tip diameter of about 600 mm. Another limitation of the measured stages is, of course, the mass flow provided by the compressor.

The ETH Zürich solution can best serve as inspiration for a similar concept - one dynamometer for two stages with the same speed. In addition, very detailed information about the construction is available thanks to the description in the dissertation [11]. The key element of the design is the separate measurement of torques of both stages by means of capacitive measurement of the torsion of two shafts (rotation of the shafts at 70% of the torque is 4 °).

The requirement for equipment adaptability (enabling easy exchange of test stage models) is a key requirement distinguishing VZLÚ equipment from most two-stage turbine test equipment operated in the world. However, this requirement significantly complicates the construction and operability of the machine. Due to the requirement of frequent replacement of test stages, configuration with mounting the test stages on the overhanging end of the shaft was used. The bearing at this end of the shaft is relatively high loaded, and this conception is more sensitive to vibrations. This led to increased demands on the rigidity of the device. The final design of the facility is shown in Fig. 3. Summary of the new test rig parameters compared with the original facility is written in Tab. 1. A more detailed description follows.

Table 1. Summary of the new test rig parameters

Parameter	Original Facility	New Facility
Tip Diameter (m)	0.60	0.64
Hub Diameter (m)	0.20	0.16
Number of stages	1.5	2 (2.5 possibly in future)
Max rotational speed (RPM)	10000 (limited due to the critical speed)	14000

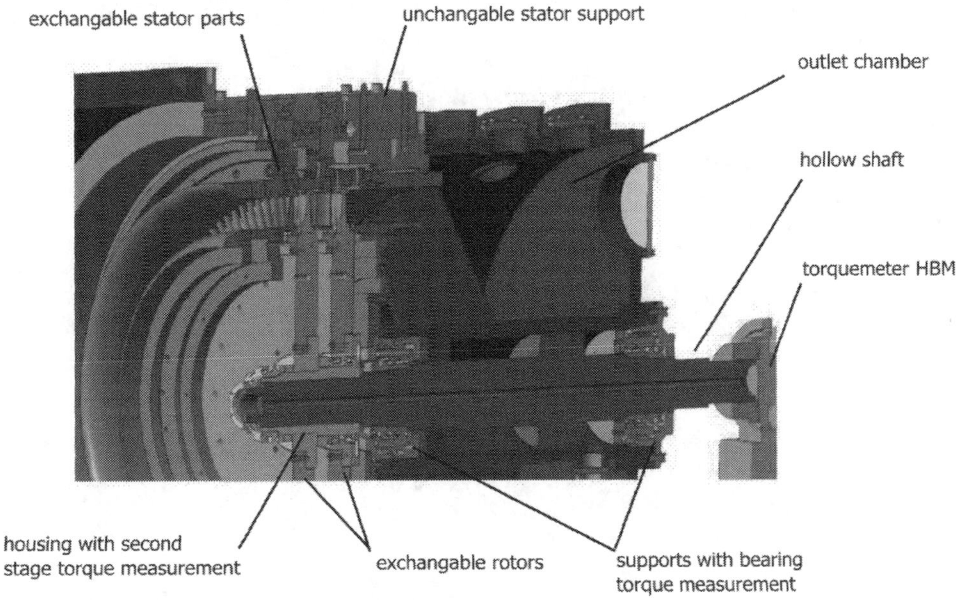

Fig. 3. The new test turbine – two-stage configuration

3 Design of individual KEY elements

3.1 Shaft bearing

The shaft (Fig. 4) is mounted on angular contact ball bearings, always in tandem on both supports. The doubling of the bearings was performed mainly to increase the rigidity of the supports; at the same time, this solution also brings a longer service life of the bearings due to a lower loading. The design of both bearing supports is unified and consists of the bearings themselves, the bearing housing assembly, and the outer bearing for the rotary bearing housing. This allows the loss torque of the bearing to be measured by a force sensor. Furthermore, the bearing housing is equipped with a thermocouple for monitoring the bearing temperature and with accelerometers for vibration monitoring. The housing design is such that the housing can be rotated about the shaft axis to prevent the outer bearing ring from squeezing. It is assumed to be rotated at least four times during the life of the main shaft bearings. The service life of the bearings should be at least 7000 h.

The shaft bearing is designed so that the front support absorbs the axial force in the flow direction (positive). The negative force is eventually absorbed by the rear bearing. The device is not expected to be operated continuously with negative force. The direction and magnitude of the force will be controlled by adjusting the pressure in front of and behind the balance piston. The front support is equipped with force sensors to measure the total axial force. The MST wave spring is used in the assembly for the permanent preload of the bearings.

The shaft assembly is inserted in a housing weldment in which the shaft assembly is supported by five ribs. The housing box also serves as an outlet chamber directing the flow from the turbine to the return path of the closed-loop wind tunnel.

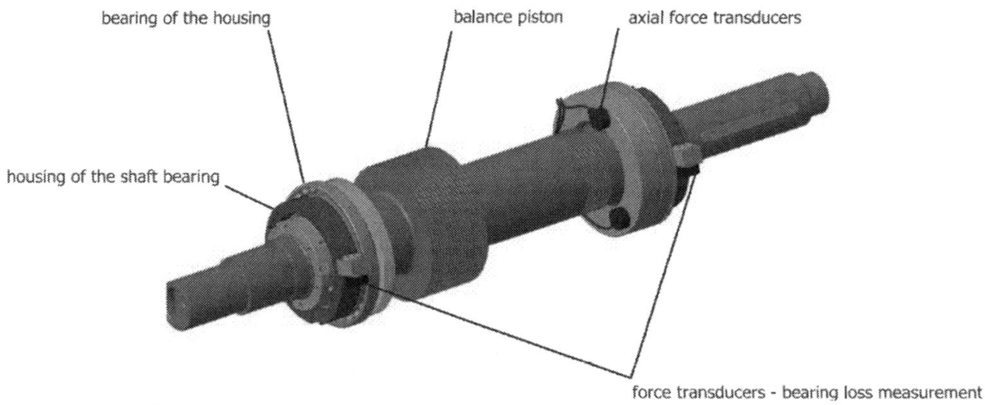

Fig. 4. The shaft assembly

3.2 Measurement of bearing losses

The measurement of the overall efficiency of the tested stage is negatively affected by the loss caused by friction in the bearings and the so-called ventilation loss, i.e. the friction of rotating parts by air (especially the disk of the tested stage).

The bearing losses in the original facility were included in the calculation of the stage overall efficiency by an empirical formula based on the measurement of bearing warming. However, the correction determined in this way may not correspond exactly to the actual losses.

For the newly built device, it was decided to use a direct measurement of the bearing response by placing the shaft bearings in housings rotatably mounted relative to the fixed frame. The torque loss is then measured using a force sensor that connects the bearing housing to the frame in a tangential direction – depicted in Fig. 4. The loss moment of each bearing is measured separately, thus the condition of both supports can be easily determined by the force measurement. The support state is monitored also by the thermocouples and accelerometers as was mentioned before.

3.3 Torque measurement

For the two-stage turbine, it is necessary to determine the torque of the two stages separately. The torque of both stages together is measured by torque meter HBM which is used to connect the turbine shaft and the dynamometer shaft. A Novel torque meter (Fig. 5) was designed and manufactured to measure the second stage torque. In fact, it determines the differential torque between the first and the second stage. Therefore, the second stage is mounted rotatably and the torque is captured by the measurement element which is instrumented by strain gages. A mechanical fuse is implemented in the assembly for safety reasons. The fuse limits the deformation of the measurement element and takes over the transmission of torque in case of a failure of the element. The power supply of the strain gauges and the signal transmission are provided by a one-channel wireless system produced by KMT Telemetry.

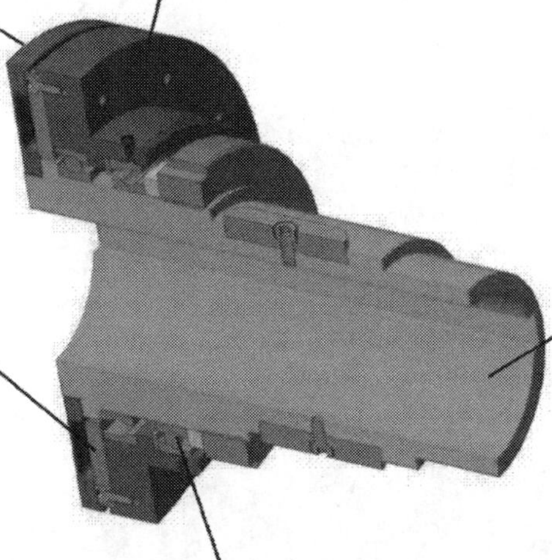

Fig. 5. The second rotor torque meter

3.4 Concept of rotating stator

During the implementation of the turbine concept, it was decided to use the concept of linear traverses and a rotating stator. The reasons for the use of this concept were mainly: the reliability of the seal and the possibility to measure the losses of the stage more reliably; and the possibility of individual circumferential adjustment of the stator vanes – the so-called indexing. A significant advantage of this solution is the simplification of the traversing device with the probe, which thus degrades only to a linear displacement with axial rotation – probe routing. This traversing device can then also be easily dismantled from the turbine rig with the probe and used, for example, in the test section for calibrating the probe.

The assembly of the stator holder is shown in Fig. 6, description follows. The rigid structure is consisted of three supports which connects the inlet and outlet consoles. The stators are mounted in holders (each stator has own holder), which are rotatably fitted in the basic structure to allow the flow field to be traversed circumferentially. The supports are fitted with bushings equipped with bearings, which ensure the rotation of the stator with the least possible resistance. This bearing concept is based on the assumption that the force acting on the stators will always act towards the outlet chamber, i.e. in the direction of flow. Therefore, the support bushes are always firmly mounted in this direction (rear), the front bearings are spring-loaded by means of MST wave springs. Similarly, radial bearings are firmly mounted on the lower supports with thrust bolts for precise alignment with the shaft. The upper support has spring-loaded bearings. Fixed support bushes in the axial direction are uniform, pre-sprung cylindrical bushes are used identically for axial and radial supports.

On the outer shell of the stator housing, there are beds for the connection of traversers with measuring probes enabling their linear radial displacement along the height of the blade. These beds are placed circumferentially by 30° relative to each other in a total of six positions. The beds allow the traversers to be adjusted in the axial direction to different measuring planes along with the entire two-stage arrangement.

A drive composed of a stepper motor, a gearbox, and a ring gear is used to rotate the stators. The stators can thus be rotated in a circumferential manner of max. 80° relative to the positions of the measuring probes. The relative angular position of the two stators can be changed manually by a simple mechanism. The clocking can be simply changed also during the facility operation.

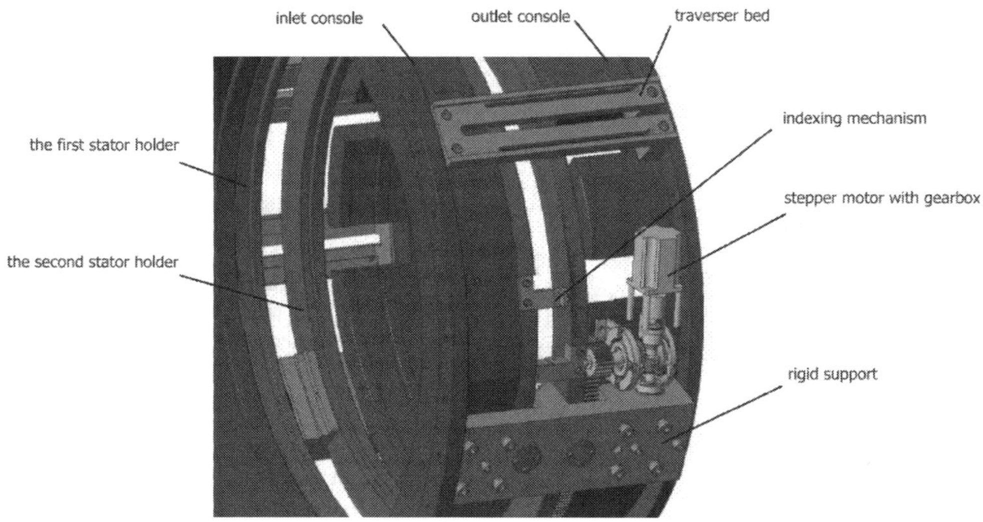

Fig. 6. The stator holder assembly

4 Trial operation results

The initial tests on the new test equipment were focused primarily on verifying the vibration behavior and the new approach to measuring the loss of torque, which plays an important role in refining the overall efficiency of the tested turbine stage. The initial testing was performed with a high-speed one-stage axial turbine. Advantageously, the stage was already measured in detail on the original turbine test rig.

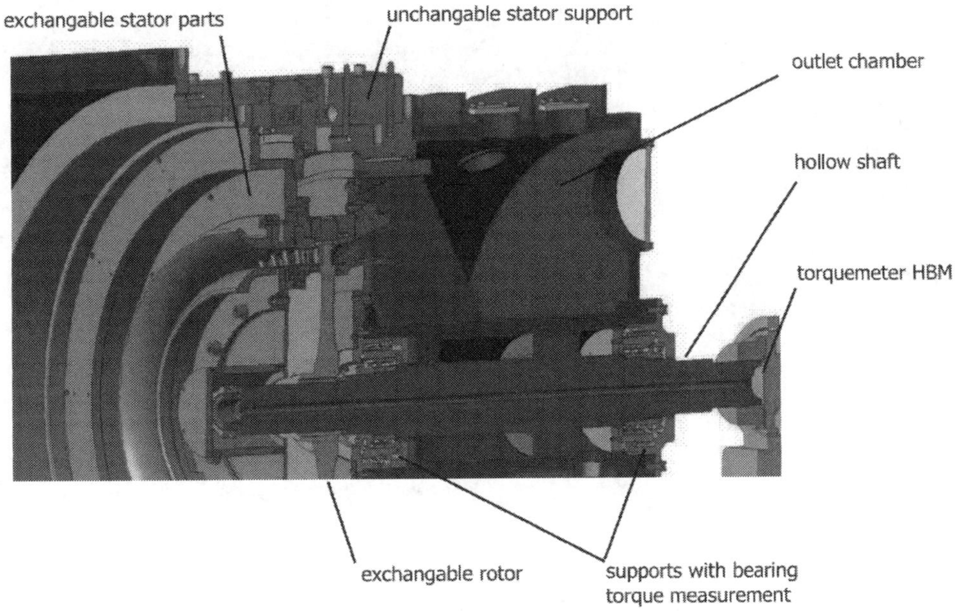

Fig. 7. The new test turbine – one-stage configuration during the first tests

4.1 Vibrations

The rotational speed range up to 14000 rpm could be tested with the particular stage. The vibrations of the front bearing and the dynamometer bearing are shown in Fig 8. These results show that the rigidity of the new base frame and the turbine body is sufficient. There are two peaks in the dynamometer vibrations characteristic (4500 and 9000 rpm). In the case of the novel facility, the peaks are kept sufficiently low thanks to the rigidity. The turbine front bearing characteristic had no significant peaks. Just slightly higher vibrations occurred in the range of the rotational speed from 10000 to 11500 rpm. Above 13000 rpm, the vibrations grew in all measured places. This is probably caused by the dynamometer behavior.

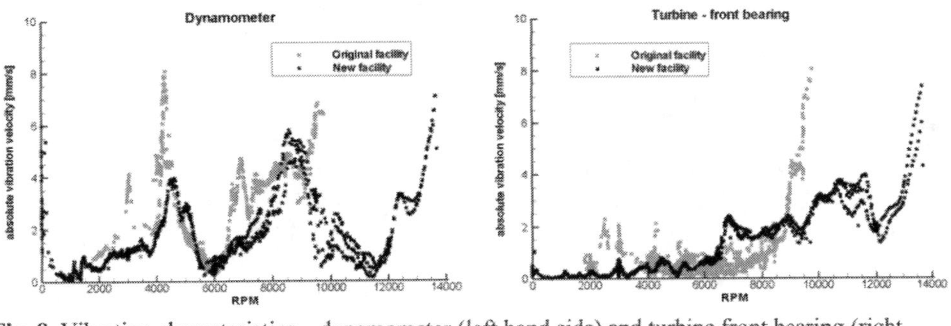

Fig. 8. Vibration characteristics – dynamometer (left hand side) and turbine front bearing (right hand side)

4.2 Bearing Losses

Figure 9 shows a bearings loss as a function of the shaft rotational speed. There is the loss computed by an SKF tool [12] for the original test facility compared with the loss estimated by the bearing's temperature measurements. The losses computed based on the housing torque measurement in the new facility are also plotted here. One can see the higher loss of bearing in the range up to 6000 RPM. These higher losses were present mainly at the beginning of the operation and in some atypical operational conditions.

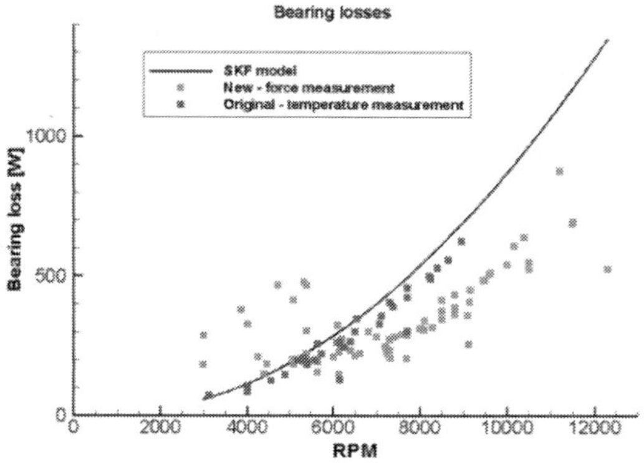

Fig. 9. Bearing losses as a function of the rotational speed

4.3 The Stage Efficiency

Overall efficiency measurement is based on the measurement of the torque, mass flow, and pressure ratio. The advantage of the facility reconstruction was, that the key technologies remained the same. The same torque meter was used, and the same configuration of the mass flow measurement was kept. Also, the shaft geometry is very similar, which is important in the ventilation losses determination. The stage tested was exactly the same including the hub and tip channel geometry parts, which was important for the comparison of the results as it influences the pressure ratio determination. Ultimately, the bearings behaviors were the most important difference between the original and novel test facility results.

The comparison of the mass flow characteristic is shown in Fig. 10. It documents that the pressure ratio and the mass flow are measured correctly, and the assumption based on the same configuration is met.

The influence of the determination of the bearing losses on the final efficiency is shown in Fig 11. The SKF model was used for both the original and the new facility, whereas the bearing-torque measurement was used only for the new one.

It shows that the results obtained in the original and in the new facility are well comparable. The effect of the determination of the bearing losses is not so noticeable on the maximum efficiency, but it rather changes the position of the optimum on the individual pressure ratios. The position of the optimum is nearly independent of the pressure ratio in the case of the bearing loss determined with the SKF model. A lower pressure ratio is characterized by a lower optimum speed ratio u/c in the case of the bearing loss measured. This result is consistent with the change of the stage reaction, which increases from 0.42 (pressure ratio 1.2) to 0.54 (pressure ratio 1.9), and the measurement of the bearing losses seems more reliable.

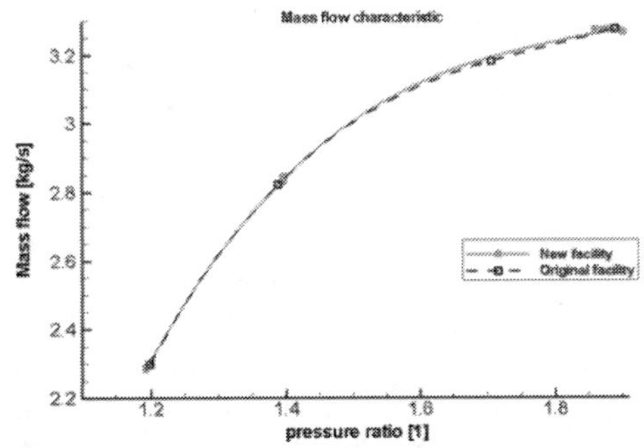

Fig. 10. Mass flow as a function of the pressure ratio – comparison of the original and the new facility

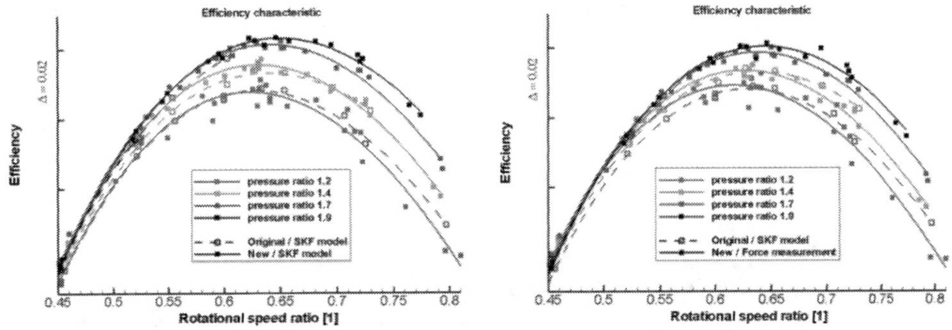

Fig. 11. The stage efficiency characteristics – comparison of the original and the new facility

5 Conclusion

This paper summarized the key consideration during the design of the new test rig for aerodynamic research on two-stage axial turbines. The main limitations of the original facility were identified as insufficient rigidity leading to a low critical speed, inability to directly measure bearing losses, and the double-skin arrangement.

The literature review revealed a relative lack of relevant sources since the major bulk of experimental facilities focus on double-shaft aircraft turbines. Therefore, the design of the new facility had to mainly stem from the lessons learned during the long-term operation of the original facility.

The design of the key elements reflecting the specific needs (especially the one of high adaptability) were described. The mounting of the shaft bearings, the measurements of torque and bearing losses, and the concept of the rotating stator were successively discussed. It can be concluded that the proposed design, which meets all the requirements, is manageable using available technical means and resources.

The trial operation of the new facility was successfully performed. The most important quantities were presented and compared with the results coming from the original facility. The initial runs proved that the rigidity of the facility is sufficient, and the rotational speed range is well extended. It is important namely for future experiments focused on turbine stages with high values of the pressure ratio.

This result was achieved with the help of institutional funding of the Czech Ministry of Industry and Trade in the program dedicated to the development of research organizations.

References

1. T. Jelínek and M. Němec, "Investigation of Unsteady Flow in Axial Turbine Stage", Experimental Fluid Mechanics 2011, pp.706-716, 2011
2. M. Němec, T. Jelínek and M. Babák, "In-Scale And Up-Scale Full Turbine Stage Measurements As A Support For Small Turbine Units Development", Proceedings of ASME Turbo Expo 2012, Copenhagen, GT2012-69693
3. J. Uher, P. Milcak, R. Skach, D. Fenderl, P. Zitek and M. Klimko, "Experimental and Numerical Evaluation of Losses From Turbine Hub Clearance Flow", In. Proceedings of the ASME Turbo Expo 2019: Turbomachinery Technical Conference and Exposition. Volume 2B: Turbomachinery. Phoenix, Arizona, USA. June 17–21, 2019. https://doi.org/10.1115/GT2019-90726
4. M. Němec, T. Jelínek, J. Uher and P. Milcak, "Effect of Stage Reaction and Shaft Labyrinth Seal in a Stage of an Axial Steam Turbine." Proceedings of the ASME Turbo Expo 2020: Turbomachinery Technical Conference and Exposition. Volume 9: Oil and Gas Applications; Organic Rankine Cycle Power Systems; Steam Turbine. Virtual, Online. September 21–25, 2020. V009T23A014. ASME. https://doi.org/10.1115/GT2020-14741
5. M. Němec, T. Jelínek and P. Milčák, "An investigation of a flow field in one and half axial turbine stage", AIP Conference Proceedings 1889, 020025 (2017); https://doi.org/10.1063/1.5004359
6. M. Němec, T. Jelínek and P. Milčák, "Clocking of stators in one and half stage of axial steam turbine", EPJ Web of Conferences 180, 02071 (2018); https://doi.org/10.1051/epjconf/201818002071
7. http://www.dlr.de/at/en/desktopdefault.aspx/tabid-7393/12440_read-30128/
8. J-M. Massonnat and T. Coton, "Development in turbine testing at ONERA", 27th AIAA Aerodynamic Measurement Technology and Ground Testing Conference, AIAA 2010-4536, Illinois, 2010
9. J. Hubinka, C. Santner, B. Paradiso, F. Malzacher and E. Göttlich, "Design and Construction of a Two Shaft Test Turbine for Investigation of Mid Turbine Frame Flows", XIX International Symposium on Air Breathing Engines 2009, ISABE-2009-1293, Canada, 2009
10. http://turbo.nd.edu/facilities/
11. J.P. Schlienger, "Evolution of Unsteady Secondary Flows in a Multistage Shrouded Axial Turbine", Diss. ETH Nr. 15230, Zürich 2003
12. SKF SimPro Quick, https://www.skf.com/group/support/engineering-tools/simpro-quick

MATEC Web of Conferences **345**, 00023 (2021) https://doi.org/10.1051/matecconf/202134500023
Power System Engineering 2021

Measurement of pressure loss coefficient for nuclear fuel assembly

Pavel Nerud[1*]

[1]Škoda JS a.s., Orlík 266, Plzeň, 316 00, Czech Republic

Abstract. Design and operation of nuclear reactor is very difficult issue from the engineering and technical point of view. All facilities must be availably designed separately, but with respect to other facilities working in the same system too. In nuclear power plant the key safety factor is the insurance of nuclear fuel assemblies cooling. The basic requirement from this point of view is an pressure loss coefficient assessment for all possible modes of coolant flow. This article describes an experience with measurement of pressure loss coefficient characteristics in Škoda JS.

1 Introduction

The one of basic safety requirements in process of design and operation of nuclear reactor is insurance of nuclear fuel assembly cooling in all possible conditions, it means during reactor operation, during pool storage and transport in container. The sufficient cooling of the fuel assembly and all its parts is possible only, when these parts have right coefficient of pressure loss coefficient and at the same time right distribution of this coefficient between parts in fuel assembly is preserved.

2 Nuclear fuel assembly

Nuclear fuel assembly for VVER reactors (PWR eventually) contains:

- bundle of fuel pins,
- top nozzle and bottom nozzle (provide right position and attachment due to other reactor internals),
- structural grids (keep right distance between fuel pins),
- supporting parts getting mechanical rigidity of fuel assembly (in some types),
- guide tubes for absorber cluster (in fuel assembly for reactor VVER 1000 and PWR),
- outer shroud (in fuel assembly for reactor VVER 440 only).

In Fig. 1 a structure of upper part of fuel assembly for VVER 440 is shown. In this cut it is possible to see top nozzle and safety grid, which protects heads of fuel pins against foreign objects fall. Six plugs in upper parts of the top nozzle are pushed away by spring. Their

* Corresponding author: Pavel.Nerud@skoda-js.cz

function is to compensate forces produced by upper reactor internals in movement to right position in reactor and during operation. Under safety grid, there is hexagonal structural grid, which keeps fuel pins in right position and enables sufficient coolant flow. An outer shroud is protecting fuel bundle against damage and acts like flow channel for coolant. Every part has its own pressure loss coefficient, which acts against coolant flow around nuclear fuel pins. In process of shape, material and surface design it is needed to take into account this coefficient and confirm this by real measurement for every flow regimes.

Fig. 1. A cut in top nozzle and bundle of fuel pins for VVER 440 fuel assembly [1]

3 Experimental loop in Škoda JS

Large water loop is experimental loop located in experimental hall in Škoda JS company. This loop is adapted for measurement for pressure loss coefficient, critical heat flux measurement on fuel pins surface and check of reliable operation for electrical drive of regulation cluster. Experimental loop contains experimental part for testing of mechanical behaviour during coolant flow; it is made by using special instrumented diagnostics fuel assembly. The next important part of experimental loop is so-called reflooding channel, where influence of hot fuel assembly reflooding after LOCA accident is tested.

This experimental loop allows pressure loss coefficient measurement for VVER 440 and VVER 1000 in two separated experimental channels. Its maximal operation parameters are written in Table 1.

Table 1. Parameters of experimental channels in Large Water Loop

	VVER 440	VVER 1000
Maximal operation pressure [MPa]	12,5	15,7
Maximal operation temperature [°C]	300	330
Maximal coolant flow [m^3/h]	180	600
Maximal temperature increase[°C/h]	120	60
Maximal electrical power [kW]	155	120

Experimental loop is equipped:

- temperature measurement in experimental channel inlet and outlet,
- static pressure measurement in experimental channel inlet and outlet,
- coolant flow measurement by Venturi nozzle in experimental channel inlet,
- pressure differences measurement between pressure taps in experimental channel.

Fig. 2 schematically shows differential pressure transducers locations in experimental channel. Some of pressure transducers are determined for direct measurement, other are determined for checksums of pressure differences. By this way the check of correctness is performed; exceptionally exclusion of incorrectly measuring pressure transducer from calculation.

The setting of right coolant flow is carried out by operator from hydraulic control room and in the next step is checked by independent measurement system.

$$Q = \alpha_{VD} \cdot A_{VD} \cdot \sqrt{\frac{2000 \cdot dp_{VD}}{\rho}} \tag{1}$$

In equation (1), dp_{VD} responses measured pressure difference in Venturi nozzle inlet and in its smallest diameter. Physical quantity α_{VD} is flow coefficient of Venturi nozzle, which is determined during calibration in certified laboratory. A_{VD} is cross section of Venturi nozzle in the smallest diameter. Coolant density ρ is determined during individual measurement as a function of measured temperature and pressure of coolant. Q is description for real flow rate.

It is needed to enumerate next values of physical quantities for gaining of pressure loss coefficient.

$$Re_{FA} = \frac{Q \cdot d_{eFA}}{A_{FA} \cdot \nu} \tag{2}$$

Equation (2) is enumeration of Reynolds number, which describes rate of coolant flow tubulisation in fuel assembly. This number is dependent on flow velocity, what is expressed by ratio of coolant flow rate Q and cross section area A_{FA}. Kinematic viscosity ν is property of coolant and characterizes its cohesiveness; it means a change of velocity on distance of adjacent layers during flow. Quantity d_{eFA} is equivalent hydraulic diameter for bundle of fuel pins.

Enumeration of pressure loss coefficient generally can be expressed by equation

$$\xi_{XY} = \frac{\Delta p_{XY}}{p_{dyn}} + \left(\frac{A_X}{A_{FA}}\right)^2 - \left(\frac{A_Y}{A_{FA}}\right)^2 - \lambda \cdot \frac{l_{XY}}{d_{eFA}} \tag{3}$$

Where

$$p_{dyn} = \frac{\rho \cdot Q^2}{2 \cdot {A_{FA}}^2} \tag{4}$$

Enumeration of coefficient ξ for one structural grid number 3 (by Fig. 2) is written as a sample.

$$\xi_{SG3} = \frac{p_H - p_L}{p_{dyn}} - \lambda \cdot \frac{l_{LH}}{d_{eFA}} \tag{5}$$

The first element in equation (5) is ratio of measured differential pressure between pressure taps L and H and enumerated dynamic pressure p_{dyn} arising by coolant flow. Dynamic pressure is the same for all measured parts of fuel assembly. The second and the third element in equation (3) are used for correction of flow cross section area. These elements are not used in equation (5), because flow cross section is not changing between pressure taps L and H. It is needed for using this correction for measurement on top or bottom nozzle of fuel assembly. Relative cross section area A_{FA} under fraction line is equal to flow area in this case. The last element in equation (3) expresses influence of longitudinal friction on bundle of fuel pins. Quantity l_{LH} is distance between pressure taps L and H reduced by height of measured structural grid. Quantity λ is longitudinal friction coefficient and it depends on flow regime and roughness of fuel pins roughness.

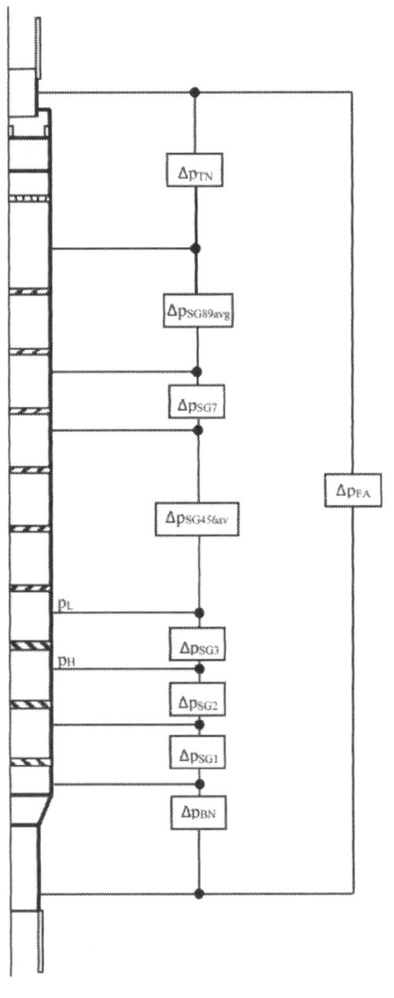

Fig. 2. Scheme of pressure differences measurement

Fig. 3. Stand for measurement static pressure and pressure differences

The result of pressure differences measurement and enumeration of pressure loss coefficients is graph, which describes development of pressure loss during change of flow velocity. This change is written by Reynolds number. Simulated sample of graph for measurement of structural grid, bundle of fuel pins and fuel assembly is given in Fig 4. In this graph appears cloud of points, which responses every single measured part. Because measured values have decreasing nonlinear character, the best interpolation is power curve. The distribution of measured points is not strictly on the line, it is caused by change of measuring error in every single measured point.

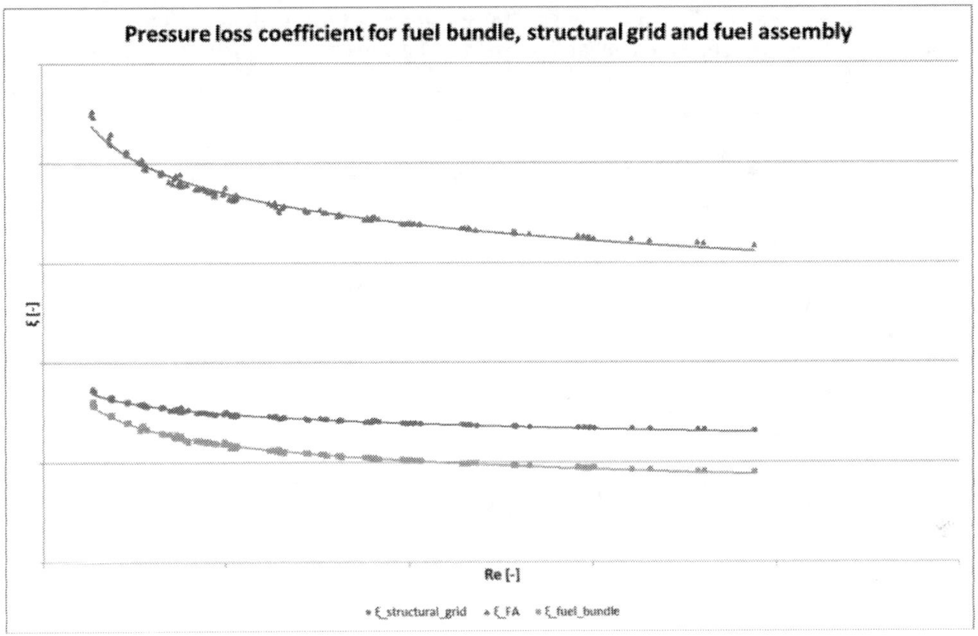

Fig. 4. Result of pressure loss coefficient measurement

4 Conclusion

Company Škoda JS built the experimental facility for pressure loss measurement for different types of fuel assemblies VVER. This facility allows performing tests in full range of pressure, temperature and flow rates, the same way like in nuclear reactor in operation. This is important advantage due to other commercial companies, which often measure in low Reynolds numbers region and in the next step extrapolate power curve to operation range of the nuclear reactor.

Experimental channel is made for different types of VVER fuel assemblies, but after some modification it can be used for measurement of western reactor fuel PWR. In present time a part of Large Water Loop is going through modernization for critical heat flux experiments.

Performed measurements of nuclear fuels from different suppliers for nuclear power plant Temelín and Dukovany prove rich experience and personal preparedness. It is supported by measurement of imitators for nuclear power plant Mochovce. The next using of technology can be waited by requirements from the Czech Republic, Poland, Slovakia, Hungary, and so on.

References

1. A. Fořtová, R. Škoda, M. Zeman, *TEPLATOR - dostupné teplárenství 21. století.* All for Power, AF POWER agency a.s. (2020). https://www.ciirc.cvut.cz/cs/teplator-dostupne-teplarenstvi-21-stoleti/
2. Nuclear Power, *Pressure Drop – Fuel Assembly.* https://www.nuclear-power.com/nuclear-power-plant/nuclear-fuel/pressure-drop-fuel-assembly/
3. Shin, ChangHwan & Chun, Tae-Hyun & Oh, Dong-Seok & In, Wang-Kee. (2012). *Evaluation of loss coefficient for an end plug with side holes in dual-cooled annular nuclear fuel.* Journal of Mechanical Science and Technology. 26. 10.1007/s12206-012-0818-4. https://www.researchgate.net/publication/257775051_Evaluation_of_loss_coefficient_for_an_end_plug_with_side_holes_in_dual-cooled_annular_nuclear_fuel
4. Nam, Ho & Kim, Jong-Man & Seo, Kyung-Won & Choi, Seok-Ki. (2005). *Development of an Experimental Correlation for a Pressure Loss at a Side Orifice.* Journal of Fluids Engineering-transactions of The Asme - J FLUID ENG. 127. 10.1115/1.1881694. https://www.researchgate.net/figure/Effect-of-the-Reynolds-number-on-the-pressure-loss-coefficient-of-the-side-orifices-Fig_fig5_245356796

Applicability of the PIV system for velocity field measurement inside the ultrasonic flowmeter

Jan Novosád[1*], *Jaroslav* Pulec[1], and *Petra* Dančová[1]

[1]Technical University of Liberec, Studentská 2, Liberec, Czech Republic

Abstract. Ultrasonic (US) flow meters are devices for a fluid flow measurement using the ultrasonic principle. Current research project dealing with developing the ultrasonic flowmeter with an innovative channel shape. Previously the work was aimed to the topology optimization using the CFD methods. The final prototype has to be analyzed experimentally to verify the velocity profiles in the US measurement section. The obtained velocity profiles in this region are necessary as the input for the US sensor data evaluation. A particle image velocimetry (PIV) was selected as a potentially suitable technique for the velocity profiles measurement. This paper presents the development and initial testing of a new experimental setup including the experimental track and PIV system. Several modifications were realized during the preparation of the experimental setup. Finally, measurements were performed with a simplified model and the results were evaluated in the form of velocity profiles in the area of interest. The most important results are the knowledge about the requirements on the model preparation. The future work will be aimed to the prototype of the ultrasonic flowmeter.

1 Ultrasonic flowmeter development

Ultrasonic (US) flow meters are devices for a fluid flow measurement using the ultrasonic principle. The flow rate is related to the fluid velocity. When the pressure disturbances propagate at the speed of sound, and the fluid has also momentum, then the absolute velocity of the pressure disturbance propagation is the algebraic sum. This principle works in several ways as an operating principle of an ultrasonic flow meter. The term ultrasonic means that the pressure disturbances are short bursts of sine waves at frequencies above 20 kHz, typically.

The newly designed flowmeters have to be tested regarding the requirements of OIML R137 [1]. This recommendation applies to gas meters intended to measure quantities of gaseous fuels or other gases. The most important test is the flow disturbance test. Flow disturbance tests shall be executed using the applicable piping configurations presented in the OIML R137, mounted upstream of the meter.

Our team works on the research project dealing with developing the ultrasonic flowmeter with an innovative channel shape. A previously published article [2] described the virtual flow disturbance test development using the CFD (Computational Fluid Dynamics) methods.

* Corresponding author: jan.novosad@tul.cz

Figure 1 illustrates the 3D CAD model of the innovative design US flowmeter DN80 fluid volume.

The flowmeter assembly is based on the two straight pipes connected together and with the input/output with several elbow segments. This composition leads to the phenomenas described by Idelchik [3] as the variation of the stream direction in curved conduits. The passage of a stream from the straight pipe to the curved section is accompanied by an increase/decrease of the pressure at the outer/inner wall. Then the diffuser phenomena lead to a separation of the stream as illustrated in Figure 2.

The previous study [2] aimed for the CFD analysis of the fluid flow field inside the ultrasonic gas meter. The numerical analysis for the sensibility of the disturbance mounted upstream of the gas was carried out. The study investigated velocity profiles in the ultrasonic measurement zone. The profiles evaluation aimed to confirm that the velocity field (profile) is flow independent. Furthermore, it was necessary to prove insensitivity to input disturbance. Based on the results, the profiles can be considered invariant when the flow rate changes. Comparison of the straight pipe and the disturbance piping mounted upstream of the gas meter indicates that the disturbance does not affect the velocity field in the measurement area. In general, the analyzed design seems to be suitable for stabilizing the velocity field for ultrasonic measurements. Finally, the research continues with some design modifications required by the reduction of the flowmeter casing. The final prototype has to be analyzed experimentally to verify the velocity profiles in the US section. The real velocity profiles are necessary as the input for the US sensor data evaluation.

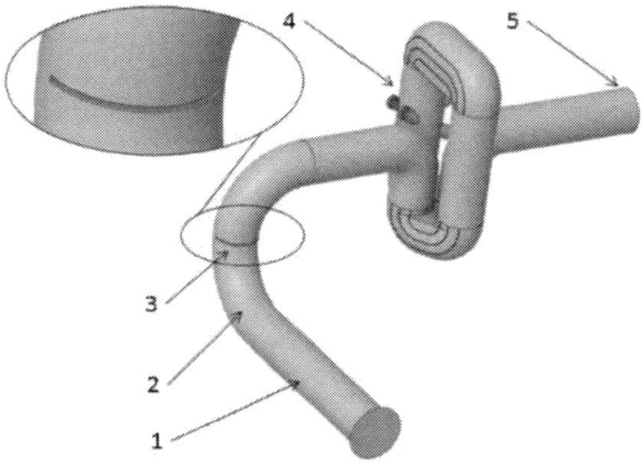

Fig. 1. US flowmeter fluid volume – 3D CAD model with the double bend disturbance [2]: 1 – inlet pipe, 2 – double bend, 3 – half-moon plate, 4 – measuring section with ultrasonic sensors, 5 – outlet pipe.

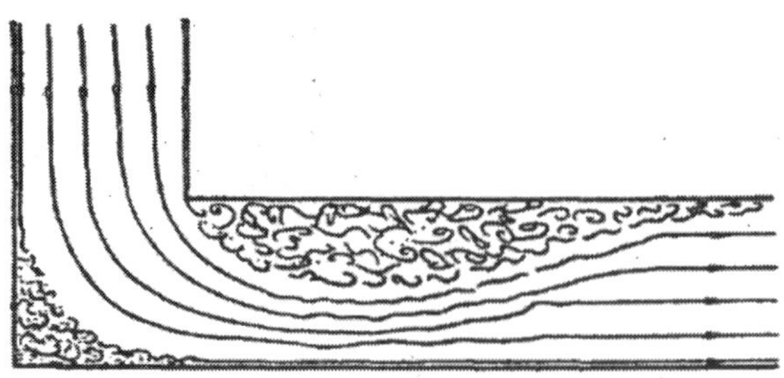

Fig. 2. Stream pattern in a 90°elbow [3]

2 Particle image velocimetry

In the field of fluid flow measurement, particle image velocimetry (PIV) is an emerging technique. It has greater reliability and convenience, particularly for micro-sized applications than other conventional methodologies that have been applied traditionally in this field of research, such as hot-wire anemometer and Pitot tube that require probes to be present inside the flow to measure. On the other hand, PIV is a non-intrusive measurement technique, which applies laser and light techniques to find the trace of the flow of particles inside the tubes or ducts. Therefore, in this technique, the flow is not likely to be impacted or deformed, enabling the increased possibility for accurate measurements. [4]

PIV is a flow diagnostic technique based on the imaging of the fluid flow, including the seeding particles. The basic underlying image processing technology is the tracking of particles or groups of particles [5]. The basic principle is to record the displacement of illuminated microparticles perfectly dispersed in the fluid flow by a digital image sensor and its subsequent digital evaluation. Laser light is illuminating the targeted region. A system of cylindrical lenses forms the laser beam into the light plane (2D and stereoscopic PIV) or into the light volume (tomographic PIV). The particles are illuminated by two laser pulses with a clearly defined time interval, and their image is recorded with photographic film or digital cameras at two different time points in a specified time interval. A well-developed computer-intensive PIV image processing procedure is employed to determine the displacements pertaining to each tracer particle or, usually, groups of tracer particles. The resulting images are then evaluated using the basic equation $v = \Delta s / \Delta t$, where v ($m \cdot s^{-1}$) is the velocity of the particle, Δs (m) is its displacement between two images, and Δt (s) is the time step between images (laser pulses). It is assumed that the tracer particles move with identical velocity as local working fluids when deducing the working fluid's velocity PIV techniques for the velocity measurement. The principle of the PIV method is illustrated in Figure 3. Detailed descriptions about the seeding particles were presented by Melling [7]. Additionally, Westerweel [8] described the fundamentals of digital PIV aimed at the mathematical apparatus for the particles velocity evaluation.

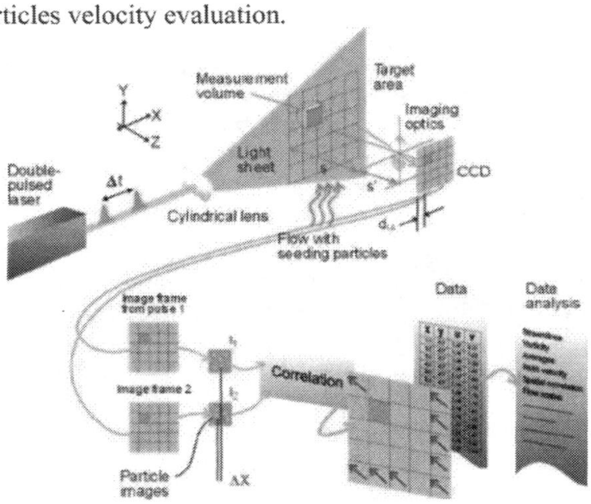

Fig. 3. Experimental arrangement and image acquisition for digital PIV [6].

3 Experimental setup

The PIV system in the laboratory at the Department of Power Engineering Equipment at the Technical University of Liberec consists of several main components. The whole system is manufactured by the LaVision company [9] with the following specification:

- FlowMaster system from LaVision for 2D, stereo-PIV, and tomo-PIV.
- 4 pcs sCMOS 5.5 MPx camera, resolution 2560 × 2160 pixels, 16 bit, a minimum time interval of two frames 120 ns, exposure time 15 μs to 100 ms, spectral range 370 nm to 1100 nm, maximum frame rate 50 Hz.
- Nd: YAG double-cavity laser, 200 mJ per pulse, wavelength 532 nm, maximum repetition frequency 15 Hz, pulse duration up to 10 nm, class 4 laser.
- 1 PC for recording and evaluation, one independent PC for evaluation, Windows 10, DaVis 10.0 software.

3.1 Measurement of velocity field inside flowmeter model

The flow field measurement inside the US flowmeter, especially in the region around the ultrasonic probes called measuring area, is the primary goal of our research. The design of the experiment is illustrated in Figure 3. The target area is composed of a transparent pipe with dimensions corresponded to the US flowmeter DN80. In this section, the flow field will be investigated by the PIV method, so the two-section planes for laser sheets are defined (see Figure 3a).

Firstly, the PIV system should be tested and verified for the possibility of the measurement inside and for finding the requirements on the experimental model design and preparation. The simplified physical model was designed and prepared, as shown in Figure 3b. The main goal of the measurement with this simplified model is to verify the ability of the used materials and experimental setup for PIV measurement in this type of task. Accordingly to the real US flow meter, two vertical pipes DN80 are mounted between the bends with $R/D = 1$. Pipes are made of 5 mm thick plexiglass. The rest of the model components are made of sheet metal with 2 mm thickness. As illustrated in Figure 3B, the air is sucked from the surroundings into the straight pipe, flows through a first transparent pipe, labeled as "Pipe 1". Then, two elbows lead the air into the second transparent pipe, labeled as "Pipe 2". For the benchmarking case, the experiment is designed as a 2D PIV setup. For this case, one camera located orthogonally to the laser sheet is used for image recording. The laser sheet is aligned with the axis of transparent pipes.

The fluid flow through the pipe elements and devices is significantly influenced by the flow rate. That's why several flow rates were chosen as the operating conditions for the experiment. The flow control is performed by the suction fan at the end of the pipe track. Table 1 notes the flow rates at the inlet section and corresponding velocities and Reynolds number, which is calculated for diameter $d = 80$ mm and the viscosity of air $1.56 \cdot 10^{-5}$ m^2/s at 25° [10].

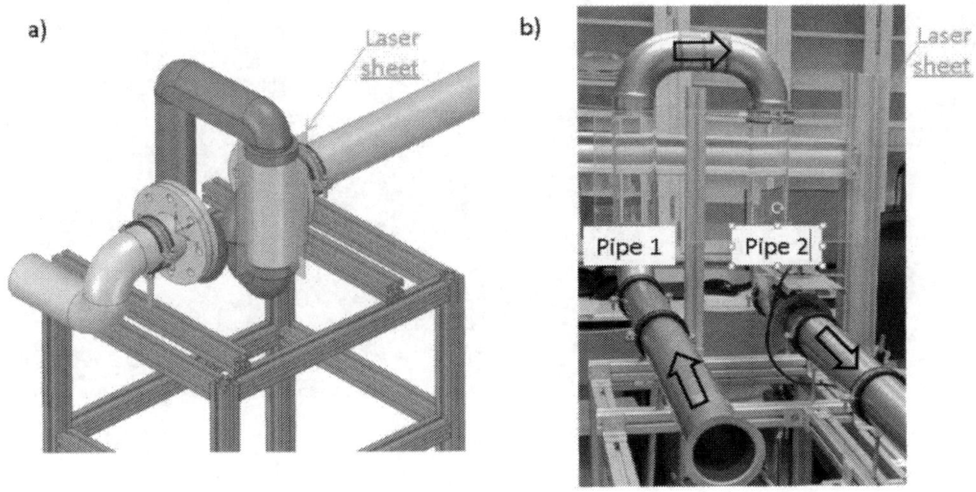

Fig. 4. Experimental setup: a) Design of the experiment for the US flowmeter, b) simplified model for PIV system verification

Table 1. Experimental setup – operating conditions (4 variants).

Quantity	V01	V02	V03	V04
Flow rate (m^3/h)	8	26	34	69
Mean velocity (m/s)	0.44	1.44	1.88	3.82
Reynolds number (1)	2254	7375	9629	19565

The preparation of the system and initial measurement shows several problems. The most difficult case is the reflection and of laser light inside the rounded plexiglas pipe wall. The troubleshooting is the black cover of a part of the pipe wall (see Target areas in Figure 4). Then the 532 nm filter is mounted on the camera lens to avoid the environmental light. Additionally, the adjustable aperture with a narrow gap was installed between the camera and target area to limit the recorded area out of the most significant reflection zones. Available calibration plates are not suitable for this task because of the dimensions. The self-printed calibration target was used inside the pipes.

The laser intensity was tuned, and the level at 14% of the maximum power was finally used. The time between the two images was 330μs, and the exposures time was set up to 15 μs. The frequency of the laser pulse was 15 Hz. Totally 300 double images were captured for each flow rate.

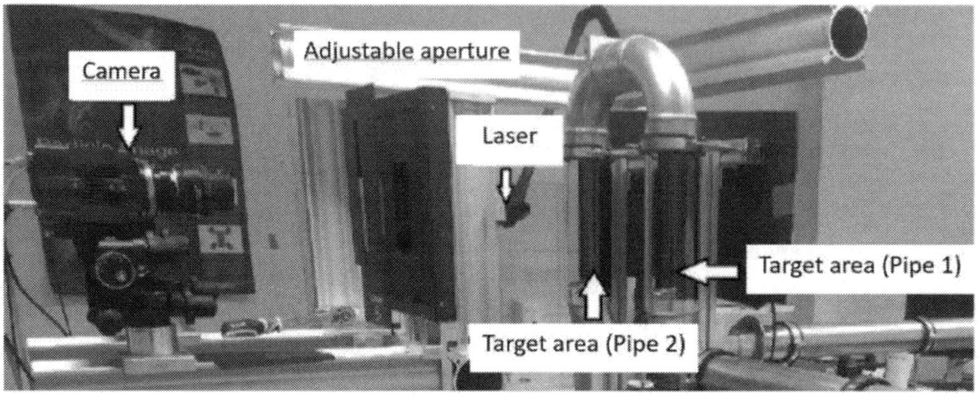

Fig. 5. Experimental setup with troubleshooting (black foil in pipes and adjustable aperture).

4 Results

The velocity fields are evaluated in the DaVis software. The postprocessing of the captured images was tuned to reach good results. The main parts of the process were the particle image velocimetry, including the image preparation, vector calculation, and finally, wrong vectors filtering. The results are presented as the selected color maps for target area in pipe 2 are shown in Figure 6.

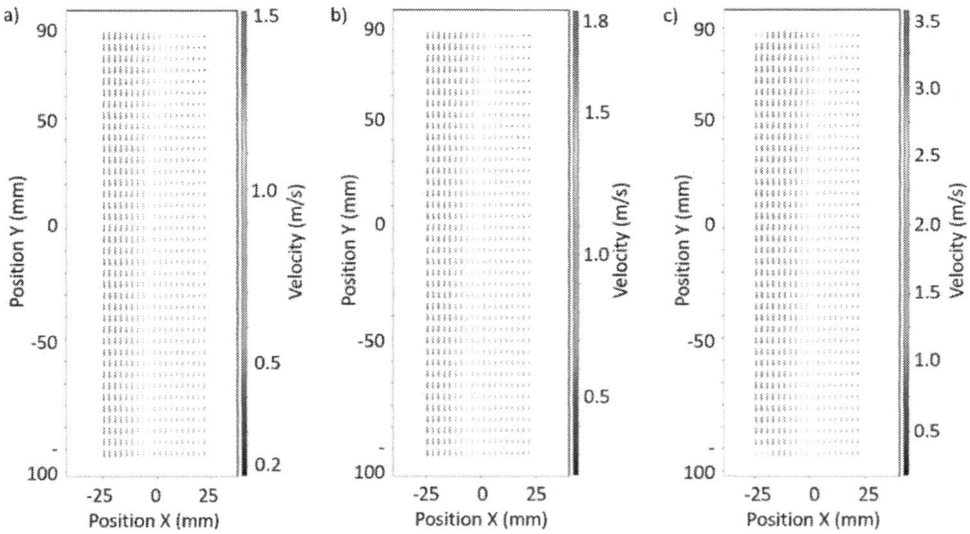

Fig. 6. Experimental results (colormap) – velocity fields, pipe 2.

The velocity profiles were evaluated from the experimental data at vertical position $Y = \{-40, 0, 40 \text{ and } 80\}$ mm. Dimensionless position X (-1,1) corresponds to the range (-25, 25) mm. The areas close to the wall was not recorded because of the reflection reduction. The dimensionless velocity $U'(1)$ is given as U/U_{max}, where U_{max} (m/s) is the maximum velocity for each variant. The results for pipe 1 and pipe 2 are shown in Figure 7 and Figure 8. The presented curves show the dependency of velocity on the flow rate. Then, the area with low velocities (a wake upstream the elbow) is decreased with higher flow rates, i.e. the flow is better distributed inside the straight pipe.

Comparison of results shows higher maximum velocity values in pipe 2 which is caused by the wakes after elbows. This corresponds to larger effective flow cross section in pipe 1. For the measurement the seeding of particles in the first pipe is worse, because of less mixing of the air in the previous section. It leads to some error points in the profile which weren't filtered by the automatic postprocessing tools.

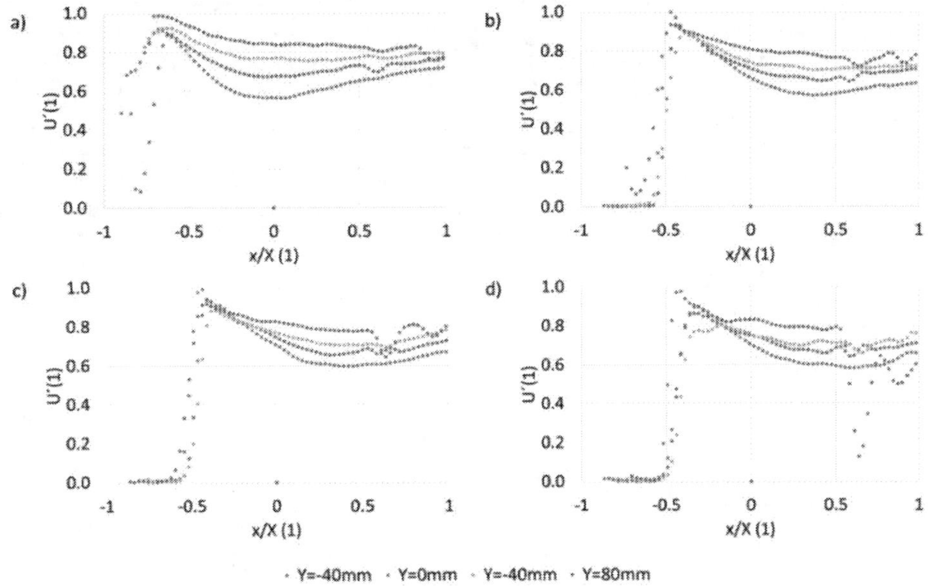

Fig. 7. Velocity profiles for pipe 1. U_{max}: a) 0.31 m/s, b) 1.16 m/s, c) 1.45 m/s, d) 2.76 m/s.

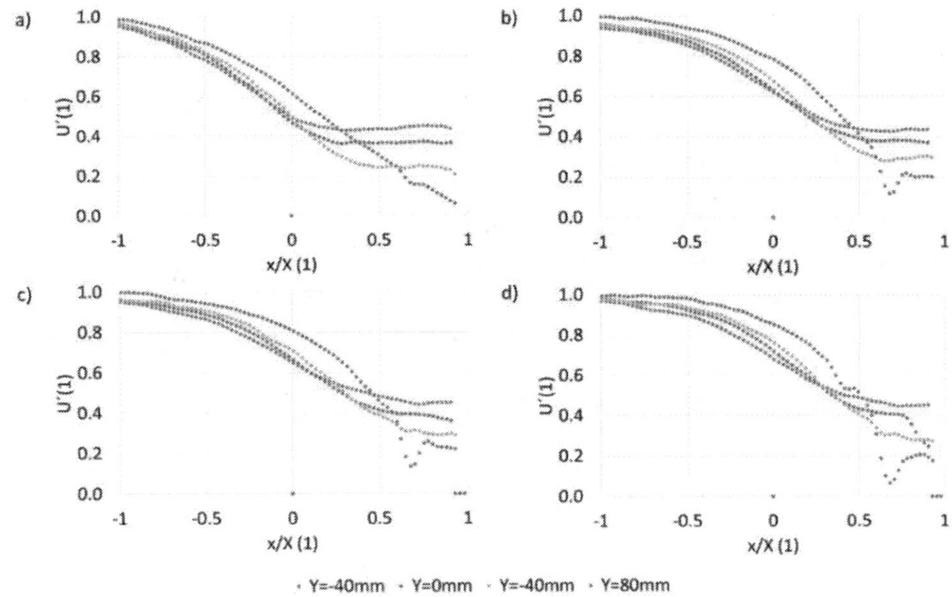

Fig. 8. Velocity profiles for pipe 2. U_{max}: a) 0.41 m/s, b) 1.53 m/s, c) 1.86 m/s, d) 3.58 m/s.

5 Conclusion

In this paper, the experimental work on the PIV system was presented. The simplified model of the newly developed ultrasonic flowmeter was made and tested with the PIV measurement. During the preparation of the experimental setup, several important requirements have been identified. The reflection and refraction caused by the cylindrical transparent model have to be avoided by installing the black foils, camera filters, and adjustable apertures. This leads to the limitations of the measurement, especially for the limit of evaluated data. The results do not describe the areas close to the walls of the target region. On the other hand, the results show the dependency of the flow field inside the channel on the flow rate. The obtained velocity profiles will be useful in the ultrasonic flowmeter development to verify the numerical simulation of the flow field inside the US flow meter.

For future work, some changes are recommended. The experimental setup and methodology of results assessment will be used for the flowmeter prototype. The inner channel shape of the prototype will be more complicated, so the manufacturing of the model should be based on a special technology for making transparent parts. The use of silicone or resin is expected. Choose of material should respect the optical properties to decrease reflections. Additionally, the seeding of particles could be improved to be more homogeneous at the inlet, which should lead to fewer errors in the pipe one region.

This publication was written at the Technical University of Liberec, Faculty of Mechanical Engineering, with financial support from the Czech Technological Agency under the project TACR FW02020048. Part of this work was realized as a part of the project "SGS-2019-5073" with the support of the Specific University Research Grant, as provided by the Ministry of Education, Youth and Sports of the Czech Republic in the year 2021.

References

1. International organization of legal metrology. Gas meters. OIML R 137-1 & 2 Edition 2012.
2. T. Vít, J. Novosád, P. Dančová. Int J. Eng Tech & Inf. (2021);2(1):01–03. DOI: 10.51626/ijeti.2021.02.00003.
3. I. E. Idelchik. Handbook of hydraulic resistance. Research Institute for Gas Purification, Moscow, Russia (1960).
4. M. R. Abdulwahab et al. Journal of Advanced Research in Fluid Mechanics and Thermal Sciences, Volume 65, Issue 2 (2020) pp. 213-229.
5. Wieneke, B. PIV Uncertainty Quantification and Beyond. (2017). DOI:10.4233/uuid:4ca8c0b8-0835-47c3-8523-12fc356768f3.
6. M. Jahanmiri. Particle image velocimetry: Fundamentals and its applications. Chalmers University of Technology, 2011.
7. Melling, A. (1997). Tracer particles and seeding for particle image velocimetry. Measurement Science and Technology, 8, pp. 1406-1416.
8. J. Westerweel. Meas. Sci. Technol. 8 (1997) pp. 1379–1392.
9. LaVision GmBH. www.lavision.de.
10. K. Kadoy, N. Matsunaga and A. Nagashima. J. Phys. Chem. Ref. Data, Vol. 14, No. 4 (1985) pp. 947-970.

Issue of heat mirror films overheating

Václav Pospíšilík[1]*, and *Stanislav* Honus[1]

[1]VSB - Technical University of Ostrava, 17. listopadu 2172/15, Ostrava-Poruba, Czech Republic

Abstract. The study focuses on the issue of coated thin plastic films overheating due to the absorption of solar radiation. These films are suspended within the sealed space of dual-pane insulating glass units (IGU). Films are deformed due to thermal stress and the glazing ceases to carry out its function. The surface temperature of the glazing system was evaluated from energy balance during a sunny day. It was supposed the intensity of solar radiation 1000 W/m2. Solar radiation incidents perpendicular to the surface of the glazing. Spectroscopic measurement was performed and optical properties of glasses and films were evaluated. Finally, suitable glazing was designed, which partially prevents transmission of the solar radiation into the inner space of glazing and the thin plastic films are not significantly overheated.

1 Introduction

Heat mirror coating finds use in many technical fields, mainly with applications in windows [1] and solar collectors [2]. The main goal of heat mirror coating is to create ideal optical properties in terms of light transmission and at the same time prevent higher heat losses from the building than necessary. Selkowitz [3] claims that thanks to the heat mirror glazing system, it is possible to save 25–75% of energy compared to ordinary glazing. The main feature of the "heat mirror" technology is high transmittance and high reflectance in the visible region of the spectrum, low emissivity in the IR region of the spectrum, respectively.

The solar optical properties of thin metal layers applied to glass or film are very well studied. Valkonen et al. [4] analyzed the optical properties of the metal layer (Cu, Ag, Au, Cr, Fe, Co, Ni, Al) depending on the metal film thickness. The mentioned study showed excellent results for noble metals - Cu, Ag, Au, especially. Its optical properties depending on the metal film thickness are analyzed in the paper [5]. Silver layer transmits the maximum amount of energy in the region $\lambda = 500$ nm, while gold layer has a maximum at $\lambda = 350$ nm, respectively. This maximum is constant and does not change with the thickness of the applied layer of material.

This study focuses on heat mirror in multiple glazing with an inner plastic sheet. This polymer film is stretched inside the glazing and divides the gas space into two. The advantage of these polymer films is their flexibility, low weight, and high impact strength [6]. The disadvantage of heat mirror films is high solar absorbance. On sunny days, these foils can significantly overheat. Subsequently, they may be permanently deformed and lose overall integrity.

* Corresponding author: vaclav.pospisilik@vsb.cz

The task of this study is to determine the film surface temperature in thermodynamic equilibrium under given ambient conditions. This approach can demonstrate how heat mirror films are thermally stressed by intense sunlight. At the end of this study, several types of glazing with coated or uncoated sheets (glass, polymer film) are compared with respect to their thermal insulation and optical properties.

When solving the heat transfer by radiation between surfaces of glasses or polymer films, it is necessary to consider their wavelength-dependent optical properties. If these surfaces are optically smooth, i.e. their average length scale of surface roughness is much smaller than the wavelength of the incident electromagnetic wave, the relations from electromagnetic wave theory can be used, even for very short wavelengths [7]. For calculation surface temperature of polymer film exposed to direct solar radiation, it will be necessary to determine the amount of absorbed energy based on the relationships from geometric optics and ray tracing.

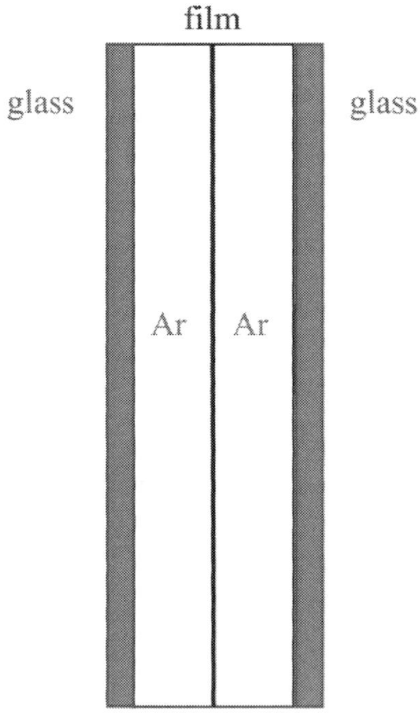

Fig. 1. Polymer film stretched inside the interglazing space, which is filled with inert gas (argon)

2 Theory

Glass and other amorphous solids tend to have extremely smooth surfaces, allowing for accurate predictions of optical properties from electromagnetic wave theory [7]. It is assumed that solar radiation with an intensity of 1000 W.m^{-2} incidents perpendicular to the external surface of the glazing (see FIG. 2). The incident radiation is partially reflected into the atmosphere, or passes into the interior, or is absorbed by the glazing.

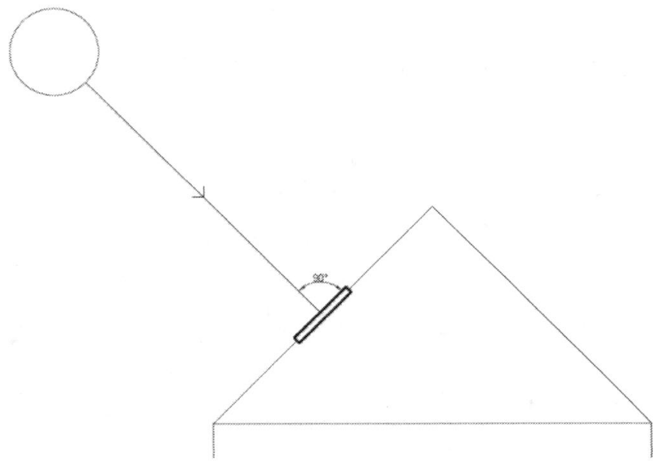

Fig. 2. Solar radiation incidents on the glazing at an angle of 90 °

The absorbed radiation energy is determined based on the consideration of FIGURE 3. The flux reflected by the opposite glass (through the inner film) e.g., $T_{s1}T_{sf}R_{s2}T_{sf}A_{s1}$ is neglected, because their value is very small. The absorbed energy of the polymer film is divided into two parts. The first of them (superscript a) is the sum of all contributions coming from the top glass. The second of them (superscript b) is the sum of the contributions coming from the lower glass.

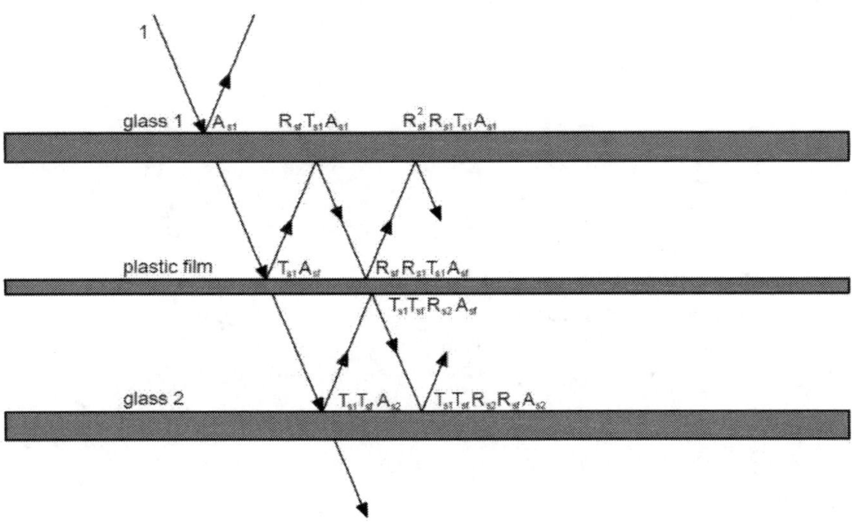

Fig. 3. Determination of absorbed energy of glass and plastic film in glazing

Spectral absorbance of glass 1:

$$A_{ABS,1} = A_{S1} + T_{S1} \cdot R_{sf} \cdot A_{S1} + T_{S1} \cdot R_{sf}^2 \cdot R_{S1} \cdot A_{S1} + T_{S1} \cdot R_{sf}^3 \cdot R_{S1}^2 \cdot A_{S1} + \cdots$$

$$A_{ABS,1} = A_{S1} + T_{S1} \cdot R_{sf} \cdot A_{S1}\left(1 + R_{sf} \cdot R_{S1} + R_{sf}^2 \cdot R_{S1}^2 + \cdots\right)$$

$$\boldsymbol{A_{ABS,1,\lambda} = A_{S1,\lambda} + \frac{T_{s1,\lambda} \cdot R_{sf,\lambda} \cdot A_{S1,\lambda}}{1 - R_{sf,\lambda} \cdot R_{s1,\lambda}}} \quad (1)$$

Spectral absorbance of film:

$$A_{ABS,f,\lambda}{}^{a} = \mathrm{T}_{s1} \cdot A_{Sf} + \mathrm{T}_{s1} \cdot \mathrm{R}_{sf} \cdot \mathrm{R}_{s1} \cdot A_{Sf} + \mathrm{T}_{s1} \cdot \mathrm{R}_{sf}^{2} \cdot \mathrm{R}_{s1} \cdot A_{Sf} + \cdots$$

$$A_{ABS,f,\lambda}{}^{a} = T_{s1} \cdot A_{Sf}\left(1 + R_{sf} \cdot R_{s1} + R_{sf}^{2} \cdot R_{s1}^{2} + \cdots\right)$$

$$A_{ABS,f,\lambda}{}^{a} = \frac{T_{s1} \cdot A_{Sf}}{1 - R_{sf} \cdot R_{s1}}$$

$$A_{ABS,f,\lambda}{}^{b} = T_{s1} \cdot T_{Sf} \cdot R_{s2} \cdot A_{Sf} + T_{s1} \cdot T_{Sf} \cdot R_{s2}^{2} \cdot R_{sf} \cdot A_{Sf} + \cdots$$

$$A_{ABS,f,\lambda}{}^{b} = T_{s1} \cdot T_{Sf} \cdot R_{s2} \cdot A_{Sf}\left(1 + R_{sf} \cdot R_{s2} + R_{sf}^{2} \cdot R_{s2}^{2} + \cdots\right)$$

$$A_{ABS,f,\lambda}{}^{b} = \frac{T_{s1} \cdot T_{Sf} \cdot R_{s2} \cdot A_{Sf}}{1 - R_{sf} \cdot R_{s2}}$$

$$\boldsymbol{A_{ABS,f,\lambda} = \frac{T_{s1,\lambda} \cdot A_{Sf,\lambda}}{1 - R_{sf,\lambda} \cdot R_{s1,\lambda}} + \frac{T_{s1,\lambda} \cdot T_{Sf,\lambda} \cdot R_{s2,\lambda} \cdot A_{Sf,\lambda}}{1 - R_{sf,\lambda} \cdot R_{s2,\lambda}}} \quad (2)$$

Spectral absorbance of glass 2:

$$A_{ABS,2,\lambda} = T_{s1} \cdot T_{sf} \cdot A_{S2} + T_{s1} \cdot T_{sf} \cdot R_{s2} \cdot R_{sf} \cdot A_{S2} + T_{s1} \cdot T_{sf} \cdot R_{s2}^{2} \cdot R_{sf}^{2} \cdot A_{S2} + \cdots$$

$$A_{ABS,2,\lambda} = T_{s1} \cdot T_{sf} \cdot A_{S2}\left(1 + R_{sf} \cdot R_{s2} + R_{sf}^{2} \cdot R_{s2}^{2} + \cdots\right)$$

$$\boldsymbol{A_{ABS,2,\lambda} = \frac{T_{s1,\lambda} \cdot T_{sf,\lambda} \cdot A_{S2,\lambda}}{1 - R_{sf,\lambda} \cdot R_{s2,\lambda}}} \quad (3)$$

Total absorbed radiative heat flux by $1\mathrm{m}^2$ glasses or plastic film is evaluated from the relation:

$$Q_{ABS} = I \cdot \frac{\int_{250}^{2000} E_{b,\lambda} \cdot A_{ABS,\lambda}\, d\lambda}{\int_{250}^{2000} E_{b,\lambda} d\lambda} \quad (4)$$

Where $E_{b,\lambda}$ is spectral emissive power of solar radiation at temperature of 5800 K.

Based on the energy flows from the surface and to the surface, the equations of the total heat balance are determined. In this application, thermal distribution within the sheet is unimportant. Therefore, it is considered a constant temperature throughout the cross-section of the glass or polymer film. Heat conduction in glass or polymer film is therefore not considered.

Glass 1:

$$Q_{ABS,1} - h_{out}(T_1 - T_{out}) + \frac{k}{d}(T_f - T_1) + \frac{1}{\frac{1}{\varepsilon_1} + \frac{1}{\varepsilon_f} - 1} \cdot \sigma \cdot (T_f{}^4 - T_1{}^4) = 0 \quad (5)$$

Film:

$$Q_{ABS,f} - \frac{k}{d}(T_f - T_1) - \frac{k}{d}(T_f - T_2) - \frac{1}{\frac{1}{\varepsilon_1} + \frac{1}{\varepsilon_f} - 1} \cdot \sigma \cdot (T_f{}^4 - T_1{}^4) - \frac{1}{\frac{1}{\varepsilon_2} + \frac{1}{\varepsilon_f} - 1} \cdot \sigma \cdot (T_f{}^4 - T_1{}^4) = 0 \quad (6)$$

Glass 2:

$$Q_{ABS,2} - h_{in}(T_1 - T_{out}) + \frac{k}{d}(T_f - T_2) + \frac{1}{\frac{1}{\varepsilon_2} + \frac{1}{\varepsilon_f} - 1} \cdot \sigma \cdot (T_f{}^4 - T_2{}^4) = 0 \quad (7)$$

The term for convection in the interglazing space is not considered in the energy balance equations. It was verified by calculating the Nusselt number, whether a natural flow doesn´t occur inside space. For all the glazings mentioned, a Nusselt number was equal to one.

The film, which has the highest temperature in the entire system due to the absorption of solar radiation, transfers heat by conduction surrounding the inert gas (argon) and at the same time by radiation to the inner surfaces of the glasses. The glasses are cooled mainly by external convection on both the exterior and interior sides. In table 1 are shown all considered values used in the calculation. The emissivities of the surfaces at ambient temperature were given by the manufacturer. Heat transfer coefficient in interior and exterior was determined as a recommended value according to EN 673. [8]

Table 1. Input values for the energy balance of glazing.

Quantity	Symbol	Value	Unit
Incident intensity of radiation	I	1000	$W.m^2$
Outside temperature	Tout	303	K
Inside temperature	Tin	298	K
Width of the gas space	s	12	mm
Thermal conductivity of gas	k	0,01734	$W.m^{-1}.K^{-1}$
Heat transfer coefficient in interior	hin	7,7	$W.m^{-2}.K^{-1}$
Heat transfer coefficient in exterior	hout	25	$W.m^{-2}.K^{-1}$

3 Spectroscopic measurement

Spectroscopic measurement was performed on four types of glass and two types of plastic films. The following sample names are well-known trade names of commercially manufactured products. Unfortunately, due to production secrets, we do not have enough information about the composition of the deposited films on the substrate. FLO4 is an ordinary uncoated glass that has the same optical properties on both sides. A metal layer is applied to the insulating glass 6304 on one side by cathode sputtering in a vacuum. In the case of Ipasol Platin and 2504 Ultra, these are glasses with a solar reflective coating, which prevents excessive overheating of the interior of the building. Klasik is an ordinary uncoated plastic film made of PET material. On HM44 is applied a layer of silver, which reduces the emissivity of its surface.

Measurements were performed using a Cary 7000 two-beam UV-Vis-NIR spectrophotometer from Agilent. Data were measured in the wavelength range of 250 nm to

2000 nm. This is the measuring range of this device. For the purposes of this calculation, the investigation of the optical properties of glasses and films in the narrower or longer wavelength spectrum is not essential. An Agilent Cary Universal Measurement Accessory (UMA) was used for the measurement, enabling automated measurement of reflectance and transmittance at selected angles of incidence. After measuring the background of the device (Baseline), the resulting reflectance and transmittance were obtained. Reflectance was determined as the ratio of the reflected part of incoming radiation and total incoming radiation and transmittance as the ratio of the transmitted part of incoming radiation and total incoming radiation. The minimum possible angle of incidence for measuring and determining reflectance is 6 °. Optical properties were measured for each side of the sample for s- and p-polarization (Rs, Rp, Ts, Tp). The resulting reflectance R and the transmittance T of unpolarized light were determined according to the relation:

$$R = \frac{R_s + R_p}{2}, \qquad T = \frac{T_s + T_p}{2}$$

The absorbance is calculated from:

$$A = 1 - R - T$$

The spectral optical properties of glasses and plastic films are shown in Figures 4-9. Figures 4 and 6 contain a small step at around 1100 nm. This is a systematic error of the spectrophotometer due to the rotation of the diffraction grating. This error does not significantly affect the overall calculation.

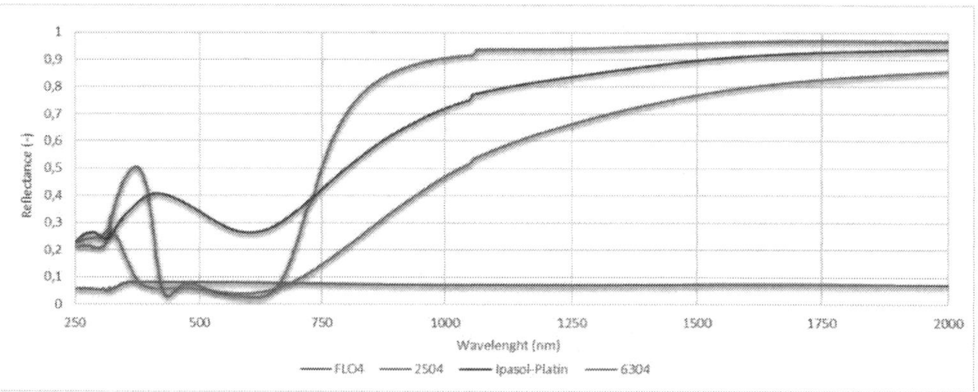

Fig. 4. Spectral reflectance of glasses

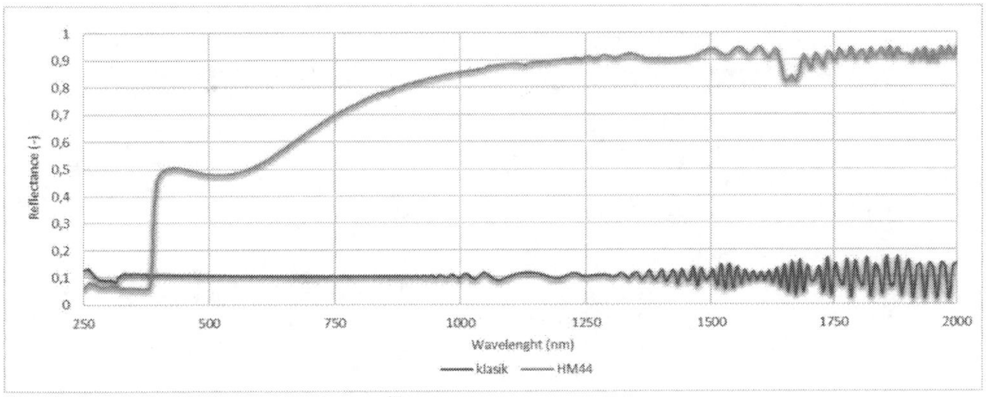

Fig. 5. Spectral reflectance of polymer films

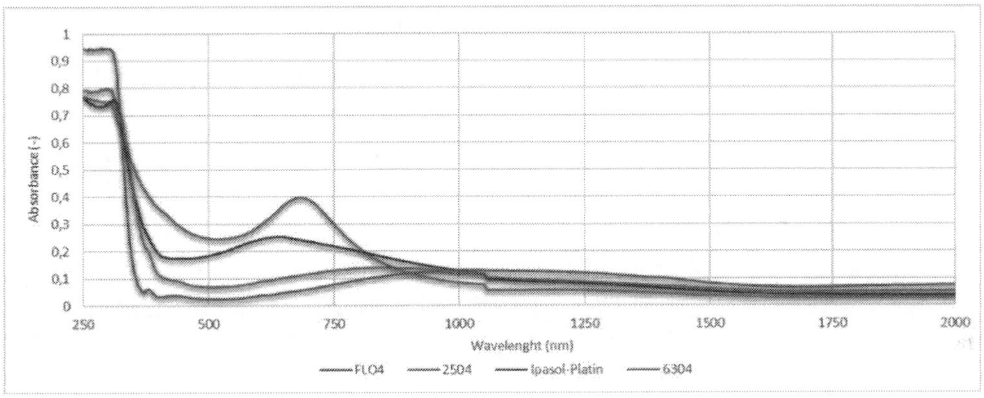

Fig. 6. Spectral absorbance of glasses

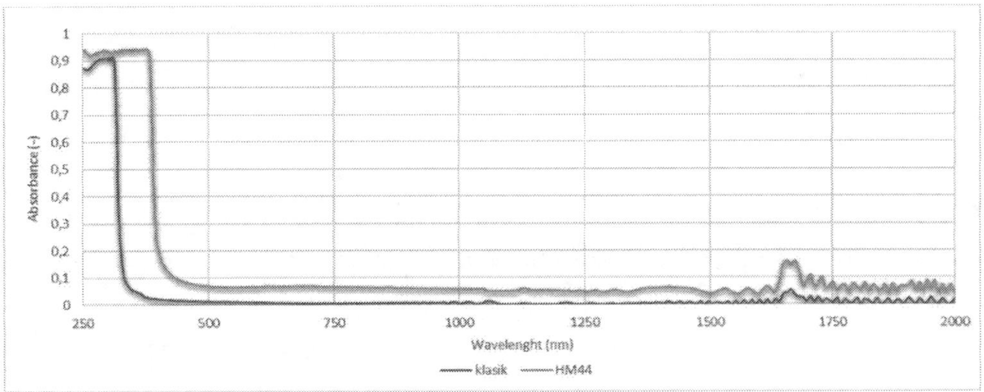

Fig. 7. Spectral absorbance of polymer films

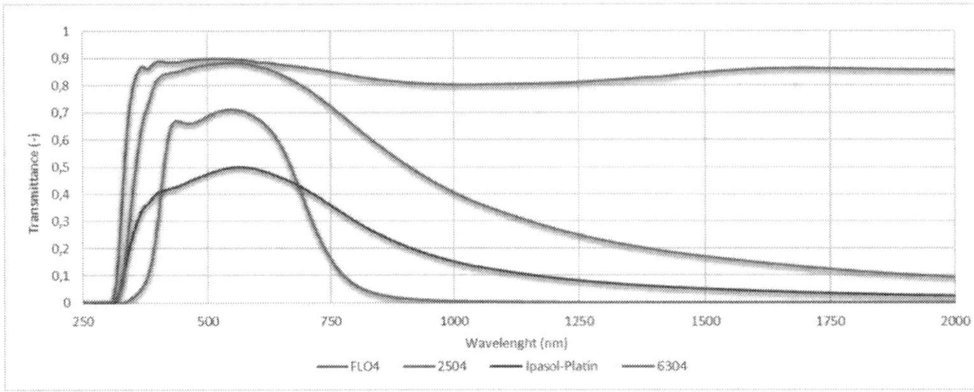

Fig. 9. Spectral transmittance of glasses

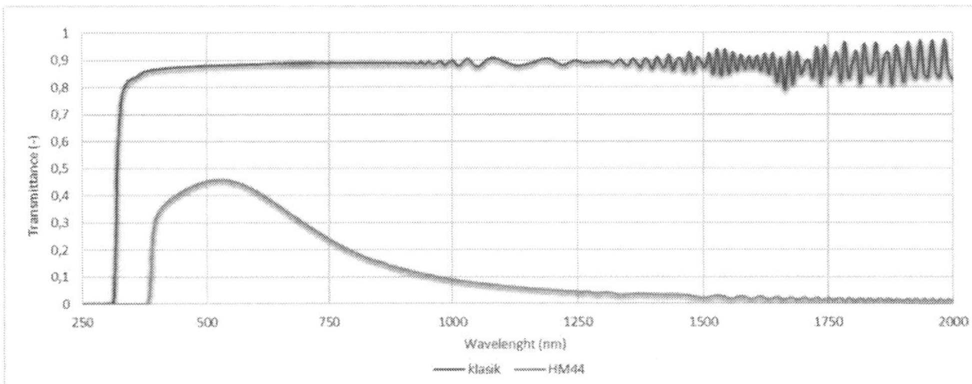

Fig. 8. Spectral transmittance of polymer films

4 Results and discussion

The surface temperature of coated and uncoated plastic film was calculated from the energy balance for seven types of glazing. Inner space inside glazing was filled with inert gas (argon). The considered glazings are shown in table 2.

Table 2. The surface temperature of the glass on the exterior side and surface temperature of plastic film inside the glazing

Glazing number	**Glass temperature t_l (°C)**	**Film temperature t_f (°C)**
FLO4 – 12 mm Ar – klasik – 12 mm Ar – FLO4	35	35,1
FLO4 – 12 mm Ar – HM 44 – 12 mm Ar – FLO4	39	75,8
6304 – 12 mm Ar – HM 44 – 12 mm Ar – 6304	40,2	68
Ipasol Platin – 12 mm Ar – klasik – 12 mm Ar – FLO4	39,3	31,8
2504 Ultra– 12 mm Ar – klasik – 12 mm Ar – FLO4	41,4	31,2
Ipasol Platin – 12 mm Ar – HM 44 – 12 mm Ar – FLO4	39,9	41,3
2504 Ultra– 12 mm Ar – HM 44 – 12 mm Ar – FLO4	41,7	36

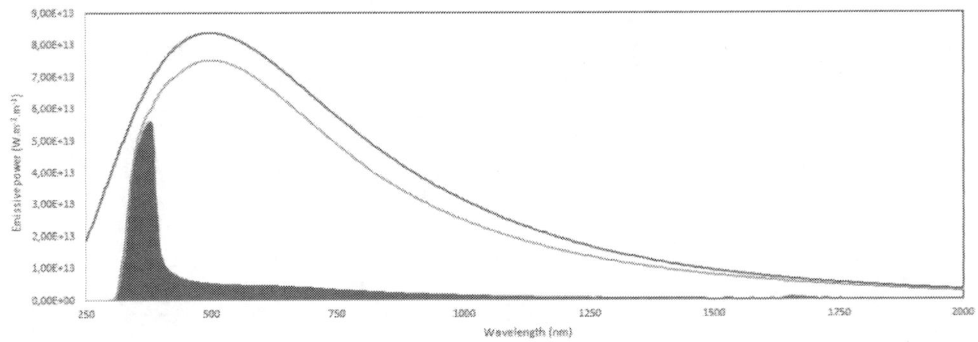

Fig. 11. Radiative heat flux absorbed by HM44 – blue area

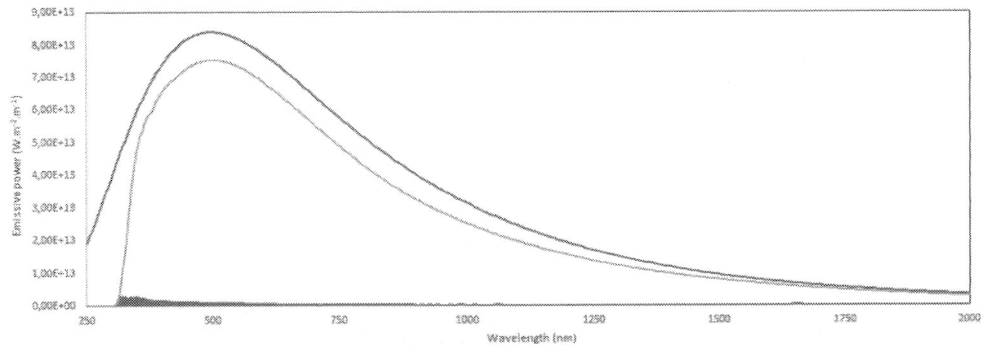

Fig. 10. Radiative heat flux absorbed by uncoated film klasik – blue area

Table 2 also shows the results of the surface temperatures of coated and uncoated glasses or films in the glazing. These glasses are the first "barrier" against the effects of solar radiation. The higher reflectance of the first glass in the UV-Visible-NIR region, the less solar radiation will be transmitted to the plastic film.

From table 2 can be seen that HM44 is exposed to significant thermal stress due to solar radiation. The triple glazing with two outer uncoated glasses and one inner film HM 44 can be heated to very high temperatures (over 75 °C). Polymer films are overheating due to poor heat dissipation from the interglazing space. The polymer film is surrounded by an inert gas with low thermal conductivity. In addition, it is desirable that gas surrounding the film doesn´t flow, thus achieving the lowest possible U-value. Plastic film transfers heat mainly by radiation to the surrounding walls. More intensive heat transfer by radiation to the surrounding walls is prevented by low emissivity in the IR region. Using coated glass (2504Ultra, Ipasol Platin) on the exterior side, which reflects part of the solar radiation to the outdoor environment, the intensity of radiation transmitted inside the glazing space is significantly reduced. Consequently, the HM 44 absorbs less energy and there is no intense overheating. This surface temperature poses a much lower risk of deforming the stretched film inside the interglazed space.

Figures 10 and 11 show the ratio between the total amount of energy incidents on the glazing – area under the red line (blackbody at temperature 5800 K), the transmitted thermal radiation through the first glass – area under the green line, and finally, the total absorbed energy of the coated or uncoated polymer film – area under blue line. It should be noted that the attenuation of solar radiation by the atmosphere is included in the calculation by the value of incident intensity of radiation 1000 W /m^2. The figures show that absorbed energy by HM44 film is many times greater than uncoated film klasik.

The combination of glasses with solar-reflective surfaces and uncoated film klasik has no significant effect on the thermal stress of the film. There is almost no thermal deformation and therefore is often used in multi-glazing. However, uncoated surface has greater emissivity in the IR region, i.e., higher heat losses through the overall glazing. Table 3 shows the calculated U-value (thermal transmittance of the glazing) according to the EN 673 [8] of all triple glazing considered above. Glazing with interglazed film klasik has U-value of 1 and 1,7 in contrast to glazing with HM 44, where the U-value is 1 and 0.7.

Table 3. U-value for several types of glazing

Glazing number	U-value
FLO4 – 12 mm Ar – klasik – 12 mm Ar – FLO4	1,7
FLO4 – 12 mm Ar – HM 44 – 12 mm Ar – FLO4	1
6304 – 12 mm Ar – HM 44 – 12 mm Ar – 6304	0,7
Ipasol Platin – 12 mm Ar – klasik – 12 mm Ar – FLO4	1
2504 Ultra– 12 mm Ar – klasik – 12 mm Ar – FLO4	1
Ipasol Platin – 12 mm Ar – HM 44 – 12 mm Ar – FLO4	0,7
2504 Ultra– 12 mm Ar – HM 44 – 12 mm Ar – FLO4	0,7

5 Conclusion

Based on the calculation of the energy balance of the sun - glazing - the surrounding environment, it was proved that there is significant overheating of coated heat mirror films located inside the interglazed space. These Heat Mirror films can be heated to very high temperatures (over 75 °C). Due to thermal expansion, the stretched film may deform, and the glazing may become unusable. It has been shown that the use of a suitable coated glass which is able to absorb or reflect a large part of the thermal radiation in short wavelengths significantly reduces the thermal loading of the films.

This study was supported by the student project No. SP2021/110 and also by doctoral grant system (CZ.02.2.69/0.0./0.0/19_073/0016945, Project DGS/TEAM/2020-030).

References

1. M. Rubin, R. Creswick a S. Selkowitz, „Transparent Heat Mirrors for Windows: thermal performance," v 5th National Passive Solar Conference, Amherst, 1980. M. P. Brown and K. Austin, Appl. Phys. Letters 85, 2503–2504 (2004).
2. S. Yoshida, „Efficiency of Drude mirror-type selective transparent filters for solar thermal conversion," Appl. Opt. 17, pp. 145-150, 1978.
3. S. Selkowitz, „Transparent heat mirrors for passive solar heating applications," v 2nd National Passive Solar, Philadelphia, 1978.
4. E. Valkonen, B. Karlsson a C.-G. Ribbing, „Solar Optical Properties of Thin Films of Cu, Ag, Au, Cr, Fe, Co, Ni and Al," Solar Energy, pp. 211-222, 1983.
5. A. Axelevitch, B. Gorenstein a G. Golan, „Investigation of Optical Transmission in Thin Metal Films," Physics Procedia 32, pp. 1-13, 2012.
6. A. Solovyev, S. Rabotkin a N. Kovsharov, „Polymer films with multilayer low-E coatings," Materials Science in Semiconductor Processing 38, pp. 373-380, 2015.
7. F. M. Modest, Radiative Heat Transfer, Boston: Academic Press, 2003.
8. DE NORMALISATION, ComitéEuropéen. EN 673: 2011, Glass in building. Determination of thermal transmittance (U value). Calculation method, 2011.

Experimental and numerical analysis of an oil-flooded air screw expander

Lukas Richter[1*], *Zdenka* Rendlova[1], *Stanislav* Planicka[1], *Jan* Seeger[2], *Marco* Taschek[2], *Michal* Volf[1], *Martin* Pelikan[1] and *Vit* Linhart[3]

[1]University of West Bohemia in Pilsen, Univerzitni 22, 306 14 Pilsen, Czech Republic.
[2]OTH Amberg-Weiden, Kaiser-Wilhelm-Ring 23, 92224 Amberg, Germany.
[3]ATMOS vyvoj s.r.o., Plzenska 149, 330 03 Chrast, Czech Republic.

Abstract. This article describes the results of a project whose task is to research and develop a screw expander through the inversion of a screw compressor. The article summarizes aspects of the construction of an experimental device that works with an oil-flooded air screw expander. The expander was subjected to experimental and numerical analysis, the results of which are presented in the article. Numerical analysis to examine the expansion process is performed both on the basis of the analytical geometric description of the working chamber of the expander and on the basis of the geometry obtained from a 3D scan of a real machine. The results of experimental and numerical analysis will be used to integrate an oil-flooded screw expander into an energy unit for the use of low-potential heat, for example in ORC systems.

1 Introduction

A screw expander is based on the inversion of a screw compressor and under certain conditions may be a suitable alternative to a turbine as a thermal cycle expansion device for the transformation of thermal energy into mechanical energy. These conditions are basically defined by the parameters of the heat source. A screw expander can be used in particular when the heat source is limited, which is typical, for example, with solar or geothermal energy and for the use of waste heat. The classical steam Rankine cycle is not suitable for the use of low-potential heat, and therefore working substances with a lower boiling point than water are usually used. If such a working substance is used, a screw expander is usually an integral part of the Organic Rankine Cycle (ORC), the binary cycle or the Kalina cycle. In the case of low-potential heat applications, a liquid phase may appear during the expansion of the low-boiling working fluid, causing erosion of the turbine blades. In such a case, a screw expander capable of working without damage should be used, even with a relatively large proportion of liquid phase. For example, the trilateral flash cycle (TFC), for which wet expansion is typical, may be suitable for the use of flow-limited heat sources.

In the case of low-potential heat applications, a screw expander is characterized by the relatively low velocity of the working medium, low speed and a lower flow rate of the

* Corresponding author: richterl@kke.zcu.cz

working medium, unlike microturbines. Microturbines for the use of low-potential heat tend to be high-speed (tens of thousands of RPM) and their phasing into the electrical grid with a frequency of 50 or 60 Hz is more demanding. Nowadays, it is possible to produce a screw expander with a maximum rotor diameter between 800 – 900 mm, which corresponds to a maximum output power of approximately 5 MW. However, for rotor diameters greater than 350 mm, production costs rise steeply, and therefore the ideal power output range for a screw expander is limited to approximately 50 kW - 500 kW. The higher the accuracy of the screw expander production, the higher the efficiency of the expander can be achieved. On modern milling and grinding machines, high precision screw rotor production can be achieved with a tolerance ranging between 3 - 5 μm [1].

The aim of our research, the outputs of which are described in the article, is to perform an analysis of the expansion process in an oil-flooded air screw expander using experimental and numerical approaches. The results are used to implement a screw expander into an energy unit for the use of low-potential waste heat. In addition, the mathematical model can be used to modify the profiles of screw bodies to optimize the expansion process.

2 Working principle of oil-flooded screw expander

Screw expanders can be made in many variants depending on the number of rotors, different profiles of the screw bodies or on the number of rotor teeth. The different configurations usually depend on the pressure conditions, lubrication or expander speed. The expander we analysed has two rotors (twin screw expander). The male rotor connected to the shaft has 4 lobes and the female rotor has 5 lobes. The rotors together with the expander casing form the working chamber, the volume of which first increases through isobaric filling up to the value defined by the geometry of the inlet port. As the rotation continues, the chamber closes and separates from the inlet port. However, the volume of the working chamber continues to grow, and therefore there is isentropic expansion and a decrease in the pressure of the working fluid. At the moment when the working chamber is exposed to the low-pressure port, the fluid is gradually discharged isobarically. Continued rotation leads to a reduction of the contact line between the teeth to the level of zero volume of the working chamber, when the fluid is completely discharged. The whole process of gradual filling, expansion and discharge is thus completed after two revolutions of the main rotor (720 °). This working principle is shown in Fig. 1 [2].

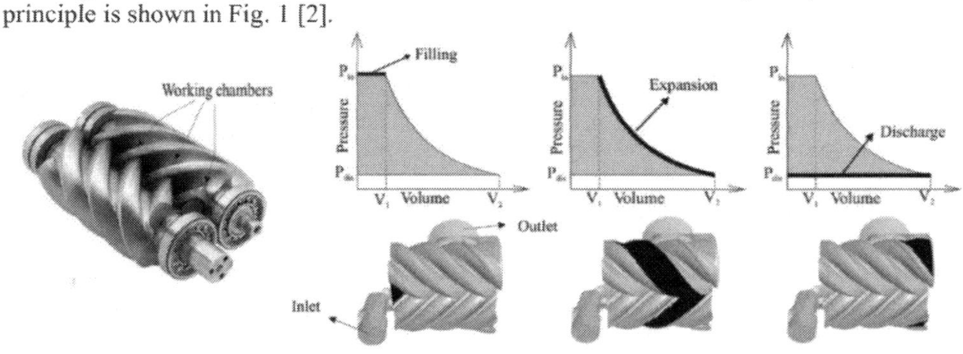

Fig. 1. Working principle of a screw expander [2].

The volume of the working chamber should be as large as possible, thus maximizing the discharge of fluid per revolution. The profiling of the rotors must be designed so that an oil sealing line is formed in each contact position both between the rotors and between the rotors and the casing, because leakage of the working fluid from the working chamber adversely affects the efficiency of the screw expander. On the other hand, the contact forces must be low so that the rotors do not deform due to friction losses and the local temperature increases.

The operation of an oil-flooded screw expander depends on a relatively large flow of lubricating oil, which is injected into the working chamber at predetermined locations. The oil is used to lubricate the rotor motion and thus to reduce friction losses, as well as to seal gaps and for cooling when the machine is operated in compressor mode. Oil-flooded machines have a simple mechanical design that requires an external shaft seal, but no internal seal. A thin oil film is formed between the rotors, which seals the inter-rotor clearance, resulting in low internal leakage. However, the viscosity of the oil limits the maximum speed at the top of the rotor. Its value can vary along the rotors depending on the temperature, but it must always comply with the specified limits. Typical values are about 50 mm^2/s, and this value decreases with increasing temperature [1].

3 Construction of test bench with oil-flooded screw expander

At first glance, it seems that the oil-lubricated screw expander is a simple inversion of a conventional oil-flooded screw compressor. A closer look shows that this is not the case at all. The main differences to be examined relate to the transmission of power between the rotors, the replacement of the self-circulating oil system by forced circulation, the regulation and control of the rotational speed, the separation of the oil from the expanded air under low pressure and high volume flow conditions, and subcooling of the working fluid during expansion (unlike the compression process, where heat is generated). Considering that in the case of expansion, lubricating oil does not act as a cooling medium (as opposed to compression), its quantity and injection method seem to be important for the generated power.

The test bench can be divided into 3 basic parts - a) compressor unit to create compressed air, b) screw expander and its oil management, c) shaft measurement and electrodynamic brake.

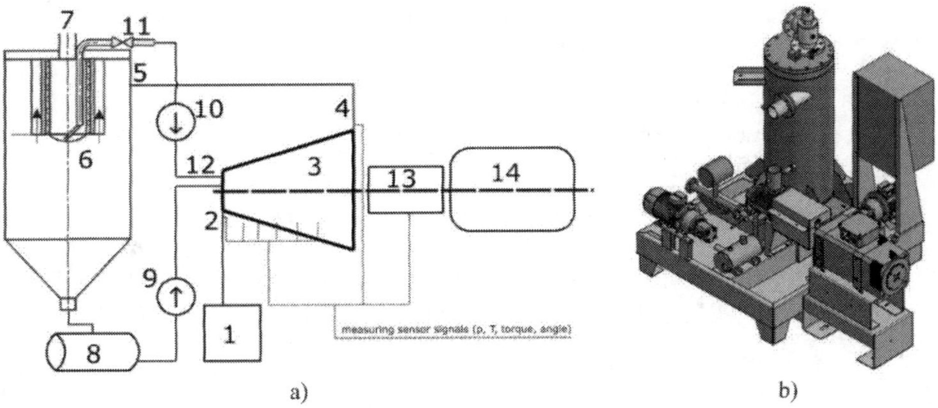

Fig. 2. a) simplified 1-D schematic of test bench, b) 3D CAD model of test bench.

The function can be described as follows according to Fig. 2(a). The compressor (1) generates compressed air which enters the expander through the high-pressure port (2). In the oil-flooded expander (3), air is mixed with lubricating oil, which enters the expander through the oil ports (12). Air expands in the expander and the entire mixture of air and oil then exits the expander through the low-pressure port (4). The amount of oil in the expanded mixture is around 5 kg per cubic metre of air. The expanded mixture then enters tangentially (5) into the oil separator (6). The oil separator vessel is designed as a cyclone separator in combination with a fine filter for separating residual oil. Based on the centrifugal principle of the cyclone, most of the oil is trapped at the bottom of the vessel. However, the air separated by the cyclone still contains a residual amount of oil (1-5 g/m^3), and therefore this mixture further passes through a filter membrane, where the residual oil is separated. Purified air with an oil content of less than 3 mg/m^3 then emerges from the vessel through the upper lid (7). Because the oil separation takes place under conditions of low pressure and relatively large volume flow, a separation vessel of relatively large dimensions was chosen (diameter 419 mm and height 1030 mm). From the bottom of the separation vessel, the oil is then drained into the oil collecting tank (8) and from there pumped (9) back into the screw block. Residual oil trapped by the fine filter is also pumped (10) back to the screw block by a scavenge line (11). The test bench is further equipped with a torque measuring shaft (13) and an electrodynamic brake (14) for controlling the speed and torque of the expander.

4 Screw expander analysis

4.1 Theoretical assumptions

The screw expander is essentially a volumetric device, in other words also a positive displacement device. From this point of view, its function can be described by means of volumetric efficiency. Volumetric efficiency is defined as the ratio of actual and theoretical volume flow, and thus can also be expressed in terms of the ratio of actual and theoretical mass flow. However, in the case of a screw expander, the actual flow may be higher than the theoretical flow, which would mean a volumetric efficiency higher than 1. For this reason, the ratio of the actual and theoretical flow is referred to as the Filling Factor (FF) [2]:

$$FF = \frac{\dot{m}_{\text{real,fill}} \cdot v\ (T_{\text{fill}}, p_{\text{fill}})}{n_{mr} \cdot z_{mr} \cdot V_{\text{fill}}} \qquad (1)$$

Where $\dot{m}_{\text{real,fill}}$ is actual mass flow of the working fluid during filling $[kg \cdot s^{-1}]$,
$v\ (T_{\text{fill}}, p_{\text{fill}})$ is specific volume as a function of filling temperature and pressure $[m^3 \cdot kg^{-1}]$,
n_{mr} is rotational speed of the male rotor $[1 \cdot s^{-1}]$,
z_{mr} is number of teeth (male rotor),
V_{fill} is the filling volume $[m^3]$.

During the operation of the expander, two main phenomena affecting the FF value occur. These are throttling losses and internal leaks. As the rotors rotate, the entire area of the inlet port is only open for a short time, and therefore throttling losses occur, which reduce the actual flow and thus the FF value. Internal leakage of working fluid occurs between the individual working chambers in the direction from the high-pressure port to the low-pressure port. As a result, additional mass is required at the inlet to the expander, which increases the FF value (possibly above 1). The value of FF varies depending on the pressure ratio and especially on the rotational speed. Higher admission pressure results in more intense internal leaks and thus a higher FF value, however, this phenomenon is also compensated by increasing pressure loss at the inlet port, which reduces the mass flow. The rotation speed has an even greater effect on the FF value. Higher speed causes an increase in throttling losses and thus a decrease in the Filling Factor, because the relative effect of internal leaks is smaller and the pressure loss at the inlet port increases. It is evident that the best results are obtained when the value of the Filling Factor is close to one [2, 3].

The significance of internal leaks decreases more or less linearly with rotational speed, while internal friction increases with the square of the speed. This partially compensates these two effects, and therefore satisfactory efficiency is achieved over a relatively wide range of speeds, unlike turbines where the slightest deviation from the nominal speed means a significant decrease in efficiency. If the operating point corresponds to a design point, screw expanders cannot in most cases compete with turbine efficiency. However, more stable performance over a wider operating range makes screw expanders more flexible and more advantageous than turbines in applications where operation over a wider range of conditions is required [1].

The disadvantage of screw expanders is their built-in volume ratio (V_I). This ratio is defined by the geometry of the expander casing. The volume ratio is defined as the ratio of gas volumes in the low pressure port (V_{LP}) and the high pressure port (V_{HP}) [4]:

$$V_I = \frac{V_{LP}}{V_{HP}} \tag{2}$$

From the adiabatic equation, it can be deduced that the built-in volume ratio also corresponds to the pressure ratio:

$$p \cdot V^k = konst \tag{3}$$

$$p_{HP} \cdot (V_{HP})^k = p_{LP} \cdot (V_{LP})^k \tag{4}$$

$$\frac{p_{HP}}{p_{LP}} = \left(\frac{V_{LP}}{V_{HP}}\right)^k \tag{5}$$

$$\frac{p_{HP}}{p_{LP}} = V_I^k \tag{6}$$

$$P_I = V_I^k \tag{7}$$

Where P_I indicates the pressure ratio and represents the discharge pressure divided by the suction pressure, i.e. the admission pressure divided by the emission pressure. The adiabatic exponent k is dependent on pressure and temperature and can be determined as the ratio of specific heat defined at constant pressure and volume ($k = c_p/c_V$). It is believed that any operation other than that corresponding to the built-in pressure ratio causes a decrease in the isentropic efficiency. At lower pressure ratios, 'over-expansion losses' occur and at higher pressure ratios, 'under-expansion losses' occur [5].

4.2 Experimental Approach

For characterization, the screw expander (ATMOS B100) was operated with air and integrated into a test rig. This rig consists of an oil supply system with high performance oil (Mobil Rarus 424), a generator and a compressor unit. Pressure sensors (Kistler 6061B, Kistler 6125C), temperature sensors (Pt100) and an angle encoder (AVL 365X) are used to perform the experiments necessary for the characterization. Five pressure sensors are positioned inside the screw expander to monitor the complete progress of the pressure. Additionally, one pressure sensor is mounted in the intake and one in the exhaust air duct. The temperature sensors measure the intake and exhaust air temperature.

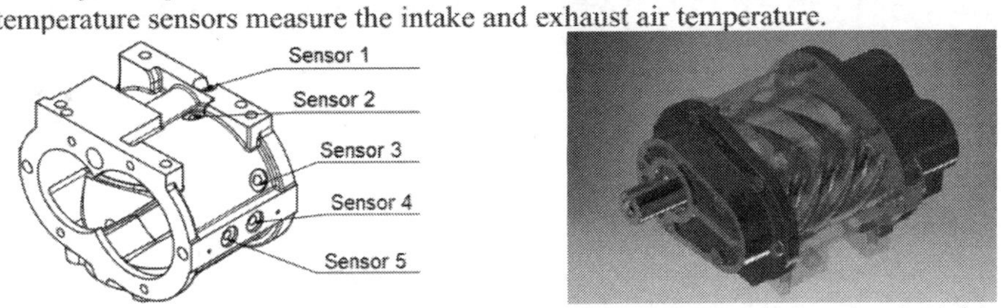

Fig. 3. Casing with drilled holes for pressure sensors (left), 3D CAD model of test bench (right).

The angle encoder is necessary to acquire the corresponding time. In addition, a torque measuring shaft (FR-A701KTR Dataflex 32/30) was positioned between the screw expander and the generator.

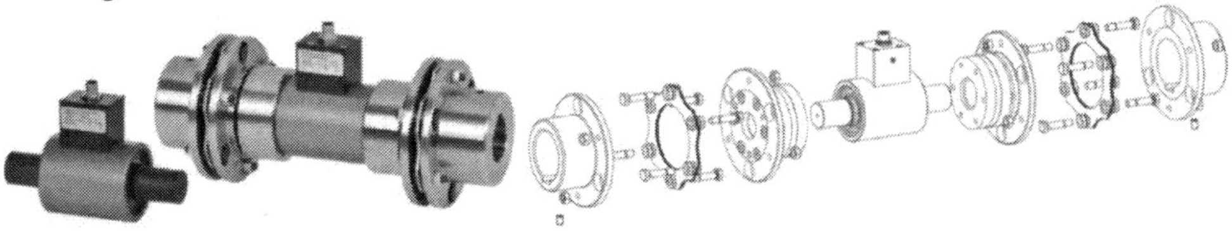

Fig. 4. Torque sensor FR-A701KTR Dataflex 32/30 [6].

During the measurements the oil temperature was 25 °C with a corresponding flow rate of 1 l/h. The air inlet temperature was adjusted to 22 °C. In Figure 5 the pressure development over time is displayed for all four cases. It can be seen that a significant pressure drop inside the screw expander is measured within the first 60° for all four cases. Then the pressure is nearly constant. More specific test conditions and the corresponding measurements are shown in Table 1. The inlet pressure and rotation speed was varied for four different test cases. All the parameters are summarized in Table 1.

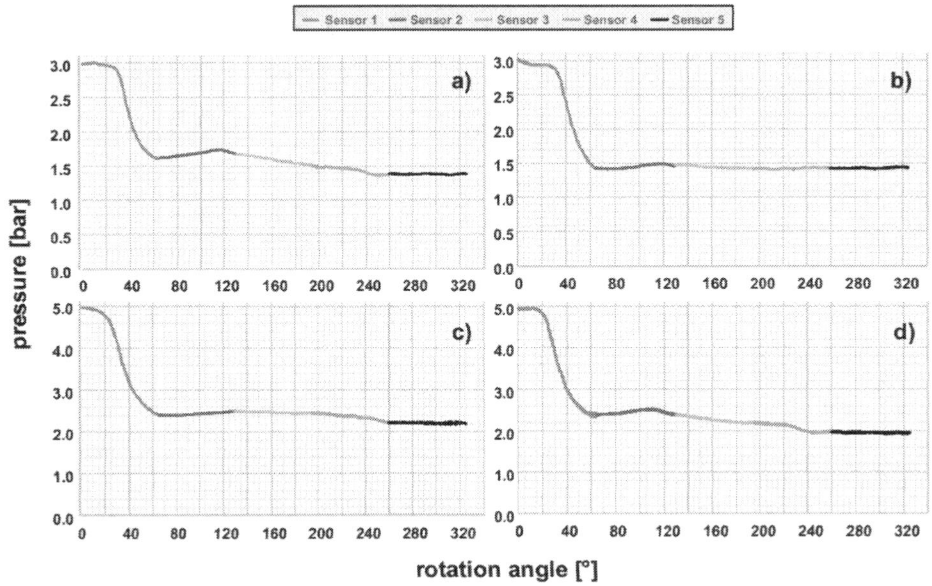

Fig. 5. Pressure development inside the screw expander.

Table 1. Parameters of the different operation points.

Measurement point	Rotation speed	Inlet pressure	Outlet temperature	Volume flow	Mechanical shaft power
	1/min	bar	°C	m³/h	W
a)	1000	3	3	72	558
b)	1500	3	3	83	609
c)	1000	5	-10	117	1563
d)	1500	5	-16	147	1987

In Figure 6a the torque of the screw expander in relation to the expander speed is displayed for an inlet pressure between 2-10 bar. Negative values indicate that energy is delivered by the screw expander. As expected, the torque rises with increasing inlet pressure. For a constant inlet pressure, the torque descends linearly due to the friction. The shaft power is shown in Figure 6b. At an inlet pressure of 2 bar nearly no shaft power is measurable. At higher inlet pressure the shaft power increases with rising expander speed. At an inlet pressure of 10 bar the compressor could supply the required amount of air only up to an expander speed of 1500 rpm and short measurement periods.

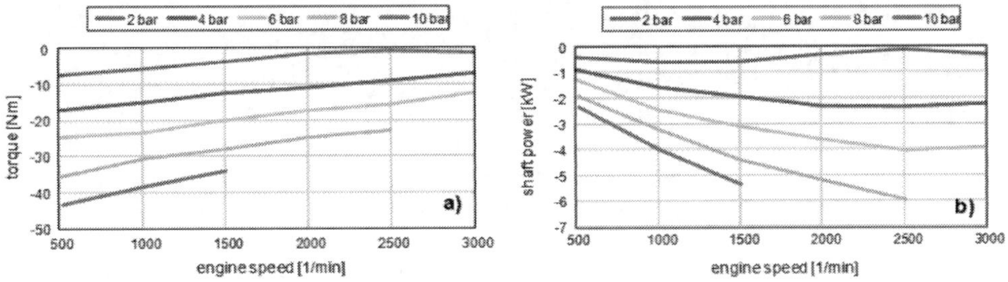

a) b)

Fig. 6. Development of torque (a), shaft power (b) and efficiency over engine speed.

4.3 Mathematical model of expansion process

The description of the working chamber shape is determined by the teeth shape of both rotors, called the 'male' and rotor and the 'female' rotor. In principle, if an imaginary section of both rotors perpendicular to the axis of rotation is made, the shapes of the rotor profiles are obtained. A precise analytical description of the female rotor profile is crucial for computer modelling the geometry. This profile can be understood as a periodic geometry that is formed by elements of the same shape (teeth) rotated by an angle of $2\pi/n$, where n is the number of teeth of the female rotor. Each tooth is described by a series of consecutive analytical curves which are most often lines and circular arcs. The tooth shape of the male rotor is derived using the envelope principle, which is mathematically given by transformation of the coordinates of the contact points of both profiles. As in the case of the female rotor, the shape of the male rotor profile is composed of a periodically repeating geometry of the tooth shape. The number of teeth of the male rotor is given by the distance of the axes of rotation of both twin rotors and by the diameters of the addendum circles that together with the teeth define the monitored working chamber area in the section, see Figure 7.

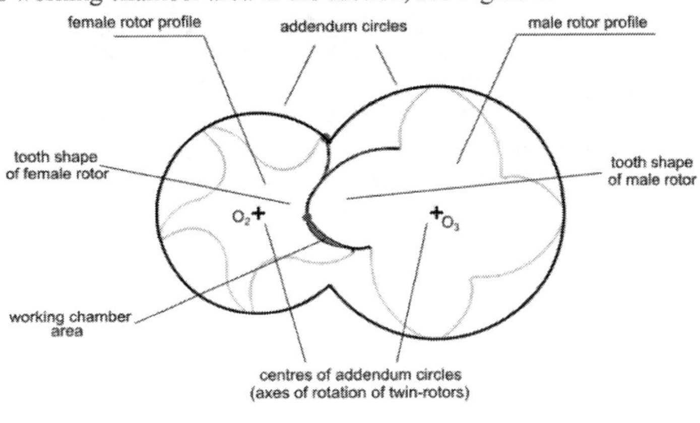

Fig. 7. Twin rotor profiles in frontal plane.

If the twin rotors are a total length l, the geometry can be represented by m parallel perpendicular slices. Taking a distance of two neighbouring $\Delta l_s, s = 1, \ldots, m-1$, then the elementary volume V_s of the working chamber is computed using

$$V_s = \frac{A_i + A_{i+1}}{2} \Delta l_s, \quad i = 1, \ldots, m \tag{8}$$

where A_i and A_{i+1} are the working chamber areas in slices i and i+1, respectively. Then, the total volume of the working chamber is the sum of the elementary volumes along the entire length of the rotor for a given angle of rotation.

For the specified angles of rotation, the working chamber is filled through the inlet port by air mixed with added oil for lubrication. The inlet port characteristics are determined by the geometry of the twin screw expander housing. Inlet port geometry is modelled using primitive curves and its position is derived with respect to the position of the twin rotors axes of rotation. The amount of supplied working medium is thus influenced not only by the geometrical characteristics of the working chamber and the angle of rotation of the rotors, but also by the size of the cross-section area of the inlet port through which the working medium can enter the working chamber. In Figure 8, the hatched area through which the working medium enters the working chamber is marked for a chosen angle of rotation.

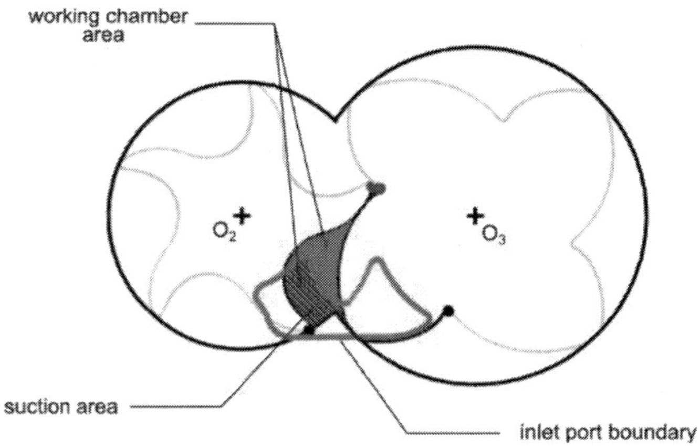

Fig. 8. Active inlet port area for a chosen angle of rotation.

The benefit of an analytical description of the geometry of the screw expander working chambers lies in its use for possible modifications to the working chamber geometry. Possible impacts of geometry changes on the expander efficiency can be evaluated, which require calculations of the expansion phase thermodynamics and especially the solution of the pressure in the working space during the expander working cycle. We developed our own simplified, computationally fast solver of the expansion phase to handle this task effectively. Our solver is based on an expansion model in an isolated single working space of the screw expander, which is derivative of the adiabatic law with the assumption of a working medium consisting of two separate phases, namely compressible gas (air) and incompressible liquid (oil). In the pure expansion phase (between injection and discharging) it can be written as

$$p(V) = p^{IN} \left(\frac{V^{IN} - V_{oil}^{IN}}{V - V_{oil}^{IN}} \right)^k \tag{9}$$

where V is the volume of the working space changing during the expansion, p is the corresponding pressure, p^{IN} is the inlet pressure, V^{IN} is the volume of the working space at the moment of closing the inlet port, V_{oil}^{IN} is the total volume of the injected incompressible oil to the working space and k is the heat capacity ratio. Detailed studies of the associated phenomena affecting the pressure in the working space (propagation of expansion waves, losses due to friction, centrifugal forces, etc.) have shown that when using air as the main working medium at the considered working pressures and temperatures (i.e. always using superheated air vapour), the adiabatic law expansion model is a sufficiently accurate approach for calculating the pressure inside the expanding working space. During the expansion, this model assumes the same pressure along the working space, which decreases over time as the working space increases its volume. This model of working medium expansion is a good approach when assuming fast processes (adiabatic expansion of gas without heat exchanges) and an ideal geometry with ideal contacts (no leakage flow losses). The proposed model based on the adiabatic law for the solution of the expansion phase is then incorporated in the solver in the form of a simple explicit algebraic equation.

Initial measurements of the expansion pressures on a test stand prototype working with a real screw expander clearly pointed out strong leakage flows. Likewise, outlet temperatures were not in accordance with the adiabatic expansion model predictions, which undoubtedly indicates a strong re-heating of the air during expansion. The solution provided by the current idealized model deviates significantly from the measured data when employing a real expander. To adjust the solutions of our solver to the measurement data, the idealized model was modified to capture these two observed phenomena.

The new solver works with all the active working spaces inside the screw expander simultaneously. The expander working cycle is discretized in time explicitly and the expansions in individual working spaces are calculated separately by the adiabatic expansion model between two consecutive time levels. Subsequently, the air leakage mass flow Q_m^{AIR} between pairs of adjacent working spaces is evaluated using the proposed flow function, which assumes only air flow through the gap between the stator and both rotors, not oil leakages

$$Q_m^{AIR} = f(\Delta p, h)\; l\; \Delta t, \tag{10}$$

where Δp is the pressure difference between adjacent working spaces, h and l are the thickness and length of the gap and Δt the time step size, the flow function is continuous and smooth and satisfies conditions $f(0,h)\ \equiv 0$ and $f(\Delta p, 0)\ \equiv 0$. The leakage flow function is a result of 140 2D simulations in Fluent for various Δp and h, starting from the main rotor geometry. The proposed leakage flow sub-model assumes equal leakages through the male and female rotor-stator gaps.

Furthermore, the heat transfer between the air and oil takes into account getting the air outlet temperature into accordance with the measurement data from the test stand. However, we still assume a fully heated expander, and therefore the heat exchange with the stator or the environment is not considered. The air and oil temperatures are thus adjusted in individual time levels according to the relations for interphase heat exchange given in [7] and the corresponding air pressure is corrected.

The expansion and the heat transfer are thus solved in all the working spaces individually, but all the active working spaces are interconnected in pairs through gaps where air leakage flow may occur. A similar solver is commonly used in some computational software e.g. [8]. The solution algorithm handles all the active working spaces together, which can generally occur in different phases of the work cycle: forming the working space, filling the chamber through the inlet port, pure expansion, forming a blow-hole and direct opening of the outlet port. During these parts of the working cycle, the volume of the working space is increasing. The thermodynamic state is calculated from the formation of the working chamber, before the inlet opening. After that, the amount of air through the inlet port with time-changing cross-section area is reduced on the basis of the theory of compressible flow in nozzles.

Finally, the calculations end with the opening of the blow-hole, when the expanding working space starts to be connected to the outlet port via the blow-hole and the older depressurized working space. We assume that the outlet pressure does not decrease further. This assumption is in good agreement with the measured data.

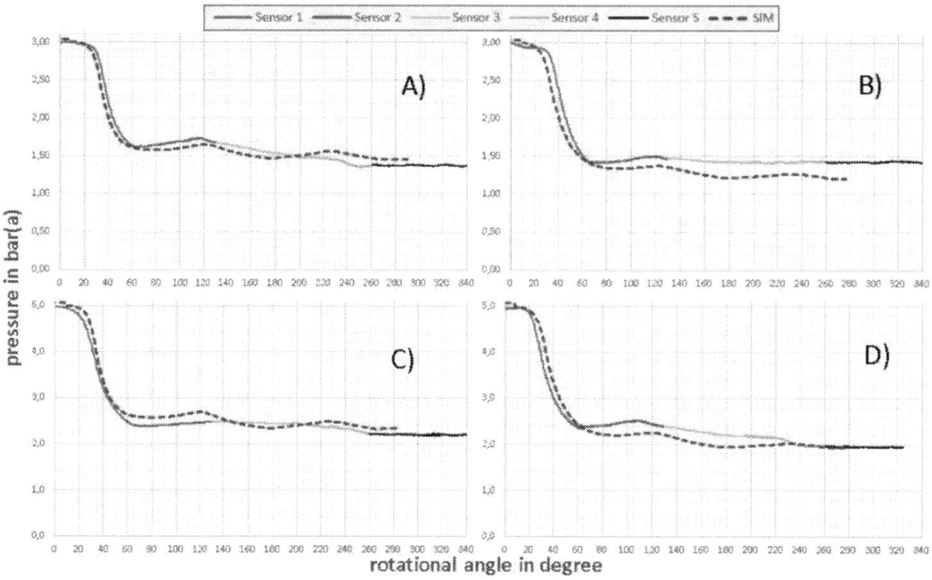

Fig. 9. Comparison of the measured pressures and the solutions of numerical simulation (dashed purple line).

Table 2. Outlet temperature and standardized volume flow rate of air calculated from numerical simulations at different operation points.

Measurement point	Rotation speed	Inlet pressure	Outlet temperature	Volume flow
	1/min	bar	°C	m^3/h
A)	1000	3	-11	70
B)	1500	3	-12	85
C)	1000	5	-25	117
D)	1500	5	-26	143

4.4 CFD Simulations of screw expander

CFD simulations should be conducted to obtain detailed information about fluid flow and the expansion process between the screw rotors. Furthermore, in the later stages, these simulations should also include oil lubrication and thus they should provide information regarding oil injection, its distribution between the rotors as well as the heat transfer between the oil and the expanded gas. As of now, simpler CFD simulations have been conducted as a proof-of-concept that the intended simulations are feasible with the computational resources and simulation software we have available (we will use ANSYS Fluent since we do not possess any single-purpose simulation software for screw machines).

Below we will briefly discuss the geometry of the screw expander which was obtained by 3D scanning the real device. Then we will discuss the approach for 3D simulations and the analyses will be presented.

4.4.1 3D Scanning of Screw Expander

The basis of each CFD simulation is the geometry (aka computational domain) which, in our case, was obtained by 3D scanning a real screw expander. A real device was disassembled, and each component was scanned using precise industrial 3D metrology, ATOS Core. It uses a stereo camera set-up working on the principle of triangulation. The result of the scanning can be seen in Fig 10. The inverse geometry of these solid components (i.e., the volume between the rotors inside the housing) is the computational domain for subsequent CFD simulations.

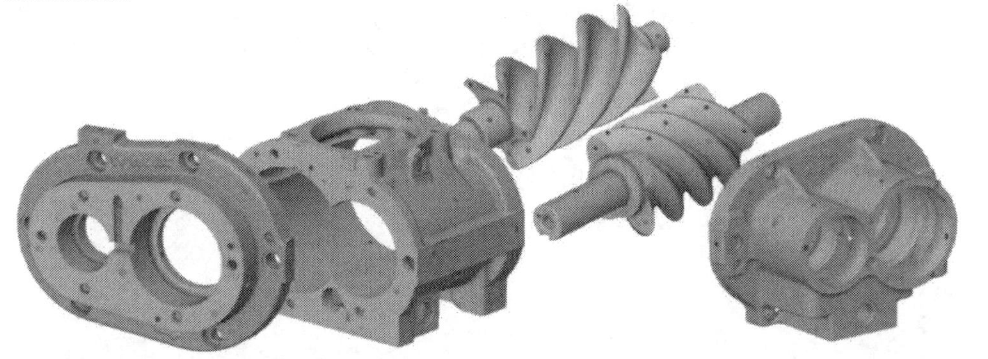

Fig. 10. 3D scan of screw expander.

4.4.2 Overview of CFD Analyses

CFD simulations of the screw expander in ANSYS Fluent can be done either using dynamic mesh or overset mesh. The former means that after each slight rotation (the size will depend on the allowed deformation of the mesh elements), the computational domain will need to be re-meshed. It follows that this procedure significantly increases the requirements on the computational resources and the overall simulation time. The latter approach does not require re-meshing as the simulation model consists of two meshes that move onto each other. However, it requires very dense meshes so that the elements overlay each other, but this approach seems to be very reliable for 3D simulations based on our previous experience.

For our simulations, we decided to first use the dynamic mesh. The main challenge in the dynamic mesh simulation lies in the proper setup of the element size and their allowed deformation before the re-meshing process. It goes without saying that several simulations must be done to verify the setup and thus a simpler domain should be utilized to decrease the computational demands. We decided to start with a 2D simulation to get the sizing of the elements and the values of the allowed deformation. Then we applied the settings to a 3D segment (created by cutting the volume inside the housing and between the rotors by two parallel planes perpendicular to the axes of rotation of the rotors) thus verifying the settings in 3D. Finally, the entire domain is simulated. This concept is shown in Fig. 11.

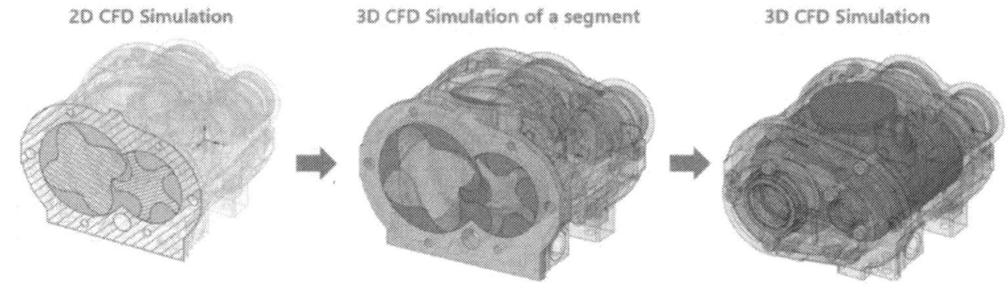

Fig. 11. Overview of CFD simulations

4.4.3 2D and 3D CFD analyses

As of now, 2D and 3D simulations of a segment have been performed. Both simulations were done with a time-step of 0.002ms (i.e., it requires 10 000 time-steps per 1 rotor revolution) and a two-equation SST k-ω turbulence model was used. The mesh is shown in Figure 12a and some missing areas can be seen in locations where the teeth are close to the housing wall, or they are close to each other. This is on purpose, as we do not intend to model the leakage in these initial simulations. Gap detection was used to remove these small gaps from the simulation.

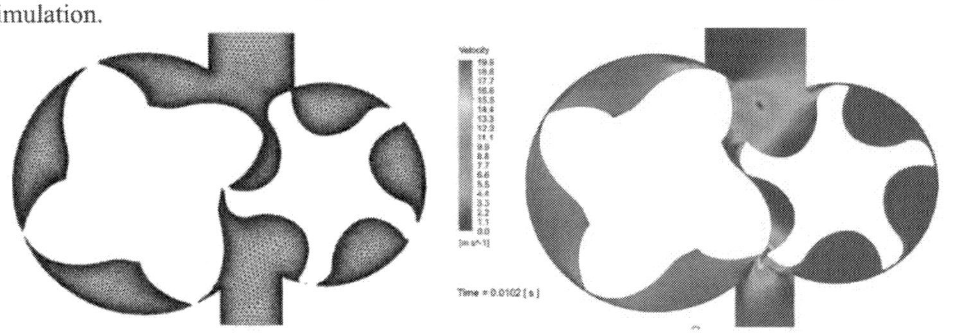

a) b)

Fig. 12. a) Computational mesh for 2D simulation. b) Results of 2D CFD simulation (velocity field).

In Figure 12b the velocity field can be seen for a specific time step. Further analysis of the results will not be presented in this section as it would be outside the scope of this article. To conclude, it should be noted that the mesh settings from the 2D analysis were applied to the 3D analysis of the segment and the simulation was stable and converged very well. However, the simulation time was not acceptable (a few days per rotor revolution), and thus further optimization is necessary before the simulation of the entire domain.

CFD simulations become much more complex when oil flow is considered. Multiphase flow requires a further increase in computing capacity. The most common approaches for solving multiphase flow using CFD solvers are the Euler-Euler (gas and liquid are interpenetrating continuous media) and Euler-Lagrangian (gas is a continuous medium and liquid is dispersed phase) approaches [9].

5 Conclusion

The joint implementation of the project by the University of West Bohemia, OTH Amberg-Weiden and ATMOS resulted in an initial analysis of the expansion process inside an oil-flooded air screw expander. Follow-up activities will use the established Design of Experiment and the developed mathematical model for further analysis of the screw expander in real conditions of the power unit. Real conditions are characterized mainly by a special working medium in the form of low-boiling organic matter, the compatibility of which with lubricating oil and sealing elements of the expander must be taken into account. The real conditions differ from the performed analysis in operating temperatures, when temperatures around 100 °C are reached at the inlet to the expander. In order to achieve satisfactory performance in real conditions, the speed of the expander must be higher, which increases the medium flow. At higher media flows, problems can occur in separating the oil from the working fluid. The next step in terms of the numerical approach will be the development of a 3D model, which is very demanding given the need to use a dynamic computing mesh.

The presented work was financially supported by the European Commission and the European Regional Development Fund. The aim of the project is to strengthen cross-border activities and cooperation in the field of research and innovation between Czech Republic and Free State of Bavaria.

References

1. Smith, I., Stosic, N., Kovacevic, A. *Power recovery from low-grade heat by means of screw expanders*. Cambridge, UK: Woodhead Publishing, [2014]. Woodhead Publishing in mechanical engineering. ISBN 978-1-78242-189-4.
2. Iva Papes, Joris Degroote, Jan Vierendeels, *New insights in twin screw expander performance for small scale ORC systems from 3D CFD analysis*, Applied Thermal Engineering, Volume 91, 2015, Pages 535-546, ISSN 1359-4311, https://doi.org/10.1016/j.applthermaleng.2015.08.034.
3. Olivier Dumont, Rémi Dickes, Vincent Lemort, *Experimental investigation of four volumetric expanders*, Energy Procedia, Volume 129, 2017, Pages 859-866, ISSN 1876-6102, https://doi.org/10.1016/j.egypro.2017.09.206.
4. Wennemar Jurgen (2009). *Dry Screw Compressor Performance And Application Range*. Texas A&M University. Turbomachinery Laboratories. Available electronically from http : / /hdl .handle .net /1969 .1 /163089.
5. Hsu, Sung-Wei & Chiang, Hsiao-Wei & Yen, Chih-Wei. (2014). *Experimental Investigation of the Performance of a Hermetic Screw-Expander Organic Rankine Cycle. Energies*. 7. 6172-6185. 10.3390/en7096172.
6. KTR-Group, *DATAFLEX.* Drehmomentmesswelle der Bauart 32/300, handbuch.
7. Brennen, C. E., *Fundamentals of Multiphase Flows*, Cambridge University Press, 2005. ISBN 0521 848040.
8. Stosic, N., Smith, I and Kovacevic, A. (2005) *Screw Compressors – Mathematical Modelling and Performance Calculation*. Springer Berlin Heidelberg New York. ISBN-10 3-540-24275-9.
9. Basha, Nausheen & Rane, Sham & Kovacevic, Ahmed. (2018). *Multiphase Flow Analysis in an Oil-injected Twin Screw Compressor*. Proceedings of the 3rd World Congress on Momentum, Heat and Mass Transfer (MHMT'18). DOI: 10.11159/icmfht18.132.

MATEC Web of Conferences **345**, 00027 (2021) https://doi.org/10.1051/matecconf/202134500027
Power System Engineering 2021

Pressure losses and oscillations in a compact valve of a steam turbine

Václav Sláma[1,2*], *David* Šimurda[3], *Lukáš* Mrózek[2], *Ladislav* Tajč[2], *Jindřich* Hála[3], and *Tomáš* Radnic[3]

[1]Doosan Skoda Power s.r.o., Tylova 1/57, 301 28, Pilsen, Czech Republic
[2]University of West Bohemia, Power System Engineering Department, Univerzitni 8, 306 14, Pilsen, Czech Republic
[3]Institute of Thermomechanics of the Czech Academy of Sciences, Dolejškova 1402/5, 182 00 Praha 8, Czech Republic

Abstract. Characteristics of a new compact valve design for steam turbines are analysed by measuring pressure losses and oscillations on the valve model. It is the model of an intercept valve of the intermediate-pressure turbine part. This valve is relatively smaller hence cheaper than usual control and intercept valves. Besides, four different valve seat angles were tested in order to investigate the valve seat angle influence. In order to further clarify measured phenomena, the wide range of numerical simulations were also carried out. Measurements were performed in the Aerodynamic laboratory of the Institute of Thermomechanics of the Czech Academy of Sciences in an air test rig installed in a modular aerodynamic tunnel. Numerical simulations were performed in the Doosan Skoda Power Company using a package of ANSYS software tools. Measurement results are compared with numerical and generalized in the form of valve characteristics and pressure oscillation maps. As a result of the pressure loss analysis, pressure losses in similar valve assemblies can be predicted with required accuracy for each new turbine where modern compact valves are used. As a result of the pressure oscillation analysis, operating conditions at which dangerous flow instabilities can occur were identified. Thanks to this, the areas of safe and dangerous operating conditions can be predicted so that the operational reliability of the valve can be guaranteed.

1 Introduction

The steam turbine governing system, which includes different types of valves, is a crucial part of any steam turbine. Therefore, the research and development of valves has a long tradition in Doosan Skoda Power Company which also cooperates with a lot of external specialist, for example from the University of West Bohemia, the Czech Academy of Sciences as well as NPO CKTI or MEI in the Russian Federation. A lot of results have been already published such as, for example, [1-8]. In this paper, a control valve of the intermediate-pressure turbine part is investigated. It is also called as an intercept valve and it

* Corresponding author: vaclav.slama@doosan.com

is shown in Fig. 1 on the left. Some results and facts can be found in [7]. Further results, especially for all tested valve seat angles (60°, 70°, 80° and 90°) are described here. The original approach is especially the depicting the intensity of pressure oscillations danger in the form of pressure oscillation maps.

In compare with control valves for high pressure (HP) turbine parts, the main disadvantage of valves for intermediate-pressure (IP) turbine parts is their greater size due to the greater volumetric mass flow. It leads to more expensive design as well as severe technological and assembly issues. For this reason, the new compact valve was designed. Compact does not mean that the whole valve is smaller than usual. The smaller is only the diameter of the control valve chamber D_k in comparison with the diameter of the valve diffuser throat D. The scheme is shown in Fig. 1 on the right, where the typical and compact valve is compared by scaling both on the same valve diffuser throat size. The dimensions are shown in Tab. 1 where D_k is the diameter of the valve chamber, D is the diameter of the diffuser throat, α is the valve seat angle which was tested (Fig. 2 on the right) and D_k/D is the diameter ratio which shows that the compact valve is approximately twice smaller than the typical valve. D_k, D and α are also shown in Fig. 2 on the left. Due to this compactness, the valve is cheaper and there are no serious technological issues. However, the pressure losses are expected to be greater.

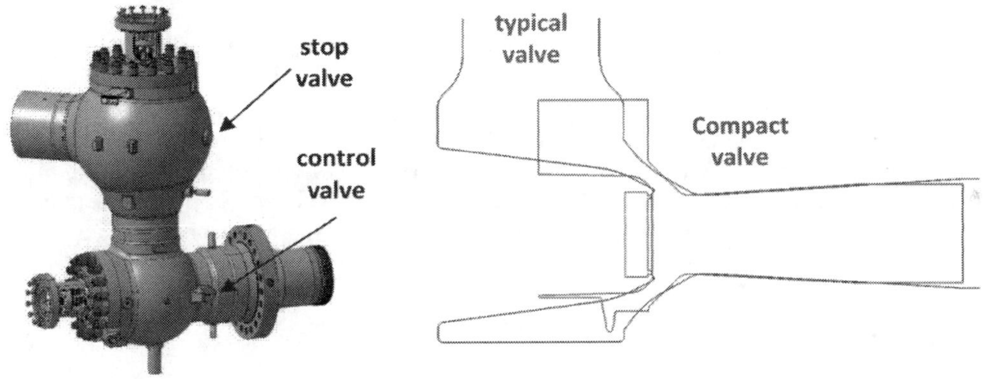

Fig. 1. Valve assembly of the intermediate-pressure turbine part (left), comparison between the valve chambers (right).

Table 1. Comparison of valve dimensions.

	compact valve (IP valve)	**typical valve (HP valve)**
D_k **[mm]**	1,100	360
D **[mm]**	420	80
α **[°]**	60, 70, 80, 90	60
D_k/D **[1]**	2.62	4.50

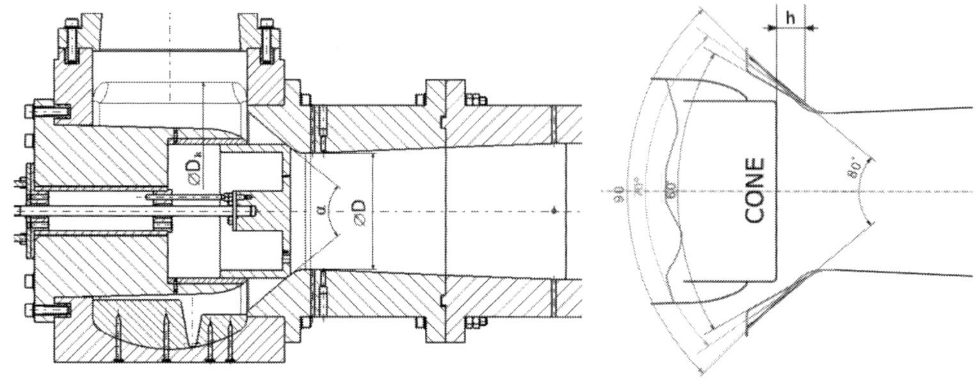

Fig. 2. Cross-section through the control valve model (left), valve seat angle α and valve cone lift h (right).

For each valve assembly there are, among others, two objectives for research and development investigation. The first is to estimate pressure losses so that the turbine efficiency can be precisely calculated for all the similar valve designs. For this reason, pressure loss models are often created [4]. The second is to avoid a danger of intensive pressure oscillations because it can negatively affect valve operational reliability. The pressure oscillations can appear under the valve cone and near the diffuser side walls or even in the outlet part of the diffuser [9-11]. The issue is that these oscillations can cause high dynamic forces which can lead to strong valve vibrations and thus valve operating reliability failure. Therefore, each valve has to be designed very carefully to eliminate potentially unstable flow field, where pressure oscillations appear, for the required valve operating range. Therefore, except the pressure losses analysis, the pressure oscillation analysis is also an important part of this presented study.

2 Experimental model

The measured model representing the real valve assembly was scaled 1:2.875. It is sufficient to satisfy laboratory capabilities and the thermodynamic similarity theory [12]. The model consists of an inlet pipeline, a stop valve, a control valve and an outlet pipeline (Fig. 3). The detail cross-section is shown in Fig. 5. In order to measure a required range of flow conditions, a modular aerodynamic air tunnel situated in the Aerodynamic laboratory of the Institute of Thermomechanics of the Czech Academy of Sciences in Nový Knín in the Czech Republic was used. The scheme with a detail description of all parts is in Fig. 4. Measured flow conditions were defined by a mass flow rate and a control valve cone lift. The control valve is the main part of the presented study. Its design is new (compact) and it has to be operated under different valve cone lifts. The stop valve, design of which had been unified, has been analyzed in previous studies. Some results were published in [4].

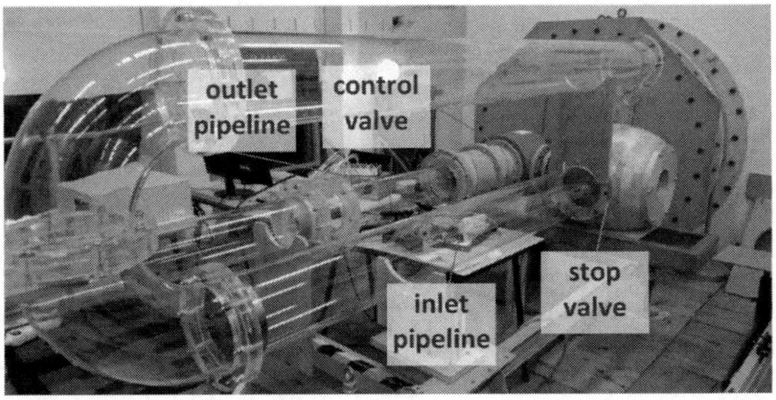

Fig 3. Model assembly in the laboratory.

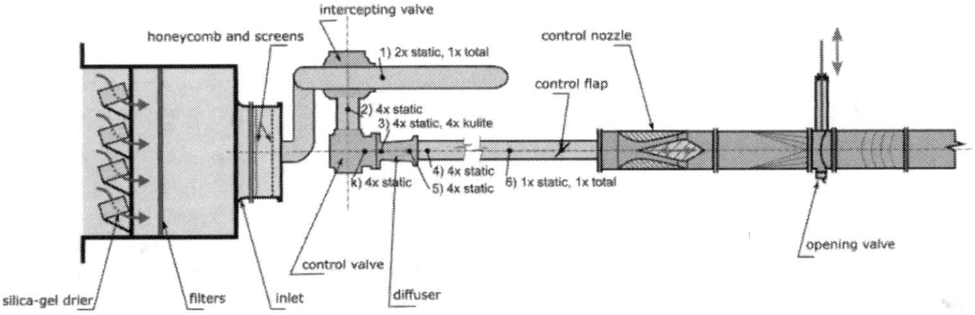

Fig. 4. Detail aerodynamic tunnel scheme.

The flow parameters were measured in different positions upstream, downstream and between the valves as it is depicted in Fig. 4 and in detail in Fig. 5. The mass flow rate was measured in the outlet pipeline using a system with Annubar probe and processing unit INMAT.. A total and static pressures upstream of the stop valve were measured by a Pitot-static tube and wall static pressure taps. At each position from p_{21} to p_{54}, there were four static pressure taps evenly distributed around the circumference [7]. These static pressures were measured by 16 channel pressure transducer NetScanner 9116. Pressure oscillations were measured by 4 fast response pressure transducers Kulite XCEL-100-1BARA near the valve diffuser throat (Fig. 5). The signal was amplified by modules DAQP-Bridge-A of Dewetron company and recorded by software Dewesoft with 14kHZ frequency. For each measured regime, the output signal was acquired for a period of 10s. From the measured time series, the root mean square (RMS) and power spectral density (PSD) were evaluated for each valve performed set of measurement using Matlab. PSD was evaluated by using Welch method with 1,024 samples in the Hann window with 50% overlap. To evaluate the intensity of the oscillations, the effective value of the RMS signal was used:

$$\mathrm{RMS} = \sqrt{\frac{1}{N}\sum_{n=0}^{N} x[n]^2} \qquad (1)$$

Relative uncertainty of the total pressure loss coefficient evaluation is better than 4% with 95% confidence interval. Because the used medium is dry air, Reynolds numbers are lower in comparison with steam, which is in real valves, approximately by two orders of magnitude.

Fortunately, previous internal studies and also publication [13] show that measurements at lower Reynolds numbers still provide accurate results. Effect of the different Reynolds number on total pressure losses will be probably the most significant in the diffuser. This may be one of the main source of differences between losses obtained experimentally and numerically.

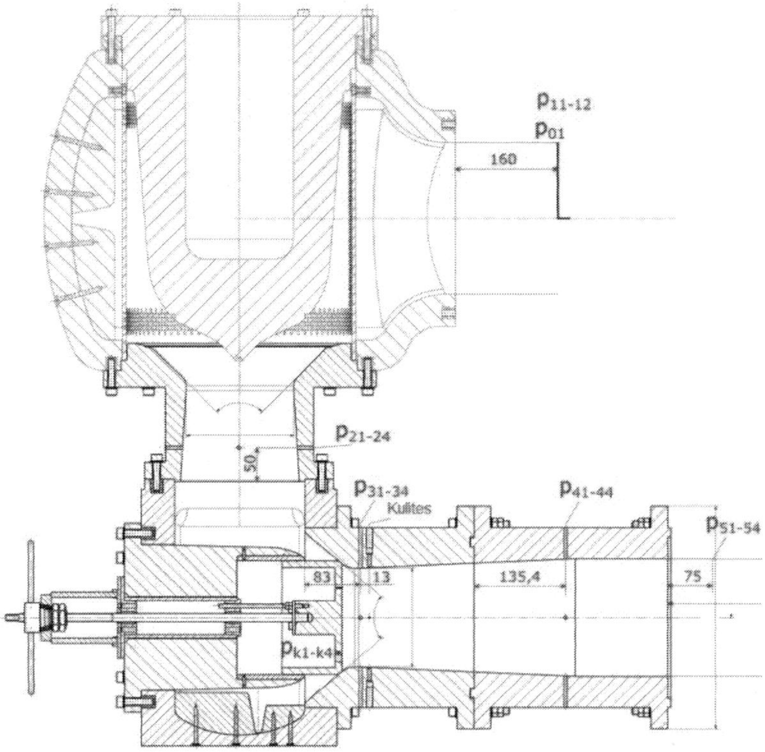

Fig. 5. Valve assembly model with measurement positions.

3 Numerical model

The numerical model (Fig. 6 on the left) was created according to the experimental one. A hexahedral as well as tetrahedral mesh (Fig. 6 on the right) was created to satisfy the typical mesh quality criteria and also the recommendation of the turbulence model to keep $y+<5$ [14]. It could be fulfilled thanks to the experiences in similar projects [4], [15]. The flowing medium was defined as dry air with thermodynamic parameters according to the experiment. A high resolution scheme was used to compute the steady state RANS solution with a compressible turbulent model. The k-ω SST turbulent model was used. Inlet and outlet boundary conditions were specified according to measured data. It means the total pressure and static temperature were defined at the model inlet and a static pressure at the model outlet in a way that static pressures at measured positions were in accordance with measured results. Simulation convergence criteria were quantified by 10^{-4} residual target value. In addition, the global mass flow rate imbalances and imbalances of required pressure losses were also monitored to be less than 0.1%. More details are in [7].

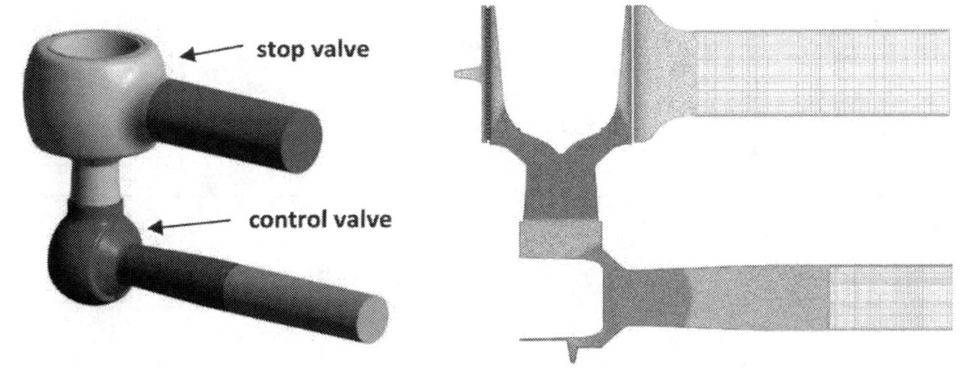

Fig. 6. Numerical model assembly (left), cross-section showing the used mesh (right).

4 Results and discussion – valve characteristic

A valve characteristic shows a dependence of three main different quantities: a mass flow, a pressure ratio and a valve cone lift. In order to generalize results, quantities in the form of m/m_{cr}, p_5/p_2 and h/D are used. m is the measured mass flow rate, m_{cr} is the critical mass flow rate, p_5 is the static pressure downstream of the control valve (Fig. 5), p_2 is the static pressure upstream of the control valve (Fig. 5), h is the valve cone lift and D is the control valve diffuser throat diameter (Fig. 2). All the values are defined or measured except the m_{cr} which is related to the valve diffuser throat and calculated as:

$$m_{cr} = A_{dif}\sqrt{\frac{p_{2t}}{v_{2t}}}\sqrt{\gamma\left(\frac{2}{\gamma+1}\right)^{\frac{\gamma+1}{\gamma-1}}} \tag{2}$$

Where A_{dif} is the cross-sectional area of the diffuser throat, p_{2t} is the total pressure at the control valve inlet, v_{2t} is the total specific volume at the control valve inlet and γ is the heat capacity ratio.

The measured results evaluated in valve characteristics graphs for all valve seat angles from 60° to 90° are shown in Fig. 7. The black line which reaches $m/m_{cr} = 0.3$ for $p_2/p_5 = 1$ represents a typical turbine characteristics and typical operating conditions of the valve. Therefore, the curves for each relative valve cone lift h/D are the most important near this line for a practical use. When comparing the results of different angles but for the same relative valve cone lift (Figs. 8 and 9), it can be seen that for the same pressure ratio the greater the valve seat angle, the greater the mass flow. Having the same valve seat angles, this could show which case is better from the pressure losses point of view. However, when valve seat angles are different, which is this case, it does not mean that the valve with $\alpha = 60°$ is the worst. The fact that the mass flow through the valve with $\alpha = 60°$ is the lowest is caused by the different mass flow. When having the same h/D, the valve with $\alpha = 60°$ has the smallest cross-section area ratio in comparison to the others (Tab. 2 on the left). In practice, the same mass flow should be achieved by the greater valve cone lift (Tab. 2 on the right). The results of $h/D = 0.144$ and $\alpha = 60°$ showed in Fig. 9 on the right are very likely to be distorted by some unexpected measurement errors. It could be improved by repeating the measurement. However, it was observed far from the common valve operating conditions.

Table 2. Different A_{seat}/A_{dif} for the same h/D (left) or different h/D for the same A_{seat}/A_{dif} (right) dependence on α.

α [°]	h/D [1]	A_{seat}/A_{dif} [1]
60	0.322	0.796
70	0.322	0.921
80	0.322	1.039
90	0.322	1.145

α [°]	A_{seat}/A_{dif} [1]	h/D [1]
60	0.800	0.294
70	0.800	0.258
80	0.800	0.231
90	0.800	0.211

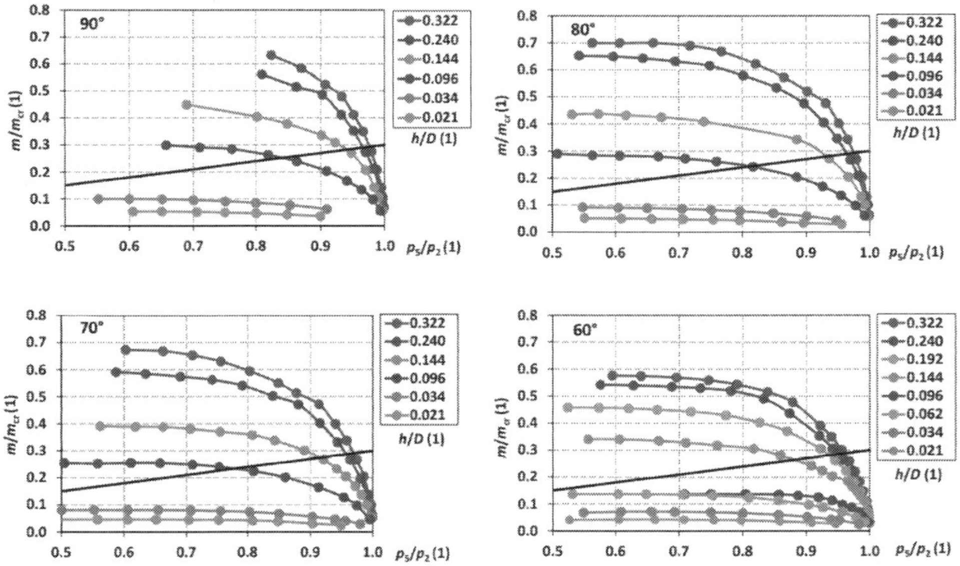

Fig. 7. Measured valve characteristics (curves for chosen relative valve cone lifts) for each valve seat angle.

The comparison of experimental and numerical data provides valuable information. Differences between experiment and CFD vary with changing operating conditions. It is shown in Fig. 10 where three different areas were identified according to results differences. The areas are roughly separated by dashed lines. The first region with the greatest differences appears for the greatest mass flows. In this region there are high velocities and both experimental and numerical inaccuracies are supposed to be greatest. Especially the mass flow measurement may be the most inaccurate there. The second is the area with greater numerical differences. They are caused by the fact that numerical simulations are carried out as steady-state, but phenomena such as flow detachment are very unsteady. This was proved by measurement as it is described in the following chapters. For this reason, the numerical results oscillate around some average value (Fig. 11 left). The results convergence was also not fully fulfilled in this area and the result is dependent on when the simulation was stopped. In the last region, differences between experiment and CFD are reasonable (Fig. 11 right). The maximal difference there is about ±10% which is a very good agreement. It is the area around and below the typical turbine characteristic line when the validity of results is the most important. Similar results were also seen for lower relative valve cone lifts.

In general, inaccuracies of both experimental and numerical modelling which cause described differences are inherent. Except the measurement uncertainties which were already mentioned there are important uncertainties of numerical simulations. There can be a difference in the flowing medium. The thermodynamic properties are defined as calorically perfect ideal gas with a constant pressure specific heat capacity and Sutherland's formula for dynamic viscosity. This may be slightly different from reality. Other differences are caused by the assumption of perfectly smooth walls, the case is defined as adiabatic and there may be inaccuracies in the method used for turbulent flow modeling [16]. Main differences between experimental and numerical data probably result from the modelling of shear layers and their development. Especially when a flow separation appears, either mass flow rate or pressure ratio will be affected depending on the boundary conditions. However, regarding the fact that the numerical simulations are used mainly to better understand qualitative features of the flow in different flow regimes, the differences shown there are not supposed to be crucial.

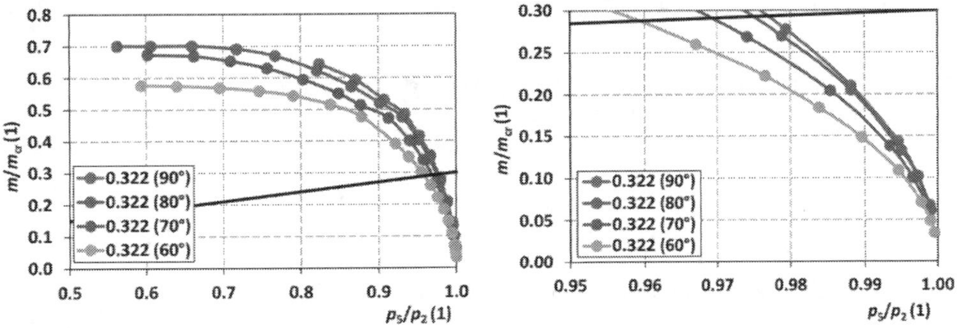

Fig. 8 Comparison of valve characteristics for $h/D = 0.322$ for different α – whole range (left), detail (right).

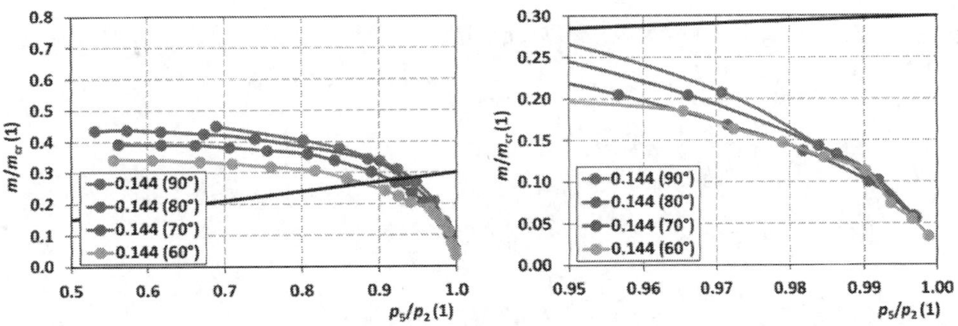

Fig. 9. Comparison of valve characteristics for $h/D = 0.144$ for different α – whole range (left), detail (right).

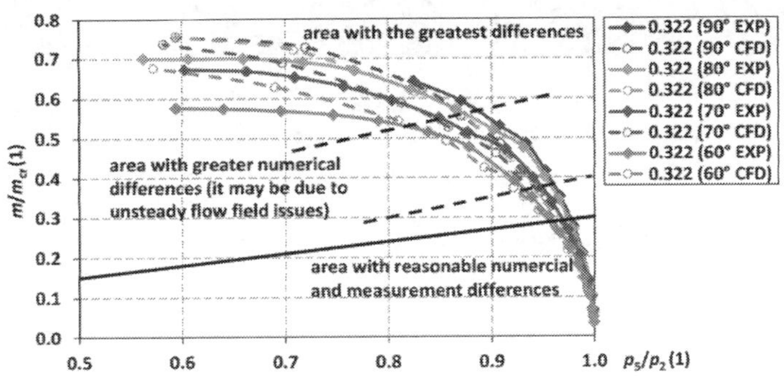

Fig. 10. Valve characteristics – comparison between experimental (EXP) and numerical (CFD) results for $h/D = 0.322$.

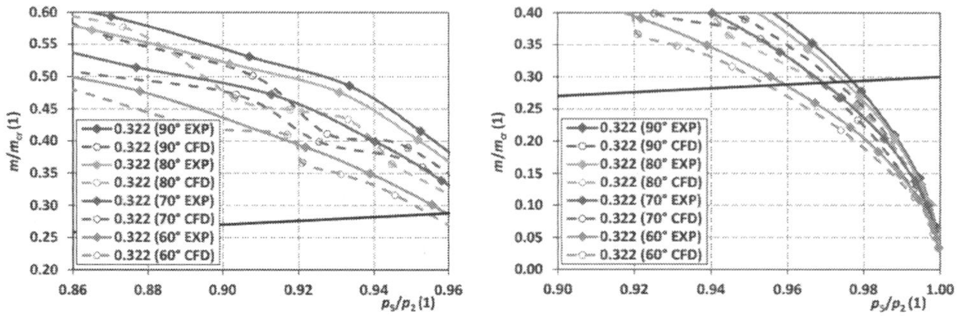

Fig. 11. Details of Fig. 10 – the area with greater (left) and reasonable differences (right).

5 Results and discussion – pressure oscillations

The pressure oscillations were evaluated as it is written in the previous chapter. An example of the raw measured data is in Fig. 12. From these raw dependences it is already evident that the amplitudes of pressure oscillations increase with decreasing pressure ratio and increasing valve cone lift. At relative valve cone lift h/D = 0.144, there are visible abrupt changes (in the read oval in Fig. 12). These abrupt changes are very probably related to the abrupt flow detachment out of the diffuser walls very near or at the valve diffuser throat. It should be also noted that the change in the nature of oscillations was audible during the measurement sets. The same abrupt changes were also visible at relative valve cone lift h/D = 0.096 (Figs. 13 and 14). For other relative valve cone lifts, this phenomenon was not observed.

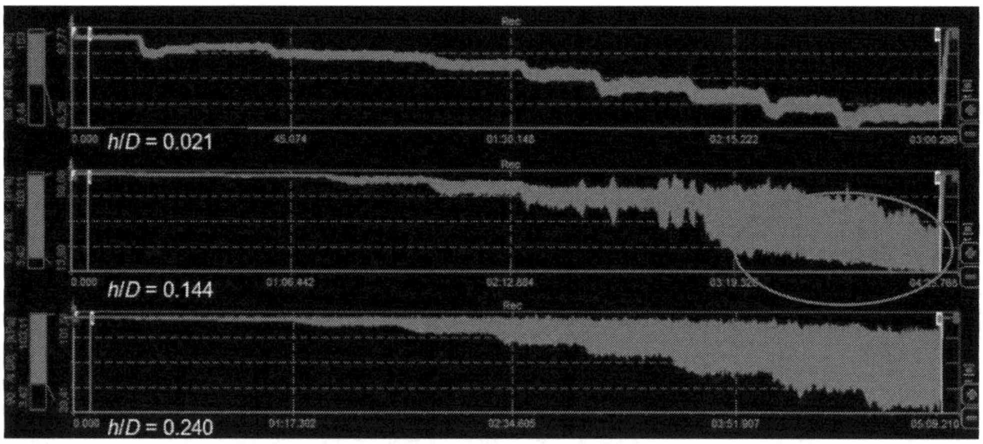

Fig. 12. Time record of pressures measured by a Kulite probe for the case with α = 80° for three different h/D with an abrupt flow detachment identification (the red oval).

All the raw data, one example of which is in Fig. 12, were further transformed to the form of pressure oscillation maps in Figs. 13 and 14. There are also shown the cases with the observed abrupt pressure changes hence the probability of an abrupt flow detachment. For the purpose of this research, the magnitude of the pressure oscillations was characterized by the RMS value according to equation (3), which was determined as the maximum value (RMS_{max}) from all four samples around the circumference of the valve diffuser throat. The results are shown in Figs. 13 and 14. There are again shown the curves for each relative valve cone lift (some of them have been already shown in Fig. 7) where each point has the RMS_{max} value.

Furthermore, the results were generalized to three modes which are shown by background colors for the cases with 60°, 70° and 80° valve seat angle. For the 90° valve seat angle there were not enough measured data for generalization hence it is not depicted. The blue color means range 0 – 2.7kPa, the yellow 2.7 – 5.5kPa and the red 5.5 – 8.0kPa. The important are again the black lines which show the typical turbine characteristic for $q_{max} = m/m_{cr} = 0.3$.

It is clear from the maps in Figs. 13 and 14 that the smaller angle of the valve seat angle has a positive effect on the magnitude of the pressure oscillations. This is mainly reflected in the interval of the greatest values of RMS_{max}, which occur in the red areas. In addition, the observed abrupt pressure changes at $h/D = 0.144$ and 0.096 are shifted to the left side of the graphs (towards lower pressure ratios) when valve seat angle decreases. From the operation reliability point of view, it is a good result because it is also far from the operating conditions where the valve is often used.

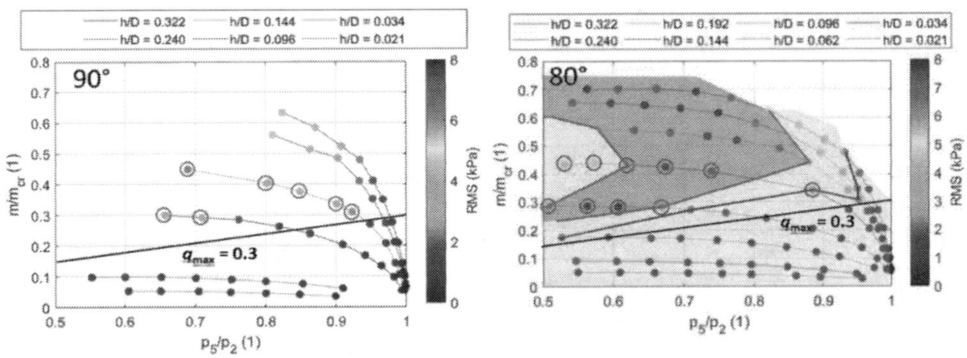

Fig. 13. Pressure oscillation maps (root mean square values) for $\alpha = 90°$ and 80° with red marked points where there is expected an abrupt flow detachment.

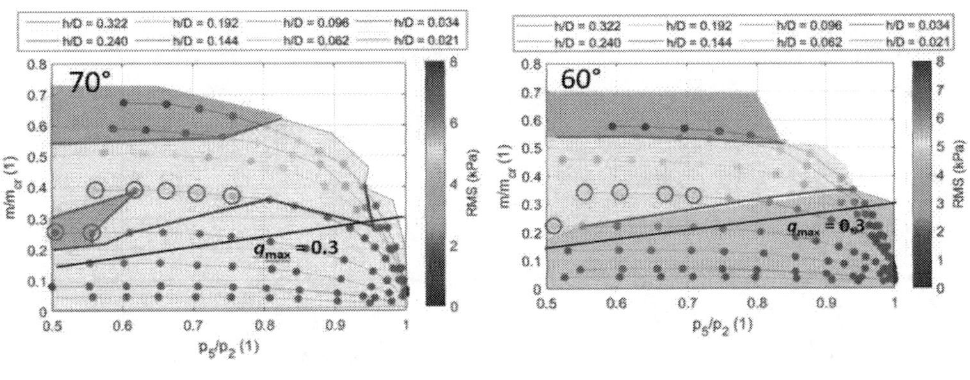

Fig. 14. Pressure oscillation maps (root mean square values) for $\alpha = 70°$ and 60° with red marked points where there is expected an abrupt flow detachment.

In order to analyze what happens inside the flow field in detail, the numerical results can be used. The observed trends are shown on the example of the case with the valve seat angle 90° for chosen mass flows and pressure ratios as well as for different relative valve cone lifts (Figs. from 15 to 17). Similar trends were visible for the other tested valve seat angles.

For the greatest valve cone lift $h/D = 0.322$, the flow field is completely asymmetric no matter whether the valve is overloaded or not (Fig. 15). The flow field itself looks similar. The only difference is in velocity magnitudes. The non-uniformities can indicate that the valve design is not fully convenient for the case where the valve is fully opened. It means that there is a potential to reduce pressure losses by optimizing the geometry. Fortunately, the measured values indicating the significant pressure oscillations which can endanger the valve reliability are not high for less overloaded cases (Fig. 13). Therefore, even if the flow field is very asymmetric, the pressure oscillations are not significant. As a result, the valve can be used from operation reliability point of view. For lower valve cone lift $h/D = 0.144$, the flow field is not too asymmetric, especially in case when the valve is less loaded (Fig. 16). From this case, it can be inferred that the lower the pressure ratio, the greater asymmetry and flow separations occur. It is in accordance with the measured pressure oscillations values which are also greater in the area with lower pressure ratios. However, it is still necessary to notice that there is the area where abrupt flow detachment can appear as shown in Figs. 13 and 14. This phenomenon cannot be appropriately simulated by the numerical approach used in this study. The flow filed for such cases may look similar as it is in Fig. 16 on the right. For the lowest lift $h/D = 0.021$, the flow field is fully symmetric no matter the pressure ratio (Fig. 17). It is also in accordance with measured results where the lowest pressure oscillations were observed.

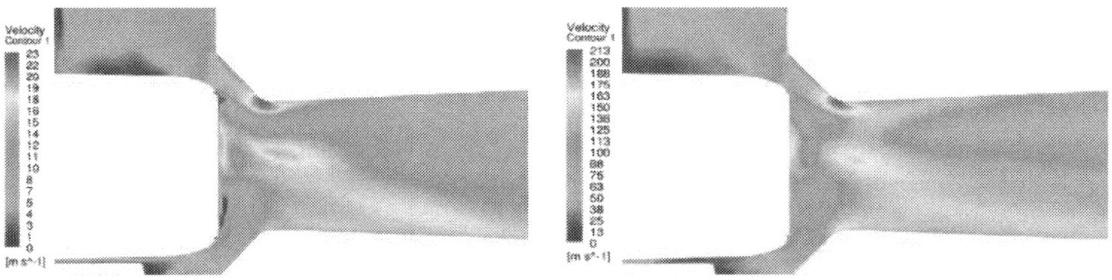

Fig. 15. Flow field in the valve cross-section for $\alpha = 90°$, $h/D = 0.322$, $p_5/p_2 = 0.999$ for the less loaded case (left) and $p_5/p_2 = 0.833$ for the more loaded case (right).

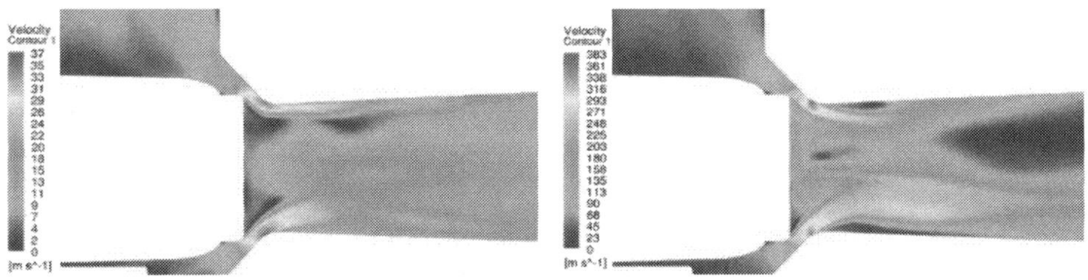

Fig. 16. Flow field in the valve cross-section for $\alpha = 90°$, $h/D = 0.144$, $p_5/p_2 = 0.997$ for the less loaded case (left) and $p_5/p_2 = 0.690$ for the more loaded case (right).

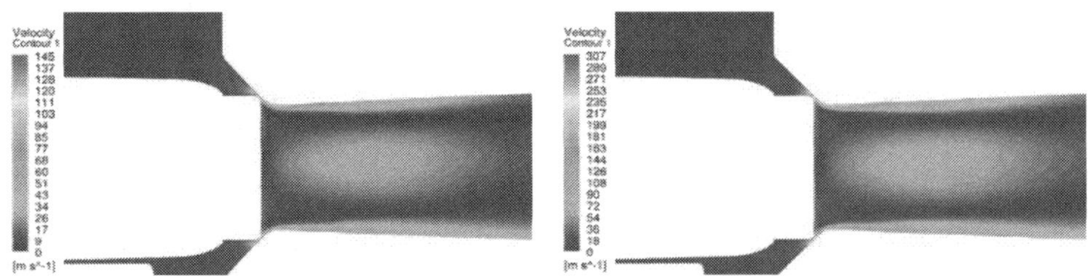

Fig. 17. Flow field in the valve cross-section for $\alpha = 90°$, $h/D = 0.021$, $p_5/p_2 = 0.899$ for the less loaded case (left) and $p_5/p_2 = 0.605$ for the more loaded case (right).

6 Conclusions

Aerodynamic measurements complemented with numerical simulations on the valve model of the compact valve assembly design for intermediate-pressure turbine parts reveal important knowledge of pressure losses which were evaluated in the form of valve characteristics and pressure oscillations which were evaluated in the form of pressure oscillation maps. It was carried out for four different valve seat angles (60°, 70°, 80°, 90°) and whole range of pressure ratios as well as valve cone lifts. As a result, the following conclusions can be derived:

- Pressure losses were evaluated and compared. The lower the valve seat angle, the lower the mass flow. It is caused by the lower mass flow area under the valve cone.
- The comparison between the numerical and experimental results were described and there were identified the different areas according to the found differences. For the most important operating conditions (around the typical turbine characteristic), the differences are reasonable.
- For relative valve cone lifts $h/D = 0.192$ or greater, the effective RMS value of pressure oscillations increases evenly with an increasing mass flow and decreasing pressure ratio. For relative valve cone lifts $h/D = 0.144$ and $h/D = 0.096$, there is an abrupt change in the RMS value of the signal at the certain mass flow and pressure ratio. This change is very probably related to the occurrence of a sudden flow detachment out of the diffuser walls very near to the valve diffuser throat.
- As the valve seat angle decreases, the area within the range of the highest RMS values decreases in the oscillating maps.
- Three different types of internal flow filed were identified. It was found out that there may be a potential to improve the valve design which would lead to pressure losses reduction.

Besides, thanks to this study, pressure losses in similar valve assemblies can be predicted with required accuracy for each new turbine and the areas of safe and dangerous operating conditions can be predicted so that the operational reliability of the valve can be guaranteed.

For the future investigation, at first, the results should be compared with other valve types, especially with those of relatively greater valve chambers in order to find out the differences in pressure losses and flow fields. As a consequence, the design of this compact valve would be optimized. Next, downstream of the valve assembly, there is a spiral inlet turbine chamber which leads steam to the turbine stages. In this turbine chamber inlet, the pressure losses have never been measured and compared with numerical results. There would be a potential to validate the pressure loss model for inlet turbine chambers as well as a possibility to optimize its design.

This research was made possible thanks to the kind support of the Technology Agency of the Czech Republic, National Center for Energetics TN01000007/05. The authors would also like to thank the Doosan Skoda Power Company and, the Institute of Thermomechanics of the Czech Academy of Sciences for giving a permission to publish this paper.

References

1. Kollross P., Tajc L.: *An Experimental Research of the DSPW Steam Turbine Control Valve*, AIP Conference Proceedings 2189, 020012, November (2019), https://doi.org/10.1063/1.5138624
2. Mrozek L., Tajc L., Vimmr J., Bublik O., Slama V., Kalista R., Dancova P.: *The computational study in the unloading slot of the control valve*, EPJ Web of Conferences 180, 02070 (2018), https://doi.org/10.1051/epjconf/201818002070

3. Panuska M., Tajc L., Kollross P., Richter L., Volf M., Zitek P.: *The effect of a strainer in the control valve of the steam turbine*, AIP Conference Proceedings 2047:1 (2018), 020015, https://doi.org/10.1063/1.5081648
4. Slama V., Mrozek L., Tajc L., Simurda D., Hala J.: *Creation and Validation of Pressure Loss Model for Steam Turbine Control Valves*, AIP Conference Proceedings 2189, 020020, November (2019), https://doi.org/10.1063/1.5138632
5. Slama V., Mrozek L., Rudas B., Simurda D., Hala J., Luxa M.: *Experimental and Numerical Study on Pressure Losses and Flow Fluctuations in a High-Pressure Valve Assembly of Steam Turbine Governing System, ASME Turbo Expo 2020*, GT2020-14474, Virtual, Online conference, September (2020), https://doi.org/10.1115/GT2020-14474
6. Slama V., Mrozek L., Tajc L., Klimko M., Zitek P.: *Flow Analysis in a Steam Turbine Control Valve with Through-Flow Valve Chamber*, Journal of Nuclear Engineering and Radiation Science, NERS-20-1141, November (2020), https://doi.org/10.1115/1.4049055
7. Slama V., Mrozek L., Rudas B., Simurda D., Hala J., Radnic T.: *Investigation of Pressure Losses and Flow Fluctuations in an Intercept Valve Assembly of an Intermediate-Pressure Turbine Part*, ASME IMECE 2020, IMECE2020-23359, Virtual, Online conference, November (2020), https://doi.org/10.1115/IMECE2020-23359
8. Mrozek L., Slama V., Sedlak K., Tajc L.: *Analysis of Losses in Steam Turbine Control Valves*, AIP Conference Proceedings 2323, 060003 (2021), March 2021, https://doi.org/10.1063/5.0042097
9. Yueyue Li, Sihua Xu, Jin He, Zhiqiang Hu, Lei Xiao: *Numerical Investigation on the Vibration of Steam Turbine Main Inlet Control Valves Considering Fluid Force Effects*, ASME Turbo Expo 2019, GT2019-90419, Phoenix, Arizona, USA, June (2019), https://doi.org/10.1115/GT2019-90419
10. Windemuth C., Lange M., Mailach R.: *Introduction of a Novel Test Rig for the Investigation of Fluid-Structure Interaction Effects in Steam Turbine Control Valves Using an Elastic Model*, 13th European Conference on Turbomachinery Fluid Dynamics & Thermodynamics, ETC2019-006, Lausanne, Switzerland, April (2019), https://www.euroturbo.eu/publications/proceedings-papers/etc2019-006/
11. Wallat S., Domnick B. C., Munch C., Brillert D.: *A Test Rig Concept to Study Fluid Structure Interactions in a Steam Turbine Valve*, ASME Turbo Expo 2018, GT2018-75094, Oslo, Norway, June (2018), https://doi.org/10.1115/GT2018-75094
12. Langhaar, H.: *Dimensional analysis and theory of models,* New York (1951)
13. Brüggeman P. et al.: *A new emergency stop and control valves design – part 1: experimental verification with scaled models*, ASME Turbo Expo 2014, GT2014-25116, Düsseldorf, Germany, June (2014), https://doi.org/10.1115/GT2014-25116
14. T. Knopp, *Model-consistent universal wall-functions for RANS turbulence modeling*, International Conference on Boundary and Interior Layers, Göttingen, Germany, January (2006).
15. Kalista R., Mrozek L., Slama V., Sedlak K.: *Steam Turbine Axial Exhaust Diffuser Investigation Using a Test Rig and its Numerical Model*, ASME IMECE 2019, IMECE2019-10267, Salt Lake City, Utah, USA, November (2019), https://doi.org/10.1115/IMECE2019-10267
16. Denton J. D.: *Some Limitations of Turbomachinery CFD*, ASME Turbo Expo (2010), Power for Land, Sea, and Air, Volume 7: Turbomachinery, Parts A, B, and C: 7

Operational and economic analysis of a Velox-type combined heat & power plant with a throttle valve

Krystian Smołka[1*], *Piotr* Wiśniewski[1], *Mirosław* Majkut[1], and *Sławomir* Dykas[1]

[1]Silesian University of Technology, Department of Power Engineering and Turbomachinery, Konarskiego 18, 44-100 Gliwice, Poland.

Abstract. For many years, the Department of Power Engineering and Turbomachinery of the Silesian University of Technology has been using a small-capacity (about 500 kW_e) steam-gas power plant. Based on many years of experience in operation of this power plant utilizing the Velox-type gas-steam system, an idea arose to modify this type of thermal power cycle to create a combined heat and power (CHP) plant of small capacity, dedicated for distributed heat and power production or production process steam with high temperature. Previous thermodynamic and economic analysis of that type of low-power CHP plant confirmed many advantages of using this type of solution in low-power engineering. The new idea is to add a throttle valve between the economizer and the evaporator for the reference cycle of the gas-steam Velox-type CHP plant. It was further assumed that water would partially be vaporized on the valve. This paper presents a thermodynamic and economic analysis of this type of a system with an added throttle valve. Downstream the valve, the evaporated water fraction is fed directly into the steam superheater. It is determined how such a retrofit affects the system thermodynamic and economic characteristics. The system is modelled in the EBSILON® Professional 14 software.

1 Introduction

Historically, the first power system using a gas-steam turbine combination was the one where a steam generator was incorporated into the gas turbine system to replace the classical combustion chamber. The concept originated in the Swiss company Brown Boveri. This Velox-type system, as it was called, was patented in 1932 and in 1939 it was used for the first time to produce electricity. Considering the amount of generated steam, the Velox steam generator has a compact design and a relatively small heat transfer surface, especially in the evaporator. This is due to the fact that the heat exchange process in the combustion chamber takes place in overpressure, which intensifies the heat transfer between exhaust gases and water and steam in the tubes of the evaporator or steam superheaters. The high velocity of exhaust gases in such a system enables effective utilization of the combustion chamber [1]. The use of such combined systems in professional power engineering contributed to the

* Corresponding author: krystian.smolka@polsl.pl

creation of high-efficient heat and electricity generation systems, including the first gas-steam cogeneration plant.

The advantages of classic gas-steam systems include the following [2, 3]:

- high efficiency of the fuel chemical energy conversion into electricity and heat,
- lower investment costs compared to conventional combined heat and power plants,
- flexible selection and use of fuels; combined systems can effectively fire natural gas, other gaseous fuels, refined oil and other oil fuels,
- high thermal flexibility from a short start-up to full-load operation; moreover, gas-steam systems maintain constant and high efficiency at loads of 70 ÷ 100% of the rated power,
- short investment completion time,
- high reliability and availability,
- relatively low maintenance and operating costs and a high degree of automation of the operating process,
- very favourable ecological characteristics, much lower emissions of harmful substances and lack of solid combustion waste; a very important advantage, especially in today's power industry, which has to cope with increasingly stringent emission standards.

On the other hand, the disadvantages of combined systems include as follows:

- higher price of combined-system products compared to conventional systems,
- high price of fuel,
- dependence of the gaseous fuel availability and price on strictly political factors; natural gas and oil, which are usually used in such systems, are strategic fuels; they are extracted in specific regions of the world and the need to build pipelines transporting the fuel over long distances substantially increases investment costs.

Considering the above, classic gas-steam systems can be used effectively in high-power engineering.

Using the experience gained so far from the operation of the Velox-type gas-steam system at the Department of Power Engineering and Turbomachinery of the Silesian University of Technology [4] and the technological progress which, among others, makes it possible to use materials enabling production of steam with much higher parameters than in the early days of this type of combined cycles, an idea originated to modify the system to create a low-power combined heat and power (CHP) plant intended for distributed heat and electricity generation or production of high-temperature process steam. Previous thermodynamic and economic analyses [5–7] of a numerical reference model of a system of a Velox-type low-power CHP plant and its comparison with a classic gas-steam cycle [3, 8] led to the following conclusions:

- a Velox-type plant can operate at much lower loads than a classic CHP plant,
- a steam-gas Velox-type power unit makes it possible to use higher temperatures of live steam,
- a Velox-type plant can be much cheaper than a classic CHP plant,
- a Velox-type system can successfully be fired not only with natural gas but also with alternative gaseous fuels, and the cost of adapting such a system for operation with other fuels is not high.

The resulting general conclusion is that the proposed Velox-type gas-steam system keeps many advantages of the classic gas-steam configuration, eliminating at the same time some of the downsides of the latter; it could well be used as a low-power system in distributed power generation.

In order to improve the efficiency of the created reference system of the gas-steam Velox-type CHP plant and to reduce the size of some of the heat exchangers necessary in the cycle, it was proposed to add a throttle valve between the economizer and the evaporator. It was further assumed that water would partially be vaporized on the valve. The paper presents an analysis of this type of a system with an added throttle valve. Downstream the valve, the evaporated water fraction is fed directly into the steam superheater. It is determined how such a retrofit affects the system thermodynamic and economic characteristics. All the analyses presented herein are performed using the EBSILON® Professional 14 software.

2 Thermal power cycles and parameters

2.1 Reference Cycle

The Velox-type CHP plant (Fig. 1) is characterized by a much higher ratio of electricity obtained from the steam turbine and the gas turbine, respectively, compared to classic gas-steam cycles which are now mostly in use. More steam with higher enthalpy is produced from the same amount of fuel [6]. The Velox-type CHP plant system was modelled in the EBSILON® Professional 14 program according to the technological structure shown in Fig. 1.

Table 1 presents the basic parameters for the nominal load of the reference cycle of the Velox-type CHP plant under consideration. It sets out all important parameters for the cycle presented above, together with the efficiency of main and auxiliary machines.

It should be noted that the net efficiency of electricity production is not high, which is expected for this type of cycles with such small power outputs, whereas the efficiency of cogeneration is in the range of many other CHP plants with different technological structures [3].

In addition, Table 2 shows the calculated parameter kA for all heat exchangers used in the reference cycle of the considered CHP plant. It can be used to estimate the heat exchangers size and their surface area.

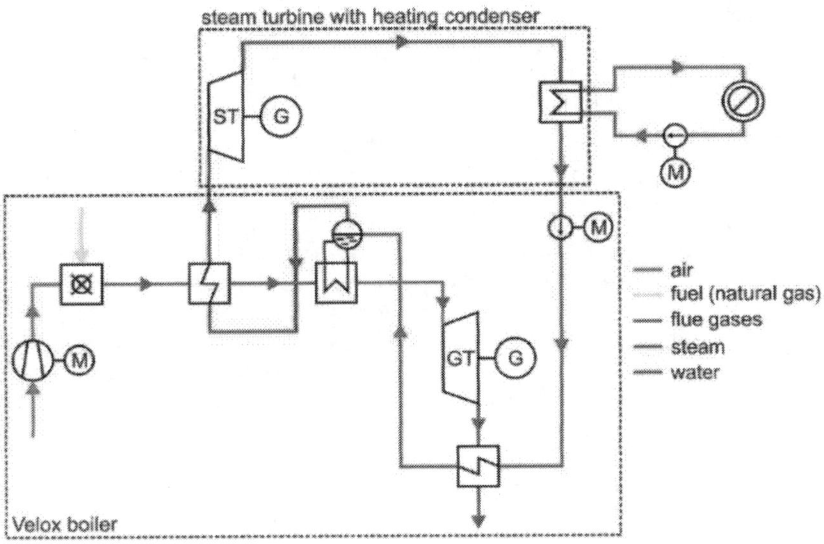

Fig. 1. Velox-type CHP plant schematic diagram – reference cycle.

Table 1. Basic operating parameters for the reference cycle of the Velox-type CHP plant under consideration (nominal load).

Parameter	Value	Unit
Fuel mass flow	0.0013	kg/s
Exhaust gas mass flow	0.101	kg/s
Exhaust gas temperature upstream the expander	333	°C
Exhaust gas pressure upstream the expander	300	kPa(a)
Exhaust gas temperature downstream the expander	192	°C
Live steam temperature	680	°C
Live steam pressure	2000	kPa(a)
Steam mass flow	0.014	kg/s
Steam temperature downstream the steam turbine	265	°C
Steam pressure downstream the steam turbine	100	kPa(a)
Power on the compressor shaft	12.989	kW
Power on the feed water pump shaft	0.034	kW
Power on the heating water circuit pump shaft	0.008	kW
Power on the gas turbine shaft	15.141	kW
Power on the steam turbine shaft	11.897	kW
Inlet temperature of the heating water circuit	30	°C
Outlet temperature of the heating water circuit	94.6	°C
Heat flux from the heating water circuit	35.547	kW
Exhaust gas temperature downstream the superheater	576	°C
Exhaust gas temperature downstream the evaporator	333	°C
Exhaust gas temperature downstream the economizer	146	°C
Water temperature upstream the economizer	99.9	°C
Water temperature downstream the economizer	184	°C
Steam temperature downstream the evaporator	213	°C
Gas turbine internal efficiency	90	%
Steam turbine internal efficiency	90	%
Efficiency of generators	98.6	%
Compressor efficiency	85	%
Efficiency of pumps	80	%
Electric efficiency of motors driving the pumps	95	%
Mechanical efficiency of motors (for pumps and compressor)	99.8	%
Net efficiency of electricity production	20.4	%
Net efficiency of heat and electricity production	76.7	%

The assumed live steam temperature of 680°C may seem extremely high for this type of cycle, especially for a steam turbine. However, technological progress in recent years has led to a significant development of nickel-based alloys that can be successfully used in the construction of turbine components. Alloys of this type have been widely used in the production of gas turbines for years, and have recently been introduced to the construction of high-power steam turbines in the search for materials for ultra-supercritical steam parameters. In the proposed cycle, there are no problems that arose in the production of large turbines for ultra-supercritical steam parameters (much lower pressure was used, and the turbine will not be large), therefore it will be possible to successfully use nickel-based alloys in its construction and experience obtained from gas turbines.

The Table 1 also shows that the assumed internal efficiency of the steam turbine is 90%. This value may seem too high for a turbine with such a low power output (~12 kW), however, the assumption was to propose a cycle for a power of up to 500 kW, hence the efficiency was adopted at a higher level than would be expected for very low power turbines. In future research, when we increase the nominal power of the entire system, this efficiency should be assumed at a level appropriate for modern steam turbines with a specific power, which will entail a slight change in the efficiency of the entire system. It should be added that the reduction of the internal efficiency of the steam turbine entails an increase in the steam temperature downstream of the turbine, and thus an increase in heat production.

Table 2. Surface area of heat exchangers (kA parameter) for the reference cycle of the Velox-type CHP plant under consideration (nominal load).

Parameter	Value	Unit
Economizer	0.230	kW/K
Evaporator	0.126	kW/K
Superheater	0.125	kW/K
Heat exchanger in the heating water circuit	1.449	kW/K
Total	1.930	kW/K

3 Cycle with a throttle valve

The proposed modification of the system consisted in increasing the feed water pressure upstream the economizer and then lowering this pressure downstream the economizer to the value required upstream the steam turbine according to the reference cycle. The lowering of the pressure between the economizer and the evaporator takes place in the throttle valve (cf. Fig. 2). Obviously, the application of such a solution in the system under consideration without any additional assumptions would only lower the efficiency of the entire system because the feed water pump would require a bigger input of work at the same generated power, both electric and thermal. Only the power of the own-needs system would be increased. For example, if the pressure upstream the steam turbine is maintained at the level of 2 MPa (reference cycle) and the pressure downstream the feed water pump is increased to 5 MPa, the efficiency of the entire system decreases by about 0.1 percentage points. It is therefore assumed that at isenthalpic throttling a small part of water has a chance to evaporate which, if this part of the evaporated water is fed directed to the steam superheater, should increase the system efficiency. An additional advantage may be a reduction in the size of the evaporator, which may translate into lower investment outlays.

Figure 2 presents a schematic diagram of the cycle described above, which – like the reference cycle – was modelled in the EBSILON® Professional 14 program. It can be seen that the saturated steam evaporated downstream the throttle valve is separated from the liquid phase in the separator and then fed directly into the evaporator. The rest of the water is traditionally directed to the evaporator and then to the superheater.

It should be noted that in order for the water downstream the throttle valve to evaporate at least in a small part, it is necessary to ensure its appropriate temperature downstream the economizer. This leads to another assumption, i.e. that the temperature of water with higher pressure (upstream the throttle valve) is equal to the saturation temperature of water with lower pressure (downstream the throttle valve). The above assumptions were taken into account to establish the input nominal parameters of the system (cf. Table 3). The internal efficiencies of individual machines and devices making up the cycle, which are not included in the table below, are assumed to be the same as in the reference cycle (cf. Table 1).

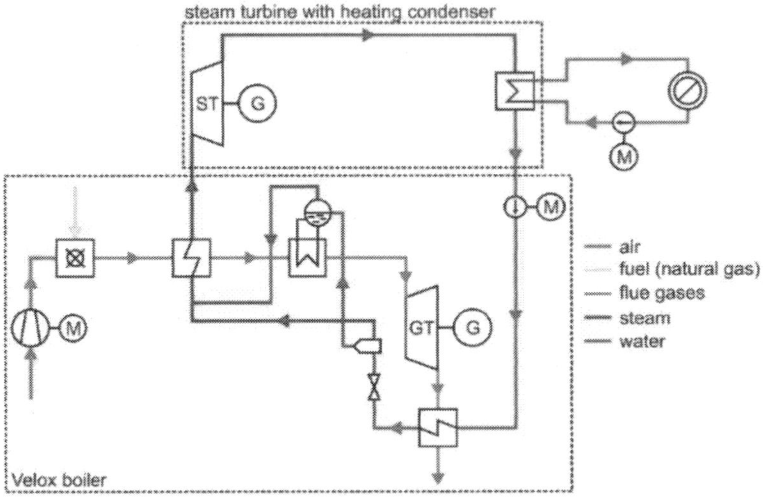

Fig. 2. Velox-type CHP plant schematic diagram – cycle with a throttle valve.

Comparing the data shown in Table 3 to the reference cycle, it can be seen that while the data for the cycle of the steam turbine with the condenser and the heating water circuit are the same for both systems, the parameters in the Velox boiler cycle are different. These differences are forced by the introduced modifications and the assumptions adopted for the cycle with the throttle valve.

Table 3. Basic operating parameters for the cycle with a throttle valve of the Velox-type CHP plant under consideration (nominal load).

Parameter	Value	Unit
Fuel mass flow	0.0013	kg/s
Exhaust gas mass flow	0.101	kg/s
Exhaust gas temperature upstream the expander	363	°C
Exhaust gas pressure upstream the expander	300	kPa(a)
Exhaust gas temperature downstream the expander	216	°C
Live steam temperature	680	°C
Live steam pressure	2000	kPa(a)
Steam mass flow	0.014	kg/s
Steam temperature downstream the steam turbine	265	°C
Steam pressure downstream the steam turbine	100	kPa(a)
Water pressure downstream the pump	5000	kPa(a)
Water temperature downstream the pump	t_s (p = 2000 kPa) = 213	°C
Inlet temperature of the heating water circuit	30	°C
Outlet temperature of the heating water circuit	94.6	°C
Exhaust gas temperature downstream the superheater	580	°C
Exhaust gas temperature downstream the evaporator	363	°C
Exhaust gas temperature downstream the economizer	156	°C
Water temperature upstream the economizer	100	°C
Water temperature downstream the economizer	213	°C
Steam temperature downstream the evaporator	213	°C

The considered system efficiency of both electricity production and combined heat and power generation depends on the adopted degree of water evaporation during throttling. For this reason, these values are not given in the table above. Their analysis will be presented further below.

4 Thermodynamic analysis of a Velox-type CHP plant with a throttle valve

As already mentioned, to model the investigated system correctly, it is assumed that some water will evaporate during isenthalpic throttling between the economizer and the evaporator. The ideal situation would be 100% water evaporation on the throttle valve, which is physically impossible. Therefore, in order to illustrate the impact of the amount of evaporated water on the system characteristics, a few evaporation levels within the range of 1 ÷ 10% are assumed. Figure 3 presents how the efficiency of the entire system, together with the amount of produced electric and thermal energy, as well as the sum of these energies in cogeneration change depending on the adopted level of evaporation.

As indicated by Fig. 3, the efficiency of the entire system increases with a rise in the level of water evaporation on the throttle valve, which is in fact expected. However, after the proposed changes are applied in the cycle, for the water evaporation level of up to 4%, the presented parameters of the system are lower or the same as for the cycle before the modifications. Only if the water evaporation level is maintained at about 5%, does the efficiency exceed the efficiency of the reference cycle. Analysing the change in energy production occurring with changes in the adopted level of evaporation, it can be observed that the level of electricity production increases very slightly, while bigger differences can be noticed for the production of heat only.

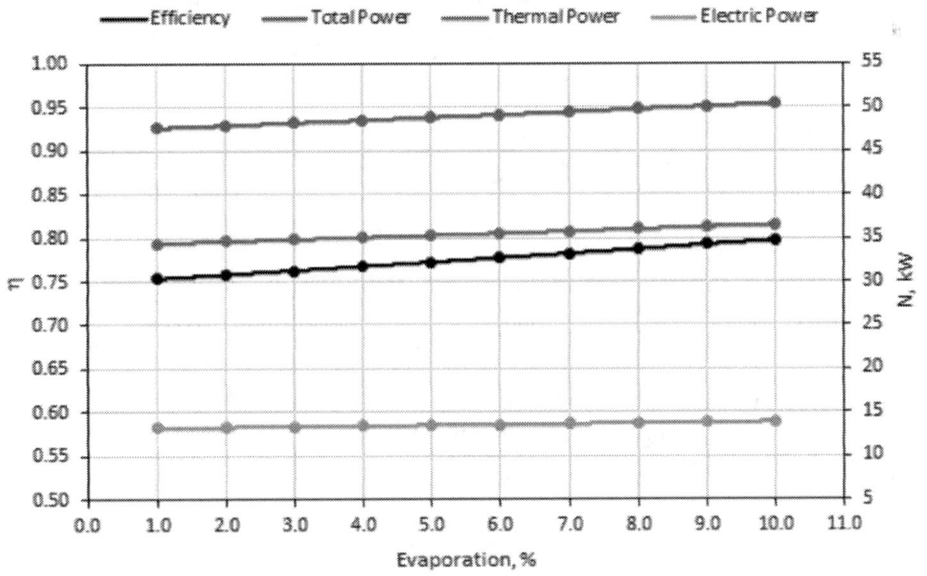

Fig. 3. Change in efficiency and energy production for a cycle with a throttle valve depending on the adopted level of evaporation.

It follows from the above that the system profitability strongly depends on the amount of water evaporated during throttling and on whether it is electricity that is produced only or whether electricity and heat are cogenerated.

For example, the cycle net efficiency at the assumed level of evaporation of 5% totals 77.2%. This corresponds to the production of 13.5 kW of electricity and 35.3 kW of thermal energy, which means the total of 48.8 kW of energy obtained in cogeneration.

Figure 4 illustrates changes in the efficiency of the entire cycle and in the heat transfer surface areas for the exchangers included in the Velox boiler cycle depending on the adopted level of evaporation. It can be seen that with a rise in the evaporation level, the heat transfer surface area of the economizer and of the superheater should be increased. It is only the evaporator heat transfer surface area that decreases slightly with an increase in the adopted level of evaporation on the valve.

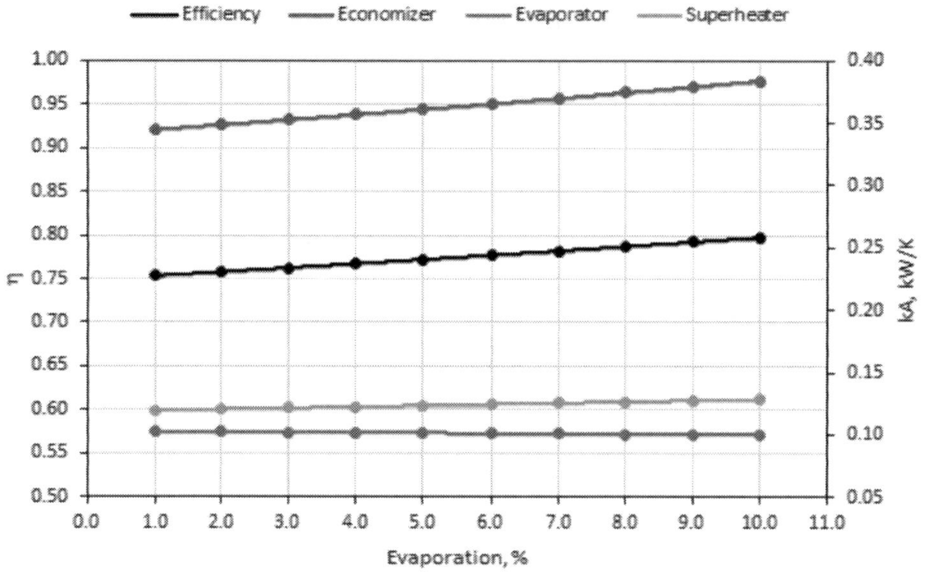

Fig. 4. Change in efficiency and in parameter kA for the exchangers included in the Velox boiler cycle for a system with a throttle valve depending on the adopted level of evaporation.

Looking at the case with the assumed evaporation level of 5% again, it can be seen that the kA parameter, through which the heat transfer surface area of individual exchangers can be estimated, totals 0.361 kW/K for the economizer, 0.101 kW/K for the evaporator and 0.124 kW/K for the superheater.

In order to better illustrate the effect of the evaporation level on the heat transfer surface area for the entire system (including the heat exchanger in the heating water circuit), this trend is shown in Fig. 5. As it can be seen here, the heat transfer surfaces in the entire range of the assumed evaporation level would have to be larger compared to the reference cycle. For example, the sum of the values of the kA parameter for all heat exchangers in the cycle at the assumed evaporation level of 5% is 2.024 kW/K, which is an increase by about 4.9% for the sum of all heat exchange surfaces in the cycle with the throttle valve compared to the reference cycle.

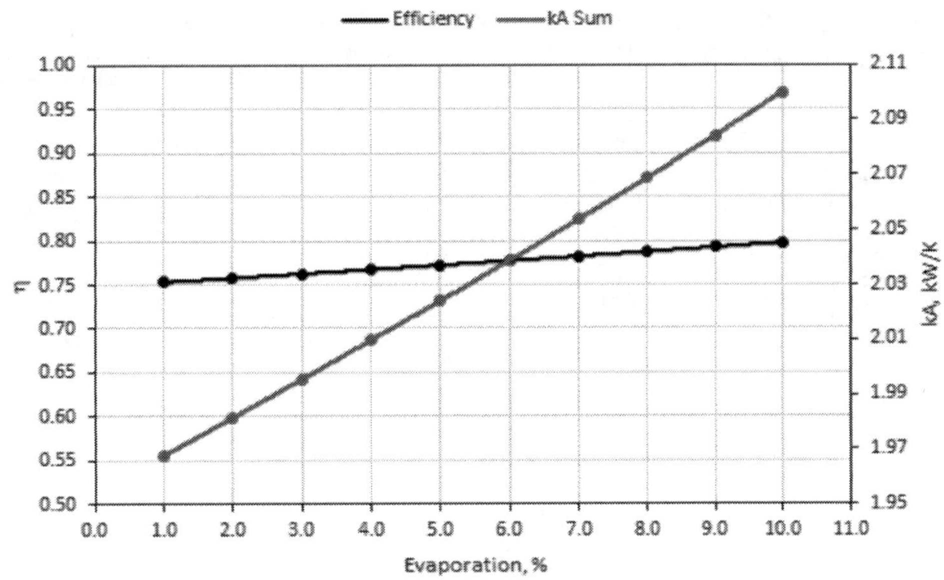

Fig. 5. Change in efficiency and in parameter kA for all the exchangers included in the cycle with a throttle valve depending on the adopted level of evaporation.

In order to analyse the operation of the cycle with the throttle valve at various loads, the previously mentioned case with the assumed evaporation level of 5% was selected. Figure 6 shows the system characteristics for operation at different loads.

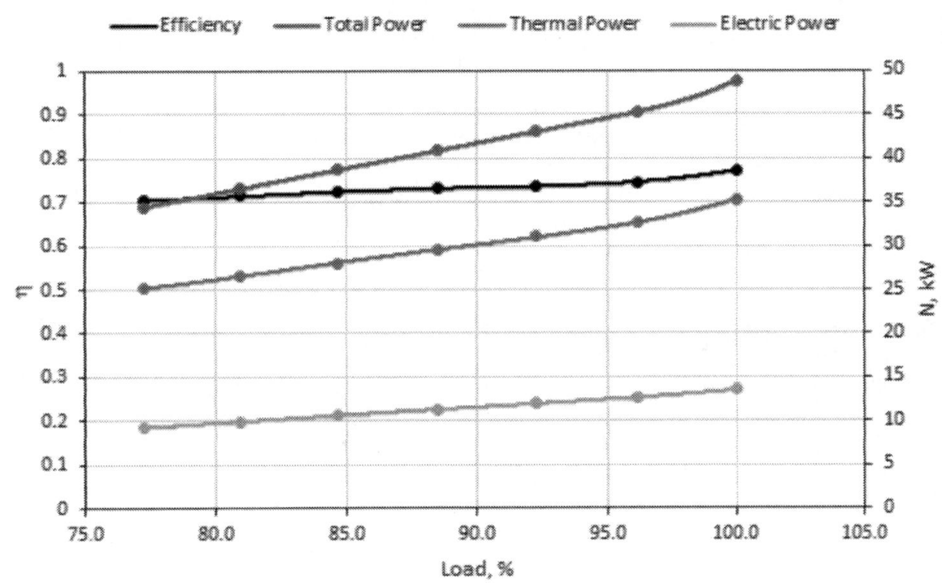

Fig. 6. Operating characteristics of the Velox-type CHP plant with a throttle valve for different loads (5% evaporation on the throttle valve).

It should be noted that this cycle is characterized by the fact that only for the nominal conditions will evaporation occur on the throttle valve because a change in load of the entire system is followed by a change in temperature downstream the economizer to values no longer corresponding to the saturation temperature for the pressure downstream the throttle valve. As shown in Fig. 6, efficiency and energy production for loads lower than 100% drop dramatically, which is to be expected considering that below the rated load, after isenthalpic throttling, there is only water that is fed directly to the evaporator. It can also be seen that the availability of the proposed cycle varies within 77 ÷ 100% of the rated load.

In order to show the temperature changes at individual characteristic points in the cycle, Fig. 7 presents t-Q diagrams for the reference cycle and the cycle with a throttle valve at the evaporation level of 5%.

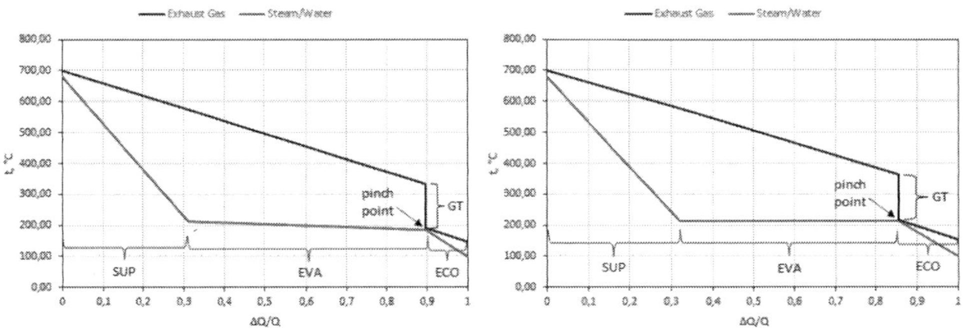

(a) (b)

Fig. 7. T-Q diagrams of the reference cycle of the Velox-type CHP plant (a) and the Velox-type CHP plant with a throttle valve (5% evaporation on the throttle valve) (b).

It can be clearly seen in Fig. 7 that for a cycle with a throttle valve, less heat is needed to evaporate the water in the evaporator (EVA). This is obviously due to some of the water evaporating in the throttle valve earlier. It can also be seen that the heat flux needed to superheat the steam (SUP) is almost identical for both cycles, and the amount of heat transferred in the economizer (ECO) is greater in the cycle with a throttle valve.

In addition, the above diagrams show the effect of the gas expander (GT) on the temperature distribution in the cycle. This effect is characterized by the fact that the exhaust gas temperature drops between the evaporator and the economizer, but no heat is transferred to the working medium in the steam part of the cycle. Hence, a significant drop in temperature at the end of the cycle is visible in both diagrams.

5 Simplified economic analysis

The costs related to the plant were analysed using the relations commonly applied in the economic analysis of gas-steam power units [3, 8–9]. The presented relations take account of only the costs of the elements included in the plant. The costs related to transport, assembly, supervision or management of the whole project are omitted. Similar economic analyses were conducted earlier in [5], where details of the calculation of the total cost of the investment can be found.

The values obtained from the relations presented in [5] and expressed in USD were updated in 2015 using the Chemical Engineering Plant Cost Index (CEPCI) and they are presented in Fig. 7 for the reference cycle of the Velox-type CHP plant and for the cycle with a throttle valve at 5% evaporation. Assuming that the cost of the reference cycle equals 100%, the obtained results indicate that the cost of the Velox-type steam-gas system with a throttle valve at 5% evaporation totals about 102% of the cost of the reference cycle. This means that the unit cost of the proposed cycle of the

Velox-type CHP plant with a throttle valve is almost the same compared to the reference cycle.

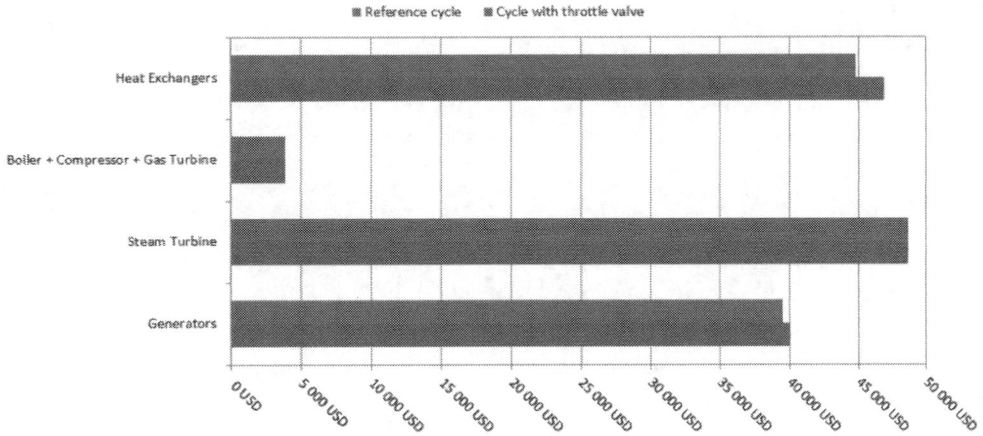

Fig. 7. Cost analysis – Velox-type CHP plant reference cycle and the cycle with a throttle valve at 5% evaporation.

In addition, it can also be noticed that the costs of the steam turbine and the gas turbine, as well as of the compressor and the combustion chamber, are identical for both the reference cycle and the cycle with the throttle valve, which results directly from the adopted input data. The higher cost of all the heat exchangers installed in the cycle is due to the fact that the cycle with the throttle valve requires a bigger surface area of the heat exchange, and the higher cost of the generators is due to the higher power generated for the nominal load of the cycle with the valve.

6 Conclusions

The following conclusions can be drawn from the analyses of the Velox-type gas-steam combined heat & power plant retrofitted with a throttle valve in the feed water circuit and from the comparison to the reference cycle:

- the cycle with the throttle valve achieves higher efficiency than the reference system assuming that the level of evaporation on the throttle valve is higher than 4%,
- the electric power generated by the two compared cycles is almost identical, but the increment in the thermal energy production is more significant for the cycle with the throttle valve,
- the sum of the surface areas of all the heat exchangers in the reference cycle is slightly smaller compared to the cycle with the throttle valve, but this does not have a significant effect on the investment outlays related to the latter cycle,
- by vaporizing some of the feed water on the added throttle valve, it is possible to improve the efficiency of the retrofitted cycle, but only if it operates at the nominal conditions; the favourable effect of the valve in terms of energy conversion will be lost if load is reduced.

Summing up, the operation of a cycle with a throttle valve is favourable only in the nominal load conditions of the proposed cycle of the combined heat and power plant. This is probably the concept biggest downside because in this way the availability of the analysed cycle is decreased. Despite the fact that the investment outlays related to the construction of the proposed cycle are only slightly different compared to the reference cycle, the operating costs of the cycle with the throttle valve may rise considerably at operation at loads lower than nominal. The cycle of a Velox-type gas-steam cogeneration plant with an added throttle valve can find application in distributed power engineering or in industrial plants where higher production of heat than electricity is required under the condition of operation close to the 100% load.

The presented work was supported by the Silesian University of Technology within Statutory Research Funds.

References

1. The Brown Boveri Review Issued by Brown, Boveri & Company, Limited, Baden (Switzerland), Vol. XXXI, January/February 1944, Nos. 1/2.
2. A. Wędzik, *Energetyka* **5**, 323–329 (2006) (in Polish).
3. Ł. Bartela, "Optymalizacja termodynamiczna oraz ekonomiczna pracy elektrociepłowni gazowo-parowej" ["Thermodynamic and economic optimization of the gas-steam CHP plant operation"] (in Polish), Ph.D. thesis, Silesian University of Technology, 2009.
4. S. Dykas., J. Kotowicz, K. Smołka, Rynek Energii **4(95)**, 47–52 (2011) (in Polish).
5. K. Smołka, S. Dykas, S. Rulik, "A comparative analysis of the performance of a steam-gas system of a Velox-type cycle and a gas-steam power station using a waste-heat boiler" in Proceedings of the 4th International Conference on Contemporary Problems of Thermal Engineering (2016), pp. 229–236.
6. K. Smołka, S. Dykas, Journal of Power Technologies **97(5)**, 343–348 (2017).
7. K. Smołka, S. Dykas, E3S Web of Conferences **137**, 01001 (2019).
8. J. Kotowicz, Ł. Bartela, Energy **35(2)**, 911–919 (2010).
9. L. Attala, B. Facchini, G. Ferrara, Energy Conversion and Management **42(18)**, 2163–2172 (2001).

MATEC Web of Conferences **345**, 00029 (2021) https://doi.org/10.1051/matecconf/202134500029
Power System Engineering 2021

An attempt to measure cycloidal rotor fan in a rectangular duct

Tomasz Staśko[1*], *Mirosław* Majkut[1], *Sławomir* Dykas[1], and *Krystian* Smołka[1]

[1]A Department of Power Engineering And Turbomachinery, Silesian University of Technology, Gliwice, Poland

Abstract. A fan with cycloidal rotor (CRF) becomes a popular idea in wide application such as aviation, HVAC (heat, ventilation and air conditioning) or marine propeller systems. This is due to advantages such as direct control of the flow direction, larger flow rates than in a conventional machines without cycloidal control. In the presented article, velocity fields of CRF placed in a rectangular channel was measured, using Laser Doppler Anemomentry (LDA) method and thermoanemometric probe (TA).

1 Introduction

Although the cycloidal rotor was invented more than a century ago [1], it is not widely used nowadays. Mostly it is used as a propulsion of small unmanned aerial vehicles in aviation, in maritime its variant in the form of a cycloidal propeller is used for maneuvering the ship [2-10]. Thanks to its advantages, such as, the possibility of a quick change of the flow direction [9-11], higher flow efficiency than in the case of machines without cycloidal regulation and the possibility of using a Gurney flap (for fans) [12]. It can also be used for HVAC. The article presents experimental measurements of the speed profiles of a radial fan with a cycloidal rotor in a rectangular duct. The aim of the research was also to determine whether the measurement of Laser Doppler Anemomentry (LDA) can be considered as a benchmark. . For this purpose, it was compared with measurements made with a thermoaneometric probe under the same flow conditions.

2 Subject of investigation

The object of measurements was a four-blade radial fan with a cycloidal gear. The blades had a chord of c = 50mm and the asymmetric CLARK Y was chosen as the profile. The rotor diameter was d = 140mm, and the total angle of deflection α_0 was controlled by a mechanical cycloidal gear.[10,11]

* Corresponding author: tomasz.stasko@polsl.pl

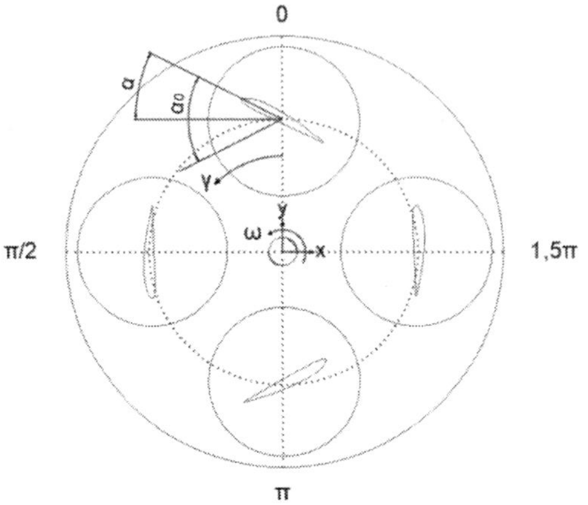

Fig. 1. Schematic drawing of a cycloidal rotor (CRF)

Fig . 1 shows analyzed cycloidal rotor in the starting position. Thanks to the use of a cycloidal gear, the angle α changes with the rotation of the rotor. The blade from the $\gamma=0$ position with a positive deflection angle α, passing to the $\gamma=\pi/2$ position, resets this angle to $\alpha=0°$, so that in the opposite position ($\gamma=\pi$) to the initial position, the same angle α is obtained only with a negative value. Then, in the $\gamma=1,5\pi$ position, the deflection angle α is again reset to zero (the angle should be the same as in $\gamma=\pi/2$) and finally the blade returns to its original position with a positive deflection angle. Each blade rotates around a middle point of camber line of an airfoil.

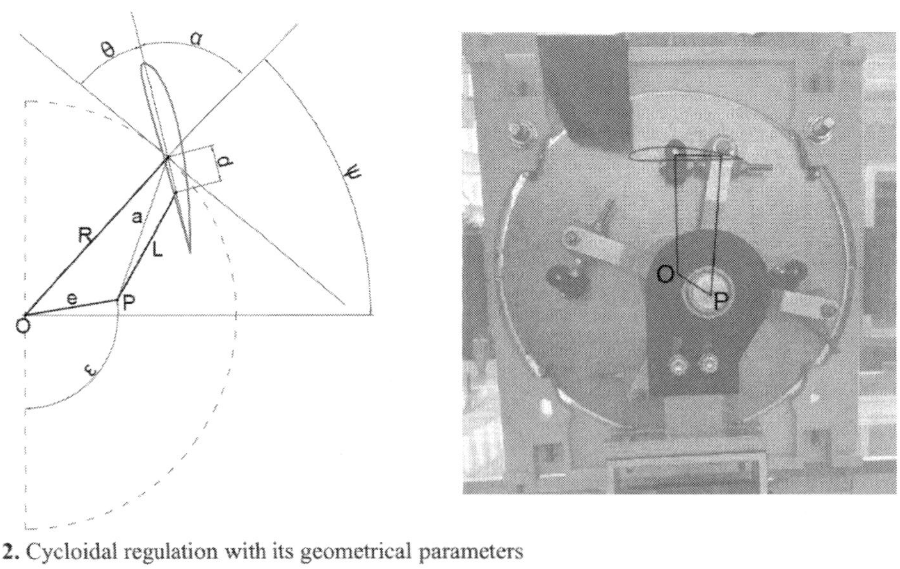

Fig 2. Cycloidal regulation with its geometrical parameters

Fig. 2 shows the cycloidal gear from the analyzed CRF with its characteristic values marked. Using ready-made equations, a program was written that allowed to determine the control parameters for optimal changes in angles and for plotting the corresponding cycloidal functions describing the movement of the blade in a circle. It was based on the following equations:[5]

$$a = \sqrt{e^2 + R^2 - 2eR\cos\left(\psi + \varepsilon + \frac{\pi}{2}\right)} \tag{1}$$

$$L = \sqrt{a^2 + d^2 - 2ad\cos(\alpha_2)} \tag{2}$$

$$\theta = \frac{\pi}{2} - \sin^{-1}\left[\frac{e}{a}\cos(\psi + \varepsilon)\right] - \cos^{-1}\left[\frac{a^2 + d^2 - L^2}{2ad}\right] \tag{3}$$

The first formula (1) determines the length of the segment a between the axis of regulation P (point P is the result of a vertical and horizontal shift of the regulation - this shift is defined as the ε angle). The second one (2) determines the length of the tendon L from the point P to the point of attachment of the blade (in this case it is its center). The third (3) formula determines the relationship between the angle of the blade θ and its location on the circle ψ. The other geometric quantities are: R – rotor diameter, e - distance shifted by the ε angle by which the regulation was shifted and d - distance of the strand hook L from the center of the blade. Using the above-mentioned program, the predicted real cycloid function, which was a sixth degree polynomial function, was plotted and compared with the ideal cycloid function, which was a *cosine* function shifted by the analyzed angle. The rotor blades and its adjustments were set in accordance with the determined parameters, and on this basis, the actual cycloidal function was obtained, characteristic for this setting.

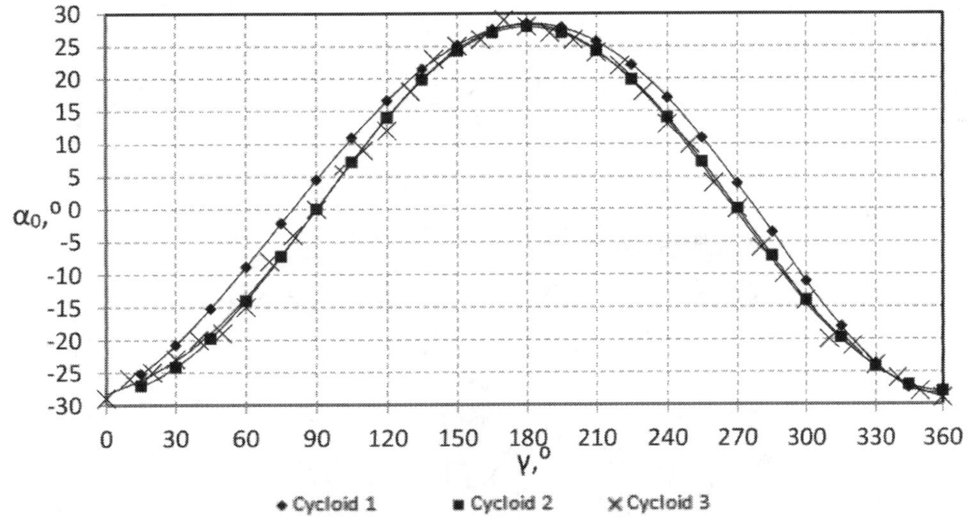

Fig 3. Comparison of cycloidal functions (*Cycloid 1* – ideal cycloidal function, *Cycloid 2* - predicted cycloid function, *Cycloid 3* – actual cycloid function)

Fig. 3 shows a comparison of 3 cycloids. For all cases, the deflection angle α_0 was checked depending on the position of the blade on the circle γ. *Cycloid 1* is an ideal cyclic function describing the change in the angle of the rotor blade from -30 to +30°. *Cycloid 2* is a function determined using a program written for this purpose, and on the basis of this function, which is a polynomial function, the geometric parameters of the cyclorotor for the analyzed range of angles were determined. *Cycloid 3* is a function determined on the basis of measurements of the actual angles of the rotor blades. It can be seen that the predicted *Cycloid 2* practically coincided with the real *Cycloid 3*, which means that the machine with the adjustment was assembled correctly. Both of the above-mentioned functions slightly differ from the ideal.

3 Measurements

The stand consisted of a rectangular duct with a fan with a cycloidal rotor. The channel was made of solid polycarbonate plates, its length was 4m, and the diameter was 250x250mm. To prevent unfavorable phenomena at the fan inlet, the CRF was placed in the middle of the duct. The cycloidal fan was measured in a closed tunnel using two methods: measuring the velocity profiles with TA and LDA. [13-15] The blade pitch angle was set to $\alpha=28°$ and the rotor speed was set to 1000rpm.

a)

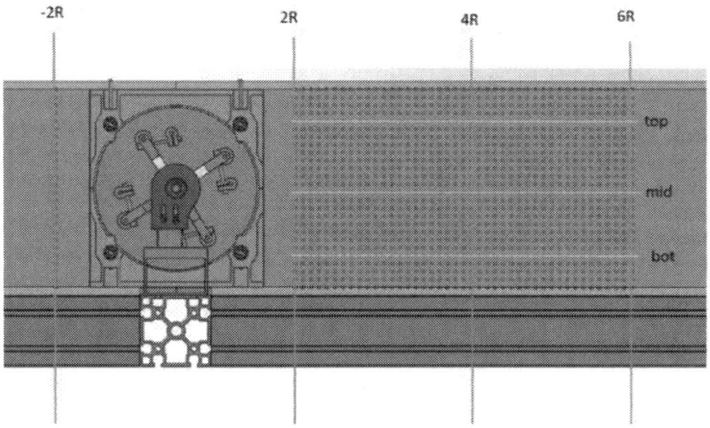

b)

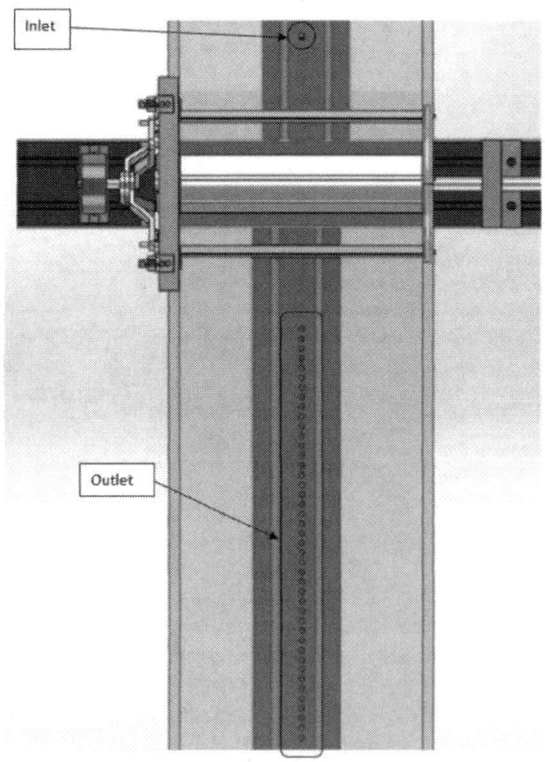

Fig. 4. Measurement areas of CRF a) plane locations b) location of measurement grid in channel

The example above (Fig. 4) shows measurement area with marked measurements points. In *Figure 4a*, horizontal (*top, mid, bot*) and vertical (*-2R, 2R, 4R and 6R*) measurement planes were showed. The horizontal planes were located to coincide with the rotor diameter. The top plane was located at the height + R from the rotor center, the lower plane was located at the same height as -R, and the middle plane along the rotor axis. The vertical planes were located at a distance R (radius) from the rotor axis, so the inlet plane was 2 times R from the axis, and the outlet plane was 2 to 6R. The measuring grid is located in the middle of the canal (Fig. 4b). Using the created mesh, a series of measurements were made using LDA and TA. The aim of the measurements was to determine the velocity profiles for the analyzed planes, to compare the results of both methods and to define the LDA method as the reference method.

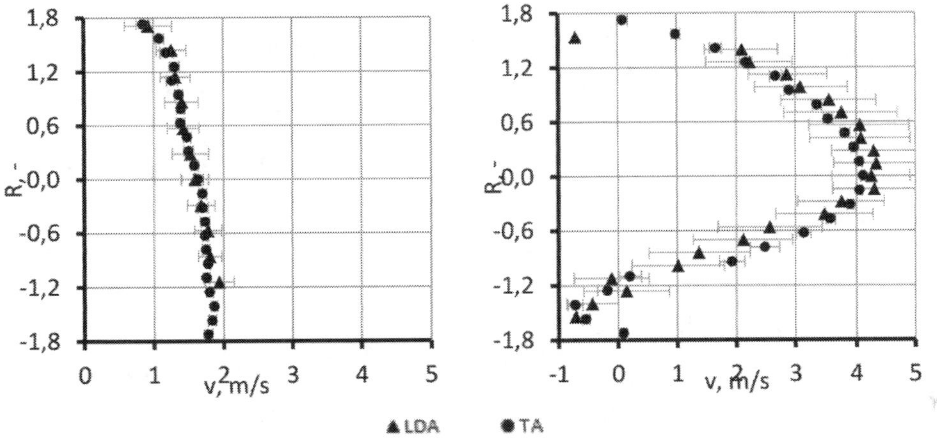

a) Velocity profile in the inlet (-2R) b) Velocity profile in the outlet (2R)

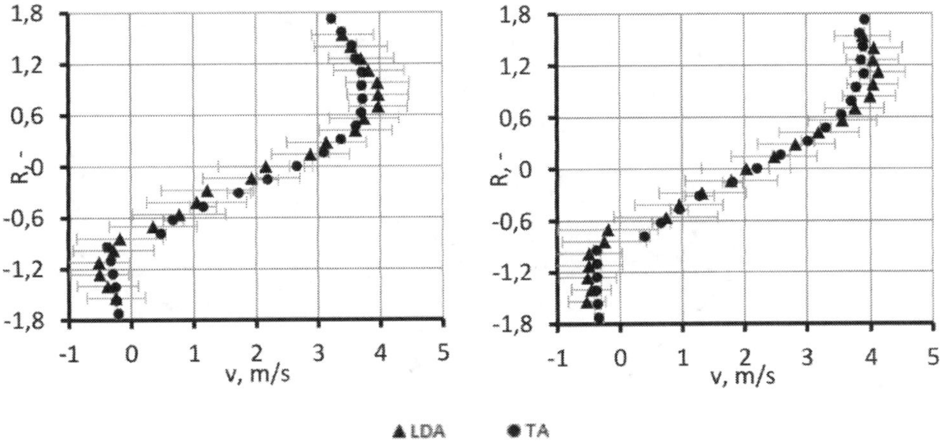

c) Velocity profile in the outlet (4R) d) Velocity profile in the outlet (6R)

Fig. 5. Comparison of the velocity profiles across the channel. Velocity profiles measured a) on inlet at *2R* b) on outlet at *2R* c) on outlet at *4R* d) on outlet at *6R*

Fig. 5 shows the velocity profiles determined with LDA and TA for different channel cross sections with standard deviation bars for each method. It can be seen that for the inlet profile Fig. 5a the velocity profile is not vertical. Increased velocity values at the bottom of the plane suggest that the rotor draws the stream more from the bottom than from entire width of the channel. In the case of the *2R* outlet, (Fig 5b) the velocity profile is already developed and its face is practically in the axis of the rotor. During the measurements, the formation of significant turbulence in the vicinity of the upper and lower part of the flow was noticed, as indicated by negative velocity values. In the case of planes *4R* and *6R*, (Fig 5c,d) the velocity profiles are very similar to each other. As the flow moves away from the rotor, it is directed more and more towards the main tunnel wall, which may cause it to rebound and slow down slightly. The formation of vortices in the lower part of the test tunnel was also observed for both profiles. Measurement errors were also determined for each velocity profile. It can be seen that for all velocity profiles, the measurements overlap satisfactorily, not exceeding the specified values of the error bars. Volumetric flow equal 0,12 m^3/s was determined with the use of LDA for inlet and inlet section.

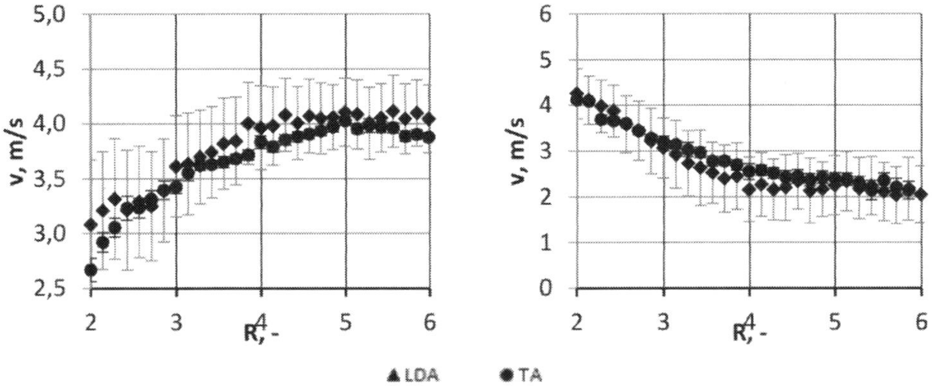

a) Velocity profile in the top plane b) Velocity profile in the center plane

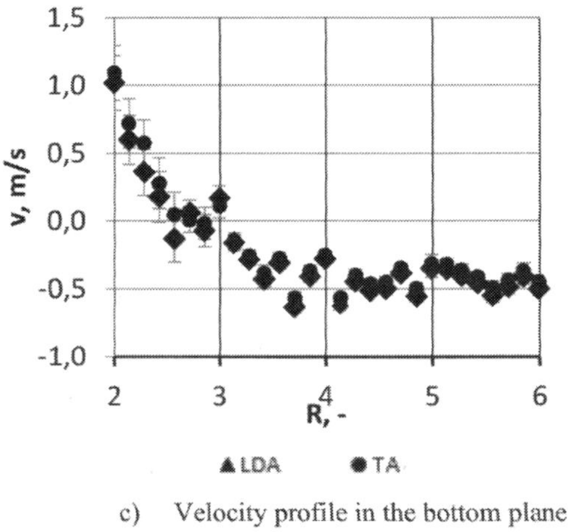

c) Velocity profile in the bottom plane

Fig. 6. Comparison of the velocity profiles across the channel. Velocity profiles measured a) on top plane b) on center plane c) on bottom plane

Fig. 6 shows the velocity profiles determined with LDA and TA for the cross section along the duct at different heights. The plane described as "top" is situated at the height of the upper edge of the rotor at a distance *R* from its center and extends from *2R* to *6R*. The "center" distance is measured along the axis of the rotor (shaft) to the same length as the "top" part. The "bottom" plane is measured at the distance *-R* and has the same length as the previous ones. In the case of the "top" plane, an increase in speed can be seen depending on the length of the exit section. The farther from the rotor, the flow velocity in the layer increases. In the "center" plane, it can be observed a decrease in the velocity value, which is associated with a change in the flow direction - the flow begins to go up in the tunnel. In the “bottom”, the velocity drops significantly to its negative values, caused by formation of vortices. As in the previous case, the compliance of the LDA and TA measurements can be considered satisfactory. As can be seen, the LDA measurement can be considered as a benchmark for further measurements and numerical calculations.

4 Conclusions

The article describes a methods of measurements and the results of a fan with a cycloidal rotor placed in a rectangular duct. The analyzed fan had a 4-blade impeller with a diameter *d* of 140mm, with blades with an asymmetrical CLARK Y profile. Measurements were carried out using two methods: one non-intrusive which was Laser Doppler Anemometry (LDA) and second intrusive - thermoanemometric probe (TA). The aim of the measurements was to determine velocity profiles for various cross-sectional areas and planes and to validate both of the measurement methods. For the given measurement regions, velocity profiles were determined along with the determination of the measurement error for both measurement methods. All the results obtained overlap to a satisfactory degree, which allows to assume that the LDA method will be the reference method and used in future measurements due to the previously obtained validation, not affecting the measurement and the possibility of visualizing the obtained results.

References

1. M. Benedict, Fundamental understanding of cycloidal - rotor concept for micro air vehicle applications. (2010).
2. J. Augusto, L. Monteiro, J. Páscoa, C. Xisto, Aerodynamic optimization of cyclorotors. Aircraft Engineering and Aerospace Technology. (2016).
3. C. Xisto, J. Páscoa, M. Trancossi, Geometrical Parameters Influencing the Aerodynamic Efficiency of a Small-Scale Self-Pitch High Solidity VAWT, April (2016), Journal of Solar Energy Engineering 138(3):pp.10
4. C. Xisto, J. Leger, J. Páscoa, L. Gagnon, P. Masarati, D. Angeli, A. Dumas, Parametric Analysis of a Large – scale Cycloidal Rotor in Hovering Conditions, July (2016), Journal of Aerospace Engineering 30(1).
5. A. Andrisani, D. Angeli, A. Dumas, Optimal pitching schedules for a cycloidal rotor in hovering August (2016), Aircraft Engineering and Aerospace Technology 88(5)
6. E. Shrestha, D. Yeo, M. Benedict, I. Chopra, Development of a Meso-scale Cycloidal-rotor Aircraft for Micro Air Vehicle Application, May (2017), International Journal of Micro Air Vehicles 9(3)
7. M. Rami and J. Pascoa, “Numerical Analysis of a Cycloidal Rotor under Diverse Operating Conditions and Altitudes,” SAE Technical Paper 2019-01-1872, (2019), doi:10.4271/2019-01-1872.
8. M. Benedict, M. Ramasamy, I. Chopra, Improving the Aerodynamic Performance of Micro-Air-Vehicle-Scale Cycloidal Rotor: An Experimental Approach, July (2010), Journal of Aircraft 47(4):1117-1125

9. E. Shrestha, D. Yeo, M. Benedict, I. Chopra, Development of a Meso-scale Cycloidal-rotor Aircraft for Micro Air Vehicle Application, May (2017), International Journal of Micro Air Vehicles 9(3):
10. S. Dykas S, M. Majkut, K. Smołka, M. Strozik, T. Chmielniak, T. Staśko, Numerical and Experimental Investigation of the Fan with Cycloidal Rotor, Mechanics and Mechanical Engineering Vol. 22, No. 2 (2018) 447–454
11. T. Staśko, S. Dykas, M. Majkut, K. Smołka, An attempt to evaluate the cycloidal rotor fan performance, Open Journal of Fluid Dynamics, (2019), 9, 292-30
12. P. Wiśniewski, F. Balduzzi, Z. Buliński, A. Bianchini, A Numerical Analysis on the Effectiveness of Gurney Flaps as Power Augumentation Devices for Airfoils Subject to a Continuous Variation of the Angle of Attack by Use od Full and Surrogate Models, Energies, vol:13,(2020)
13. C. Xisto, J. Páscoa, J. Leger, Cycloidal Rotor Propulsion System with Plasma Enhanced Aerodynamics. In ASME 2014 International Mechanical Engineering Congress & Exposition; November 14-20, (2014); Montreal, Canada.
14. F. Heitmann, M. Juling, J. Steinbock, Performance of the LDA Volumetric Flow Rate Standard Under Severely Disturbed Flow Conditions, Flow Measurement and Instrumentation, Volume 74, (2020), 101756, ISSN 0955-5986,
15. Y. Han, S.Q. Yang, N. Dharmasiri, M. Sivakumar, Effects of sample size and concentration of seeding in LDA measurements on the velocity bias in open channel flow, Flow Measurement and Instrumentation, Volume 38, (2014), Pages 92-97, ISSN 0955-5986,

Particle image velocimetry measurement of turbulent wake past circular cylinder at Reynolds numbers from $2.5 \cdot 10^3$ to $1.5 \cdot 10^4$

Ondřej Sterly[1*]

[1]University of West Bohemia in Pilsen, Univerzitní 22, Pilsen, Czech Republic

Abstract. A canonical case of air flow past a circular cylinder is studied by using Particle Image Velocimetry technique. This contribution focus to the ensemble statistics (first and second moment) of the stream-wise and transverse velocity component as well as to the in-plane vorticity component. Although the range of explored Reynolds numbers is narrow, we observe a significant shortening of recirculation bubble within this range.

1 Introduction

Many of engineering applications have to deal with interaction of flow and a solid structure [1]. This includes the effectivity of transportation devices, resistance of buildings against weather or the power generation. The last means not only the complex flow inside a closed steam turbine [2, 3] or the open wind turbines [4], but also the heat exchanger flows as well as the natural convective cooling [5, 6, 7]. Lot of experiments focusing this question is performed with cryogenic helium [8] due to its high thermal expansivity and thus the possibility of reach high Grashof numbers (in the relevant literature, the Rayleigh number is often used instead [7]). The experimental study of turbulent flows has a very wide practical uses.

The effect of an obstacle to the flow can be mainly studied within the wake [9], i.e. the part of fluid affected by the former interaction with the obstacle and carried downstream with the convective velocity component. The wake is typical by periodic oscillations [10] not only, if the obstacle motion is periodic [11, 12]. The most studied case of wake flow is the wake past a prismatic circular cylinder with axis perpendicular to the incoming flow [9, 36], which is the case studied in this contribution as well. This setup is very simplified and for a practical purposes a more complex prismatic obstacle (or profile) is used [13, 14]. Of course, a non-prismatic obstacles are studied as well [15, 16], but this approach usually needs a more complex Stereo Particle Image Velocimetry technique [17, 18], while here we use this method only in its 2D configuration, which still offers a deep insight into the physics of the flow [19].

Despite its simplicity, the experimental configuration of a prismatic cylinder placed perpendicularly to the incoming flow shows a lot of interesting physical phenomena, e. g. the flow is three dimensional, although the geometry has two-dimensional symmetry [20], the

* Corresponding author: sterlyo@students.zcu.cz

flow is periodic even in a steady incoming flow [10]. Therefore, new ideas are often tested in this simple case [21, 22]. One of the most fascinating properties of this simple canonical experiment is the wealth of regimes depending only on a single parameter: Reynolds number. Figure 1 shows the variation of base pressure coefficient C_{pb} (i.e. non-dimensionalized static pressure p_b at the trailing point of the body) with Reynolds number spanning 6 orders of magnitude as it is published by Williamson [23].

$$C_{pb} = \frac{2(p_b - p_\infty)}{\rho U_{ref}^2}$$

Where ρ is the fluid density, U_{ref} is the reference velocity and p_∞ is the static pressure far enough (at the wind tunnel wall)

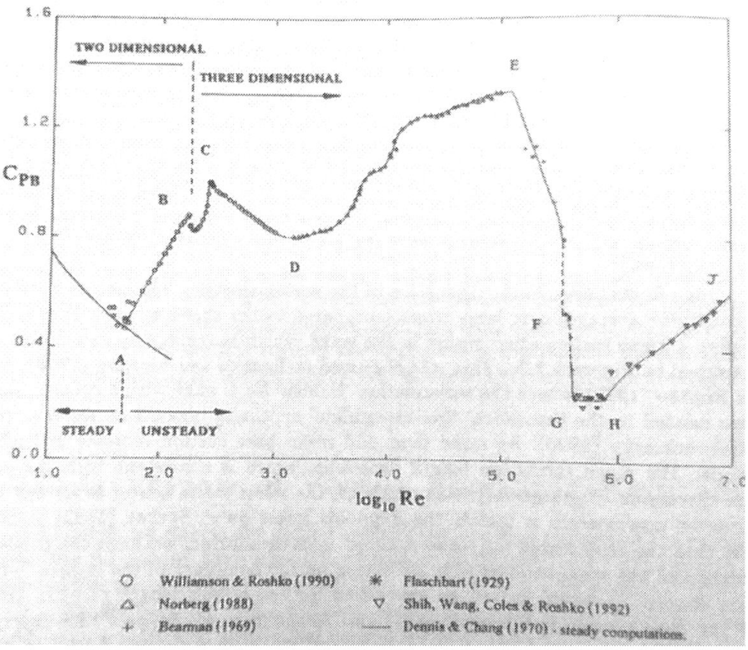

Fig. 1. Regimes of the wake past a circular cylinder shown in the terms of base pressure coefficient dependence on Reynolds number. The figure is copied from the book chapter [23] by C. H. K. Williamson. The mentioned works are references [24 - 28] except for Flaschbart (1929), which I did not find. The regimes distinguished by Williamson are denoted A – J. © Kluwer Academic Publishers.

The regimes observed in the base pressure development in Figure 1 are described in more details by Williamson [23]. The regime up to A is a laminar steady regime, the laminar vortex shedding is observed between A and B. Between B and C, there is a transition to three-dimensional wake. Up to D, the fine-scale three-dimensionalities increase their disorder. The steep decrease from E to G is caused by the asymmetric reattachment of the bubble (sometimes called critical transition). Between G and H, the reattachment become symmetrical and, finally, beyond H the boundary layer transits. More details can be found in the book chapter [23].

The present contribution shows results of a student project of measuring the wake past a circular cylinder by using Particle Image Velocimetry technique at intermediate Reynolds numbers from $2.5 \cdot 10^3$ to $1.5 \cdot 10^4$, which is very narrow range in comparison with the data in literature.

1.1 Experimental setup

The measurement has been performed inside the open wind tunnel at University of West Bohemia in Pilsen [29, 30]. The commercial Particle Image Velocimetry (PIV) system from company Dantec consists of a 4 MPix FlowSense MkII camera and a solid-state double pulse laser New Wave Solo of dominant wavelength 532 nm and pulse energy up to 500 mJ. The used system is the "slow" one with maximal repeating frequency 7.4 Hz, therefore it is not possible to measure the time development of individual vortical structures in the wake. The used lens of focal distance 60 mm together with the used camera gives the spatial resolution of acquired images 34 × 34 mm, i.e. 3.4 × 3.4 the cylinder diameter.

This PIV system has been used for many measurements in the past, for example the flow inside an axial single stage-turbine [3, 31], which was not so easy as it sounds [32]. From the laboratory experiments, this system participated in the measurement of secondary flow in a corner of channel [33], measurement of turbulent mixing past a grid [34] and past a strange-shaped counter-swirler [16] or in a turbulent jet [35].

The cylinder is made of aluminum, its diameter is 10 mm and height 125 mm filling the entire height of the wind tunnel test section. The Reynolds number has been controlled via velocity ranging from 3.8 to 23.0 m/s, which differs only 6 times, but still this range displays an interesting phenomena of shortening of the recirculation bubble. The working fluid is atmospheric air of temperature 19.1 ± 1.3 °C. The seeding particles are produced by a commercial fog generator Safex.

In the following, the stream-wise velocity is denoted u, the transverse velocity component is denoted v. The third velocity component along the cylinder axis is not measured in current configuration. The ensemble consists of around 700 snapshots. The coordinate system displayed in figures is in millimeters and it originates at the beginning of the field of view (FoV), the position of the cylinder is [-5, 16] in the coordinate system of the FoV, the cylinder radius is 5 mm, its height is 125 mm filling the entire height of the wind tunnel test section.

2 Results

The focus is given to the ensemble averaged quantities, the stream-wise velocity component u and the span-wise velocity component v. The ensemble average u of instantaneous velocity signal u_i is calculated as

$$u(\vec{x}) = \langle u_i(\vec{x}) \rangle = \frac{1}{N} \sum u_i(\vec{x})$$

And for the second velocity component v analogically. Here, $\vec{x}$ is any spatial position within the field of view, N is the number of snapshots in the ensemble, i.e. around 700 in this measurement.

Figure 2a displays the evolution of the ensemble averaged stream-wise velocity u along the wake axis. The area of negative velocity called *recirculation bubble* is formed past the body and with increasing Reynolds number it shortens significantly with length around 2.3 cylinder diameters at the lowest velocity down to 0.9 diameters at the highest velocity. Accordingly, the position of minimal velocity moves as well. The deep of the minimal velocity first slightly decreases down to -0.30 times reference velocity at Re = $5.1 \cdot 10^4$ followed by slight increase to values around -0.25 times reference velocity at the higher Reynolds numbers.

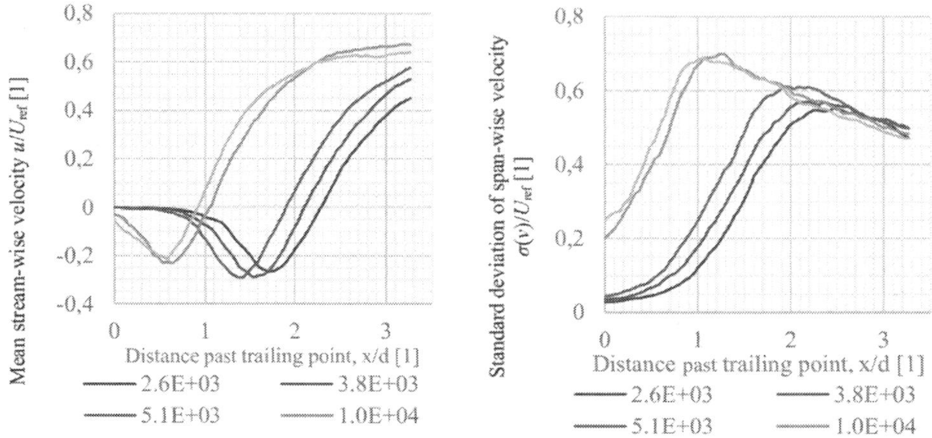

Fig. 2. (a) Ensemble average of stream-wise velocity component u as a function of distance past trailing point of the cylinder, the profile is taken at the wake axis. (b) Standard deviation of transverse velocity component v as a function of distance past trailing point. The number in the legend represents the Reynolds number with the notation that "E+n" plays for "$\cdot 10^n$", we apologize for this inconvenient notation.

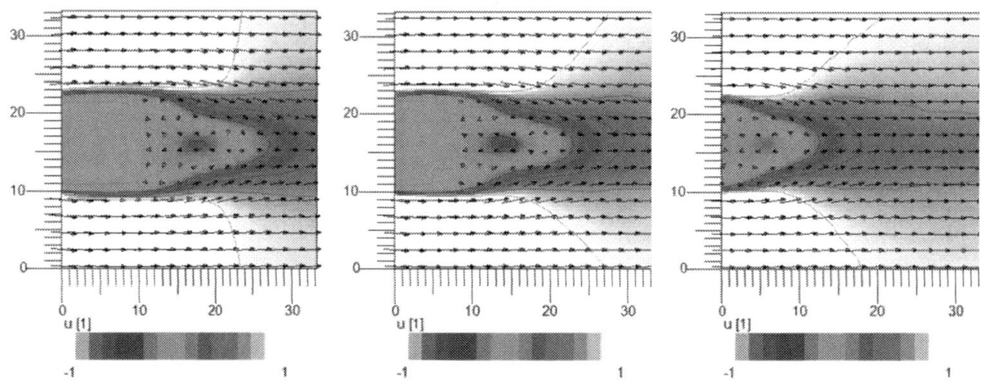

Fig. 3. Ensemble average of measured stream-wise velocity component u. The Reynolds numbers are from left: $2.55 \cdot 10^3$, $5.1 \cdot 10^3$ and $1.53 \cdot 10^4$. The coordinates are in millimeters, horizontal one represents stream-wise direction, the vertical one the transverse direction. The cylinder axis is perpendicular to the field of view and it is placed left hand side of it in a such way, that the trailing point has coordinates [0, 16] mm.

Figure 3 shows the spatial distribution of the ensemble averaged stream-wise velocity and Fig. 4 displays its standard deviation, i.e. second raw moment, which is calculated using:

$$\sigma[u](\vec{x}) = \sqrt{\langle (u_i(\vec{x}) - \langle u \rangle(\vec{x}))^2 \rangle}$$

Symbol $\langle \cdots \rangle$ means the average, u_i denotes the instantaneous velocity at certain position $\vec{x}$. The spatial distribution of $\sigma[u]$ differs significantly from that of $\sigma[v]$ displaying a pair of peaks past each side of the cylinder. Those peaks are displayed in Fig. 5. First, the peaks are sharp and probably much higher than the value measured by finite resolution of PIV system. Later downstream a sudden widening follows splashing the fluctuating fluid mainly towards the wake axis, later dissipating outside of the wake as well. This is in contrast to the distribution of fluctuations in transverse component v, Fig. 7, which has a maximum in between the peaks of stream-wise velocity fluctuations.

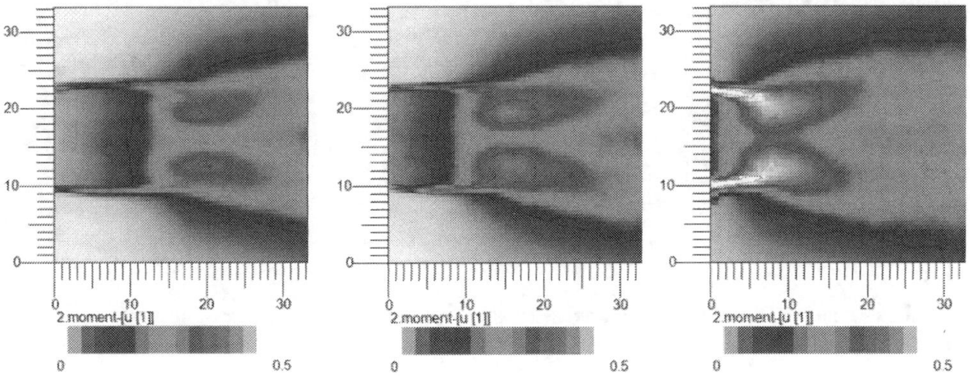

Fig. 4. Standard deviation of measured stream-wise velocity component u. The Reynolds numbers are from left: $2.55 \cdot 10^3$, $5.1 \cdot 10^3$ and $1.53 \cdot 10^4$.

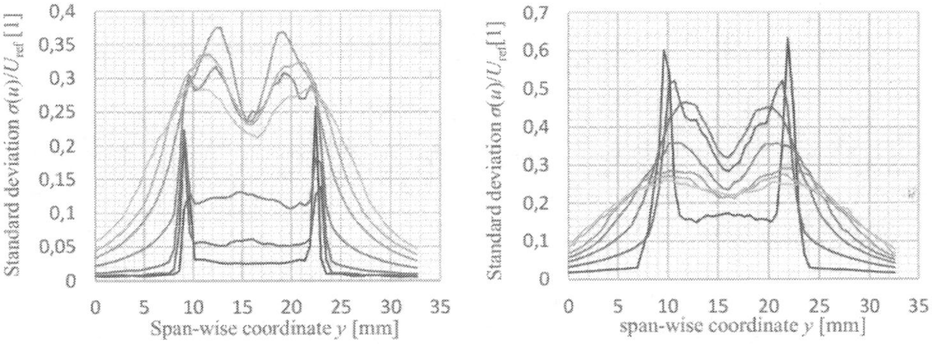

Fig. 5. Graphs showing cuts through the ensemble of standard deviations of the velocity component u. Darker color means lesser distance from measured cylinder. Values of second moment are on the y axis, on the x axis are y coordinates. The Reynolds numbers are from left: $2.55 \cdot 10^3$ and $1.53 \cdot 10^4$.

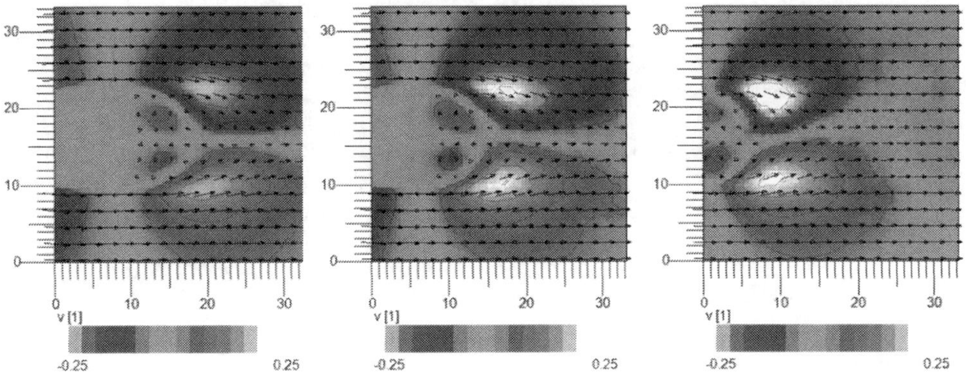

Fig. 6. Ensemble average of measured stream-wise velocity component v. The Reynolds numbers are from left: $2.55 \cdot 10^3$, $5.1 \cdot 10^3$ and $1.53 \cdot 10^4$.

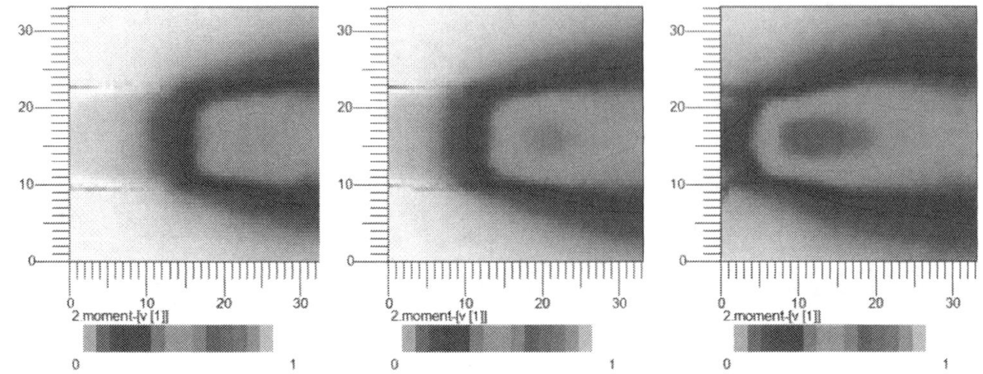

Fig. 7. Standard deviation of measured span-wise velocity component v. The Reynolds numbers are from left: $2.55 \cdot 10^3$, $5.1 \cdot 10^3$ and $1.53 \cdot 10^4$.

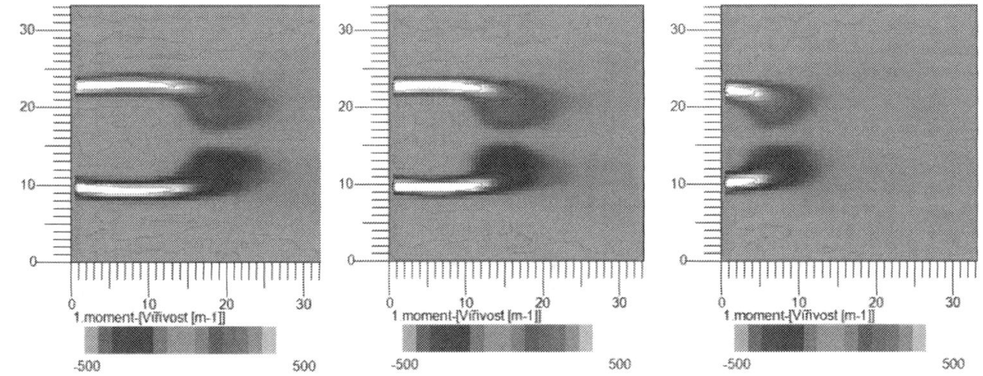

Fig. 8. Mean vorticity component parallel with the cylinder axis. The velocity field is normalized by the reference velocity U_{ref} in order to compare distributions at different velocities and thus the effective unit of vorticity changes from [s^{-1}] to [m^{-1}]. The Reynolds numbers are from left: $2.55 \cdot 10^3$, $5.1 \cdot 10^3$ and $1.53 \cdot 10^4$.

Vorticity is calculated using rotation of velocity as:

$$\vec{\omega} = \nabla \times \vec{u}$$

Note that only the component perpendicular to the field of view can be calculated as only two velocity components are measured. Thus this quantity behaves as a scalar in the data. The shear layers are characterized by strong vorticity signal, see Fig. 8, while around the stagnation point it disappears, which does not mean, that the fluid was quiet there, the opposite is true, as can be visible in the map of standard deviation (second raw moment) of vorticity in Fig. 9.

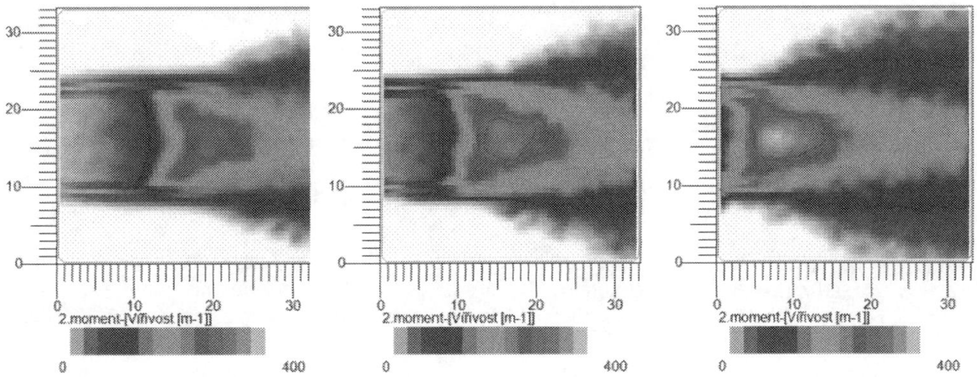

Fig. 9. Standard deviation of measured vorticity. The Reynolds numbers are from left: $2.55 \cdot 10^3$, $5.1 \cdot 10^3$ and $1.53 \cdot 10^4$.

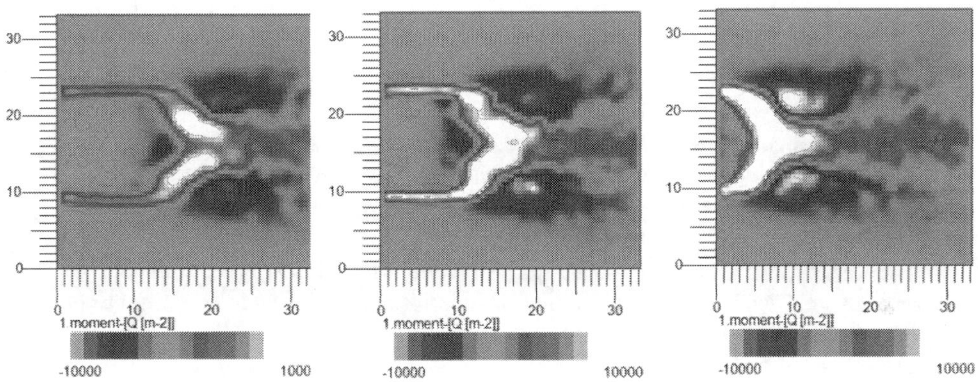

Fig. 10. Mean Q. The Reynolds numbers are from left: $2.55 \cdot 10^3$, $5.1 \cdot 10^3$ and $1.53 \cdot 10^4$. Again, the velocity field is divided by the reference velocity, thus the unit of Q parameter changes from [s^{-2}] to [m^{-2}]

Due to the nature of vorticity – it is made of two components rotation and shear – there apparently comes need to differentiate these two components of each other. Here we use the Q parameter

$$Q = \frac{1}{2}(\omega^2 - s^2)$$

this physical quantity tells us which component of vorticity is predominant. Simplified, if Q is negative, the vorticity is made mainly of shear s, $s = \frac{1}{2}\left(\frac{\partial u_i}{\partial x_j} + \frac{\partial u_j}{\partial x_i}\right)$. If Q is positive, then component of rotation is predominant. The ensemble average of Q is displayed in Fig. 10 showing a single “Y-like” structure instead of pair of opposite branches as the vorticity do (Fig. 8). This is caused by the presence of vortices around the stagnation points, but these vortices are positive and negative averaging to zero signal, while the Q ensures summing, if the structure is rotational.

3 Conclusion

We used the experimental method of Particle Image Velocimetry (PIV) for measuring the instantaneous velocity past a circular cylinder [36] of diameter 10 mm placed inside an open wind tunnel test section. We observed a significant shortening of the recirculation bubble appearing in the ensemble-averaged maps. The fluctuations differ between components – the stream-wise component fluctuates most past the boundary layers of the cylinder with a local minimum at the wake axis, while the fluctuation of span-wise velocity has maximum at the axis.

In future we will continue analyzing this data, in focus to the individual vortices and their statistics, we plan to compare this results to some numerical simulation and later we will plan better next measurement learning from the current troubles and inaccuracies.

I gratefully thank to RNDr. Daniel Duda for supervising this small student project, although the measurement was limited due to the general limitations of student works at the university due to the COVID-19 pandemic. I gratefully thank to Ing. Martin Pelikán for the opportunity to attend the conference and for paying all the conference fees. I thank to Ing. Martin Novák for the organizing the student project selecting very interesting topics.

References

1. S.B. Pope, Turbulent Flows, Cambridge University Press, (2000)
2. G. Ilieva "A deep insight to secondary flows", Defect and Diffusion Forum **379**: 83-107 (2017)
3. D. Duda, T. Jelínek, P. Milčák, M. Němec, V. Uruba, V. Yanovych, P. Žitek, "Experimental Investigation of the Unsteady Stator/Rotor Wake Characteristics Downstream of an Axial Air Turbine", Int. J. Turbomach. Propuls. Power **6**(3): 22 (2021), doi: 10.3390/ijtpp6030022
4. J. Schottler, J. Bartl, F. Mühle, L. Sætran, J. Peinke, M. Hölling, "Wind tunnel experiments on wind turbine wakes in yaw: redefining the wake width", Wind Energ. Sci. **3**: 257 – 273 (2018), doi: 10.5194/wes-3-257-2018
5. J. J. Niemela, L. Skrbek, K. R. Sreenivasan, "The wind in confined thermal convection", J. Fluid Mech. 449: 169 – 173 (2001), doi: 10.1017/S0022112001006310
6. D. Duda, M. La Mantia, M. Rotter, L. Skrbek, "On the visualization of thermal counterflow of He II past a circular cylinder", J. Low Temp. Phys. **175**: 331 – 338 (2014), doi: 10.1007/s10909-013-0961-z
7. L. Skrbek, P. Urban "Has the ultimate state of turbulent thermal convection been observed?" J. Fluid Mech. 785: 270 – 282 (2015), doi: 10.1017/jfm.2015.638
8. V. D. Arp, R. D. McCarty, D. G. Friend "Thermophysical properties of helium-4 from 0.8 to 1500 K with pressures to 2000 MPa", NIST technical note **1334** (1998)
9. C. H. K. Williamson, "Vortex Dynamics in the Cylinder Wake", Annual Review of Fluid Mechanics **28**:1, 477-539 (1996), doi: 10.1146/annurev.fl.28.010196.002401
10. V. Strouhal, "Über eine besondere Art der Tonerregung", Ann. Phys, **241**: 216 – 251 (1878), doi: 10.1002/andp.18782411005
11. D. Duda, P. Švančara, M. La Mantia, M. Rotter, L. Skrbek "Visualization of viscous and quantum flows of liquid He 4 due to an oscillating cylinder of rectangular cross section", Phys. Rev. B **92** 064519 (2015), doi: 10.1103/PhysRevB.92.064519
12. D. Duda, M. La Mantia, L. Skrbek "Streaming flow due to a quartz tuning fork oscillating in normal and superfluid He 4" Phys. Rev. B **92**: 024519 (2017), doi: 10.1103/PhysRevB.96.024519
13. V. Yanovych, D. Duda, V. Uruba "Structure turbulent flow behind a square cylinder with an angle of incidence" European Journal of Mechanics B/Fluids **85**: 110 – 123 (2021), doi: 10.1016/j.euromechflu.2020.09.003

14. D. Duda "How Manufacturing Inaccuracies Affect Vortices in an Airfoil Wake" Proceedings Topical Problems of Fluid Mechanics **2021**: 48 – 55 (2021), doi: 10.14311/TPFM.2021.007
15. W. Terra, A. Sciacchitano, F. Scarano, Aerodynamic drag of a transiting sphere by large-scale tomographic-PIV, Exp Fluids **58**:83 (2017), doi: 10.1007/s00348-017-2331-0
16. D. Duda "The PIV study of air flow past the counter-swirler 53983" MATEC Web of Conferences **168**: 05004 (2018), doi: 10.1051/matecconf/201816805004
17. C. Tropea, A. L. Yarin, and J. F. Foss, "Springer handbook of experimental fluid mechanics", Springer (2007)
18. D. Duda, V. Uruba "PIV of air flow over a step and discussion of fluctuation decompositions" AIP Conference Proceedings **2000**: 020005 (2018), doi: 10.1063/1.5049912
19. D. Duda and V. Uruba, "Spatial spectrum form Particle Image Velocimetry data", ASME J of Nuclear Rad Sci. **5** (3): 030912 (2019)
20. C. H. K. Williamson, Three-dimensional wake transition. Journal of Fluid Mechanics, **328**: 345-407. (2006), doi:10.1017/S0022112096008750
21. P. Pavlíček "Local Reynolds number" AIP Conference Proceedings **2118**, 030035 (2019); doi: 10.1063/1.5114763
22. I. K. Marchevsky, K. S. Kuzmina, E. P. Ryatina "Flow Simulation around a Circular Cylinder for Low Reynolds Numbers and Determining the Flow Separation Point in VM2D Code" Proceedings Topical Problems of Fluid Mechanics **2020**: 144 – 151 (2020), doi: 10.14311/TPFM.2020.019
23. C. H. K. Williamson "Vortex dynamics in the wake of a cylinder", chapter in S. I. Green "Fluid vortices", 155 – 234, Kluwer Academic Publishers (1995), isbn: 0-7923-3376-4
24. C. H. K. Williamson, A. Roshko "Measurement of base pressure in the wake of a cylinder at low Reynolds numbers" Zeitschrift für Flugwissenschaften und Weltraumforschung **14**: 38 – 46 (1990)
25. C. Norberg "Effect of Reynolds numbers and a low-intensity freestream turbulence on the flow around a circular cylinder" Chalmers University of Technology, publication Nr. 87/2, S-412-96 (1987)
26. P. W. Bearman "On the vortex shedding from a circular cylinder in the critical Reynolds number regime" J. Fluid Mech. **37**: 577 – 585 (1969), doi: 10.1017/S0022112069000735
27. W. C. L. Shih, C. Wang, D. Coles, A. Roshko "Experiments on flow past rough circular cylinders at large Reynolds numbers" 2nd Int. Coll. On Bluff Body Aerod. And Applic., Melbourne (1992)
28. S. C. R. Dennis, G. Z. Chang "Numerical solutions for steady flow past a circular cylinder at Reynolds numbers up to 100" J. Fluid Mech. **42**: 471 – 489 (1970), doi: 10.1017/S0022112070001428
29. V. Yanovych, D. Duda, V. Horáček, V. Uruba "Research of a wind tunnel parameters by means of cross-section analysis of air flow profiles" AIP Conference Proceedings **2189**: 020024 (2019), doi: 10.1063/1.5138636
30. V. Yanovych, D. Duda "Structural deformation of a running wind tunnel measured by optical scanning" Strojnicky Casopis 70(2): 181 – 196 (2020), doi: 10.2478/scjme-2020-0030

31. D. Duda, T. Jelínek, M. Němec, V. Uruba, V. Yanovych, P. Žitek "Observation of flow structure past a full-stage axial air turbine at the nominal and off-design states" AIP Conference Proceedings **2323**: 030004 (2020), doi: 10.1063/5.0041491
32. D. Duda, T. Jelínek, M. Němec, V. Uruba, V. Yanovych, P. Žitek "Particle image velocimetry measurement inside axial air test turbine - Effect of window" AIP Conference Proceedings **2323**: 030005 (2020), doi: 10.1063/5.0041492
33. D. Duda, J. Bém, V. Yanovych, P. Pavlíček, V. Uruba "Secondary flow of second kind in a short channel observed by PIV" European Journal of Mechanics B/ Fluids **79**: 444 – 453 (2020), doi: 10.1016/j.euromechflu.2019.10.005
34. D. Duda, V. Yanovych, V. Uruba "An experimental study of turbulent mixing in channel flow past a grid" Processes 8: 1 – 17 (2020), doi: 10.3390/pr8111355
35. D. Duda "Preliminary PIV measurement of an air jet" AIP Conference Proceedings **2047**: 020001 (2018), doi: 10.1063/1.5081634
36. V. Uruba, P. Procházka, "The Reynolds number effect on dynamics of the wake behind a circular cylinder", AIP Conference Proceedings **2189**, 020023 (2019), doi: 10.1063/1.5138635

Parameterized, numerical design of a two-wheel Curtis steam turbine for small scale WHR

Philipp Streit[1*], and *Andreas P.* Weiß[1]

[1]OTH Amberg-Weiden, Center of Excellence for Cogeneration Technologies, Kaiser-Wilhelm-Ring 23, 92224 Amberg, Germany, +49 9621 482 3531

Abstract. In contrast to the current trend of converting waste heat into electricity in the small power range below 100 kW_{el} by means of an ORC plant, the authors are pursuing the concept of a micro steam power plant equipped with a micro turbine. Water avoids many of the problems often associated with organic working fluids, such as flammability, toxicity, greenhouse gas effect and high fluid costs. However, water vapor makes turbine design more challenging. The physical reasons for this are repeated, and thereby it becomes clear why a velocity compounded two wheel Curtis turbine has been chosen. The used in-house 1D turbine design tool is briefly introduced. More focus is put on the shortcomings of the implemented 1D loss model and their negative impact on the current turbine design. Consequently, the authors continued actual turbine design by a parameterized approach in 3D CAD/CFD. This approach is explained, and finally, the CFD flow field and the performance maps of the designed turbine are discussed. The turbine is currently under construction and will be installed in 2022 in a waste heat recovery (WHR) plant in Nuremberg/Germany.

1 Introduction

Waste heat recovery (WHR) is an important building block towards an increase in energy efficiency in industry and for a successful energy transition. Pehnt et al. 2010 [1] reported that about 80 TWh of waste heat >140 °C are rejected yearly only in the German industry. Assuming a conversion efficiency of about 10%, which is realistic for small WHR units, 8 TWh per year could be extracted. This is the annual electricity production of a 1000 MW steam power plant. In cases where the waste heat temperature is below 300 – 350 °C, an ORC system is currently the preferred solution for waste heat recovery [2]. In large Combined Cycle Power Plants (CCPP), a steam turbine reuses the waste heat of the upstream gas turbine (500 - 600 °C), which leads to an overall electric efficiency of more than 60% [3]. The main advantage of an Organic Rankine cycle (ORC) is the higher vapor pressure of an organic fluid compared to water for a given heat source temperature [2]. Thanks to this, a low temperature heat source can be exploited more effectively. However, the maximum cycle temperatures of ORC are limited to 300-350 °C due to the danger of degradation of the working fluid. This issue limits the achievable maximal cycle efficiency in cases in which

* Corresponding author: ph.streit@oth-aw.de

higher waste heat temperatures are available, like in the glass, steel or cement industry or downstream of a biogas engine. Water as a working fluid does not suffer from this limitation. Furthermore, water cannot burn, is not toxic, does not spoil the atmosphere and is significantly cheaper than any organic working fluid. Thus, the research project in which the reported work has been carried out aims for a small steam Rankine cycle with about 50 kW_{el} power using a micro turbine as an expander [4]. The major goal of the current project is to reduce specific costs (€/kWel) of the existing WHR system [5].

The challenge of designing a steam turbine for small shaft power (< 100 kW_{el}) is at least twofold. As pointed out by Macchi & Astolfi [2], steam provides a very high isentropic enthalpy drop for a given maximum and minimum cycle temperature compared to any organic fluid. The specific work Δh_{is} which can be processed by one turbine stage is equal to a constant C times the square of the rotational speed u on mean diameter D, i.e proportional to u^2 (see Table 1). This is true for any turbine. The value of the constant C depends on the turbine type. Live steam parameters are 400 °C/17.5 bar in the current project. Exhaust or condensing pressure is 0.2 bar (60 °C), steam mass flow rate is about 80 g/s. So, the isentropic enthalpy drop, which must be processed by the turbine, is about 880 kJ/kg [6]. If the constant $C = 1$, what is true for a 50 % reaction turbine stage, $u = 938$ m/s (Table 1). Such a high value of u is not feasible. The wheel would burst due to centrifugal forces – no matter which material is used.

The second challenge in designing micro steam turbines is caused by the high enthalpy drop as well. The power of a turbine is equal to the mass flow rate times the enthalpy drop. Thus, the higher the enthalpy drop, the smaller the mass flow rate for a given or required power. The mass flow rate determines the wheel diameter D and the blade height h_{blade}, which are the main dimensions of the turbine. The diameter D should be as big as possible to achieve a high u combined with an acceptable angular velocity ω or rotational speed n, respectively. However, this leads to very small blade heights, which limits this approach. To sum up, the smaller the aimed power output and thus the corresponding mass flow rate, the smaller the turbine wheel diameter D and thereby the higher the angular velocity or rotational speed must be.

In small turbines for power output < 100 kW_{el}, the rotational speed is in the magnitude of several 10^4 rpm [5,7–9]. This fact makes bearing design and generator design challenging. Sophisticated aerodynamic or magnetic bearings must be applied [5,10]) and a high speed generator must be employed, which is expensive and contributes typically two third to the specific costs (€/kW) of a small turbo generator. Of course, these physical obstacles can be overcome by a multi-stage design like in big steam power plants (10 – 1000 MW). Unfortunately, this solution is too costly and too laborious for those small machines. Therefore, the authors looked for other approaches. In the history of turbomachines [11,12], various turbine architectures have been introduced and were employed till the concentration of power generation started at the beginning of the 20th century. The multi-stage axial turbine has finally prevailed in huge thermal power plants of the several 100 MW class.

Already when the steam turbine started to displace the piston steam engine at the end of the 19th century, three different main turbine concepts existed. There was the simple impulse turbine of de Laval [13] (Table 1). It can be easily derived that its stage enthalpy drop is $2 x u^2$ in such an impulse stage without pressure drop across the buckets [14]. Therefore, for the given project boundary conditions (400°C, 17.5 bar, 0.2 bar, 0.080 kg/s steam), the required circumferential speed is 663 m/s compared to 938 m/s of the 50 % reaction turbine, which was introduced originally by Parsons [15] (Table 1).

Thanks to the constant pressure over the rotor blading, impulse and Curtis stages can be designed to work with partial admission. This means that a part of the total arc of the annulus is blocked off. The small mass flow rate is concentrated on a section of the rotor wheel. Thereby, too small blade heights can be avoided [8,14].

The third well-known turbine concept is the velocity compounded Curtis stage [16]. The nozzles and the first wheel might be considered to be identical to the described impulse stage, however, the main difference is the required circumferential speed u, which is only half of that of the impulse stage. Thus, the required circumferential speed is only 331 m/s which is feasible with a steel wheel (Table 1). It has to be mentioned, that the efficiency potentials of these three turbine architectures differ significantly. Due to higher relative Mach numbers, higher necessary flow deflections compared to a reaction turbine, the impulse turbine achieves only about 80 % efficiency, the Curtis about 70 % compared to 90 % of the reaction turbine. All these turbine concepts can and have been implemented by an axial or by a radial architecture. The radial architecture might be considered to be advantageous because it supports stage enthalpy processing by the centrifugal pressure field for a radial inflow turbine [14]. In contrast, volume flow rate increase due to expansion is facilitated in a radial outflow turbine [17].

In the current project, the authors have investigated analytically and by an in-house 1D turbine design tool (1DTDT) [18] various un-common turbine architectures like multi-stage radial-outflow turbines [17], quasi-impulse cantilever turbines [19] and radial inflow Curtis-turbines for the given expansion task (400 °C/17.5 bar, 0.2 bar, about 80 g/s steam). Few uncommon turbine architectures have been promising from an aerodynamic point of view. However, mechanical pre-investigation showed that only the presently implemented architecture, a two-wheel axial Curtis-turbine [5], is actually feasible. It is equipped with a titanium rotor. Thus, the authors aim for an improved axial Curtis turbine with increased efficiency and reduced costs compared to its predecessor. In the following, the parameterized, mainly numerical design of this turbine is introduced and discussed.

Table 1 Comparison of different turbine concepts and their velocity triangles for the same stage enthalpy drop Δh_{is} (value of u for the given project boundary conditions)

	Reaction turbine	Impulse Turbine	Curtis turbine (2 wheels)
Turbine type	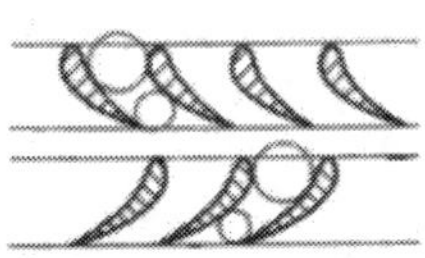	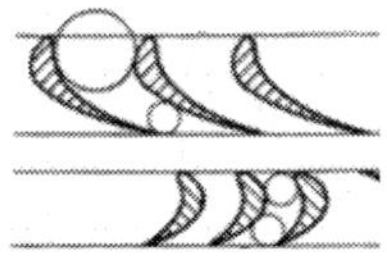	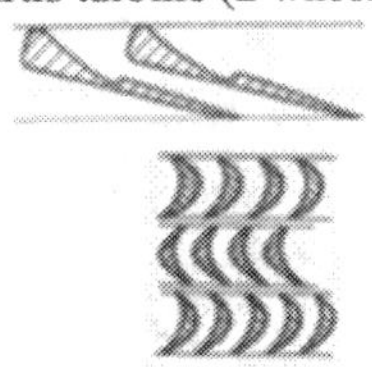
velocity triangles	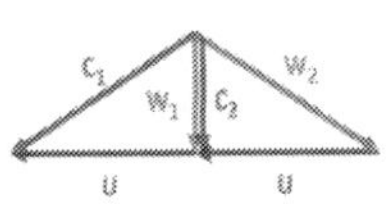	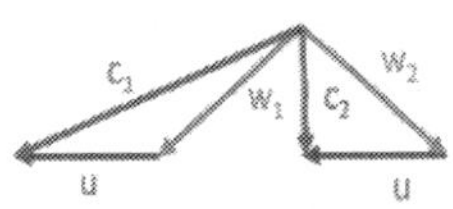	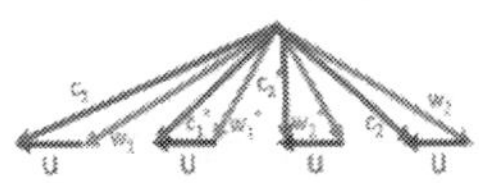
□h_{is}	$1 x u^2$	$2 x u^2$	$8 x u^2$
u_{opt}	$(\Delta h_{is})^{0.5}$	$\left(\frac{\Delta h_{is}}{2}\right)^{0.5}$	$\frac{1}{2}\left(\frac{\Delta h_{is}}{2}\right)^{0.5}$
	938 m/s	**663 m/s**	**331 m/s**
partial admission	not feasible	feasible	feasible
efficiency potential	90%	80%	70%

2 Methodology

2.1 1D Turbine design tool (1DTDT)

In order to comprehensively and reliably assess and compare different turbine concepts, a computational tool has been developed for the following considered turbine concepts:

- Axial and radial (cantilever quasi) impulse turbines
- Axial or radial two-wheel velocity compounded Curtis turbines
- Radial inflow, axial outflow reaction turbines (90° IFR)

It has been developed so that the main geometry data and the efficiency of a turbine can be assessed automatically and quickly with the following inputs to the model:

- Working fluid and mass flow rate
- Pressure at the turbine inlet and outlet and temperature at the inlet
- Wheel mean diameter and rotational speed

The tool is based on a 1D meanline model with an in-house developed appropriate simple loss model, which is described in more detail in [20,21]. The 1DTDT calculates efficiency and power output. The loss model for nozzles and buckets allows the prediction of the static thermodynamic conditions of the working fluid at the nozzle throat, nozzle exit and at the blading inlet and outlet by applying REFPROP fluid properties [6]. Hence, the velocity triangles at the mean diameter and the required cross areas can be determined in the form of nozzle angle, height and width, blade height and the blade angle at inlet and outlet. Based on these few geometry specifications, an experienced designer is able to build up the 3D model of the turbine in CAD. With the available 3D CAD model of the turbine, the next design step, the analysis and optimization of the turbine flow by means of Computational Fluid Dynamics (CFD), can be carried out. The turbine design data of the 1DTDT is shown in Table 2.

Table 2. Turbine main design data

Parameter	Unit	Curtis-Steam-Turbine
working fluid	-	steam
wheel diameter D_{mean}	mm	160
rotational speed n	rpm	36,000
degree of admission ε	%	35
design pressure ratio PR (ts)	-	87.5
degree of reaction r	%	20
nozzle exit Mach number Ma_2	-	2.22
rotor inlet relative Mach number Ma_{r2}	-	1.62
stator inlet absolute Mach number Ma_1	-	0.56
stator outlet absolute Mach number Ma_{2*}	-	1.19
rotor inlet relative Mach number Ma_{r2*}	-	0.68
expected shaft power	kW	38
predicted total to static isentropic efficiency (1D loss model) $\eta_{ts,1D}$	%	53.6

2.2 CFD Physical models and boundary conditions

To carry out these analysis and turbine optimizations, appropriate CFD software has to be chosen. For that reason, the software Fine/Turbo by NUMECA [22] which the authors already used for previous investigations [23,24], was used for the simulations. For the rotor stator interface, a full non-matching frozen rotor was used. As a turbulence model, the Spalart-Allmaras model was applied. The turbine has to be meshed 360° because due to the partial admission, it is not possible to simulate only a segment of the whole turbine. Table 3 shows the boundary conditions and physical models applied to the CFD simulations.

Table 3. Physical Models and Boundary Conditions of the CFD Simulations

Model or Condition	Parameter
Mathematical Model	Turbulent Navier Stokes
Modelling of turbulence	Spalart-Allmaras
Rotor-Stator Interface	Full Non-Matching Frozen Rotor
Efficiency Definition	Total to Static Isentropic
Fluid Model	Water (Condensable Fluid)
Inlet Boundary Condition	Absolute Total Pressure (17.5 bar)
Outlet Boundary Condition	Static Pressure (0.2 bar)

2.3 Mesh study

To assess and estimate the quality of the mesh, different grid accuracies were analysed and compared to each other. Figure 1 shows the inlet and outlet mass flow rate and the total-to-static isentropic efficiency over a varying number of grid points for the design boundary conditions. It becomes clear that for a very low number of grid points (0.7 Million), the mass flow rate extremely differs from the design mass flow rate (82 g/s). The mass flow rate course flattens out at about 81 – 82 g/s at 6.1 up to 53.8 Million grid points. The efficiency varies between 61.54 % and 61.20 % in the case of 53.8 to 6.1 Million grid points, respectively. The efficiency value for 0.7 million grid points is not conclusive due to the high mass flow rate difference. The mesh with 48 Million grid points was used as a compromise between accuracy and computing time. The still non-negligible deviation from inlet to outlet mass flow rate is very likely to be caused by high unsteady effects in the flow field due to partial admission. An unsteady time averaged CFD simulation confirmed these assumptions since a smaller deviation could be achieved (82 g/s at the inlet and 82.8 g/s at the outlet).

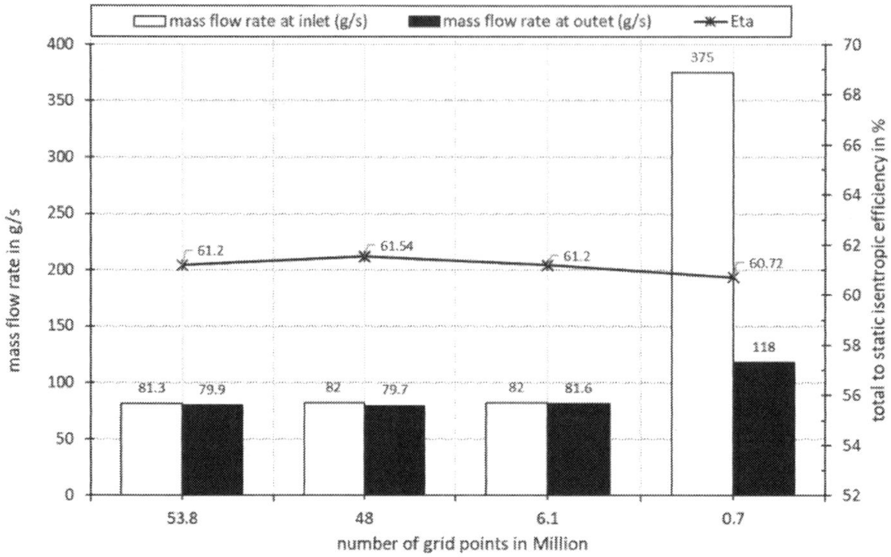

Fig. 1. Mass flow rate and total to static isentropic efficiency as a function of the number of grid points

2.4 Parameterized turbine design

In Figure 2 (top), the axial pressure distributions through the whole turbine at specific axial positions for different versions of the Curtis meridional flow channel are shown. The dotted line with the squared markers (1D-Design (15%)) and the dotted line with the triangular markers (1D-Design (20%)) show the static pressure evolution designed by means of the 1DTDT. The solid line with squared markers (CFD V1 (15%)) and the solid line with the triangular markers (CFD V16 (20%)) represent the pressure distributions achieved by 3D-CFD simulations. It has to be addressed that in the parametrization process, the reaction in the stator wheel was up scaled from 15 % to 20 % – the versions in Fig. 2 are named accordingly. If the degree of reaction is at 15%, it implies that 85 % of the entire turbine isentropic enthalpy drop is converted into kinetic energy in the supersonic nozzles and 15 % in the stator wheel – 20 % reaction follows accordingly. Both wheel passes are originally designed for constant pressure flow in the 1DTDT. The predicted blade heights (1DTDT) should ensure the aimed pressure distribution. However, it becomes clear that there is a large deviation between the designed (1D-Design (15%)) pressure distribution and the achieved (CFD V1 (15%)) pressure distribution in 3D-CFD (Fig. 2). The biggest differences occur between the inlet of the first rotor wheel and the stator wheel outlet. In contrast to the constant pressure (impulse) design, the pressure drops from about 0.5 to 0.2 bar over the first rotor wheel in V1. This pressure drop leads to a high reaction in the first wheel and increasing relative velocity over the blade channel. After this pressure drop over the first rotor wheel, the static pressure remains at the outlet conditions till the outlet area.

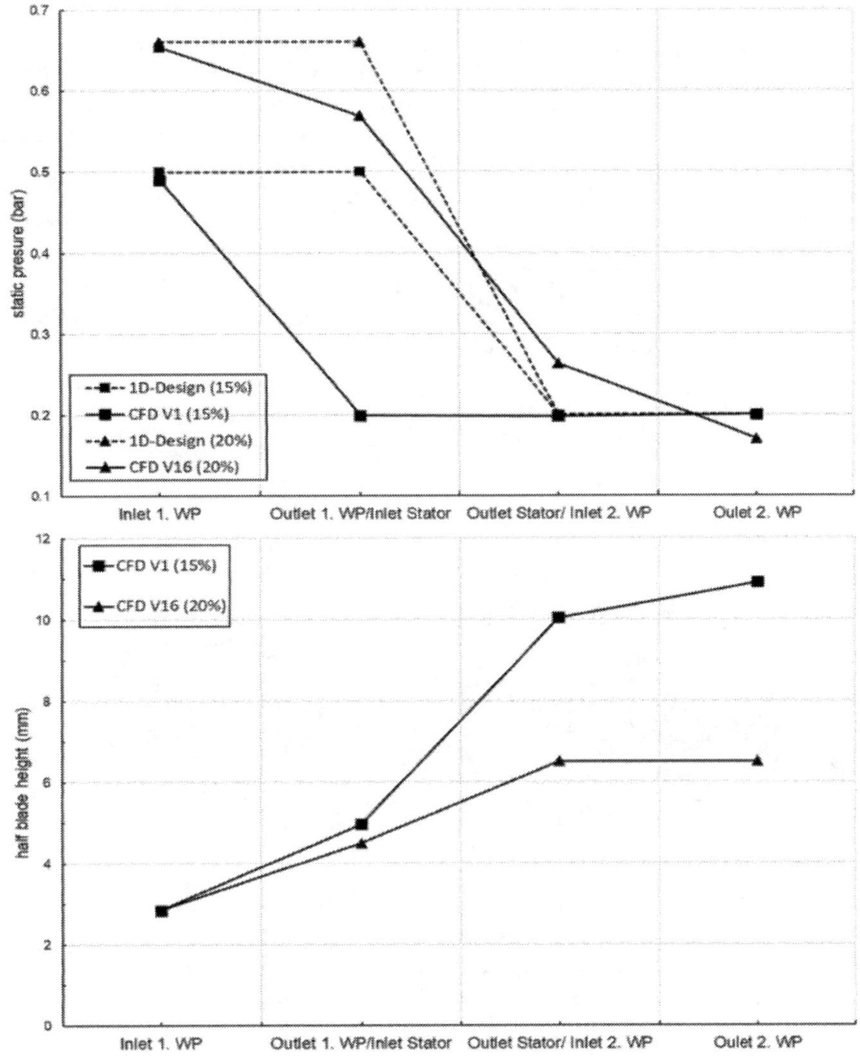

Fig. 2. Static pressure (top) and half blade height (bottom) at different positions through the turbine (mean diameter D_m = *const.*)

The origin of the significant deviation between 1D-design and 3D-CFD pressure distributions is the applied 1D loss model. The 1D loss model is not only needed to predict turbine efficiency. Furthermore, it is needed to determine the fluid properties like density at any axial position in the turbine. Thus, too high losses result in too low fluid density and thereby too big flow areas or blade heights, respectively. The loss model of the 1DTDT has mainly been developed and applied previously for single wheel axial and radial impulse turbines, working with organic fluids [21]. So far, deviations this big between 1DTDT and 3D-CFD have not occurred yet. Therefore, the authors decided to continue the optimization of the Curtis turbine directly in 3D CAD and 3D CFD, applying a parameterized design approach.

By adjusting the meridional flow channel or the blade heights (Fig. 2 bottom) step by step, a desired axial pressure distribution (CFD V16, Fig. 2 top) could be achieved. The prismatic blading remained mainly unchanged – besides rounding the leading edges of the stator and second rotor wheel. As mentioned above, reaction or enthalpy drop across the stator wheel has been increased from 15 % to 20%. The V16 stator wheel is working as a transonic nozzle. Furthermore, additionally, a slight reaction has been introduced over the two rotor wheel passes, although axial thrust forces should be avoided. This slight streamwise pressure drop should ensure that there is always a small but clearly defined axial force in the streamwise direction on the bearings. The necessary changes in blade heights are significant for the stator wheel and the second rotor wheel but much smaller for the first rotor wheel. This confirms the previous reasonable results for axial and radial impulse turbines [19] designed by the 1DTDT. Total-to-static isentropic efficiency has been increased from 57.5% (CFD V1) to 61.5% (CFD V16).

To achieve and to obtain the V16 design, it was necessary to model the turbine in a parameterized way and to iteratively exchange the information between CAD and CFD. The steps, which were necessary for this, are described below in more detail.

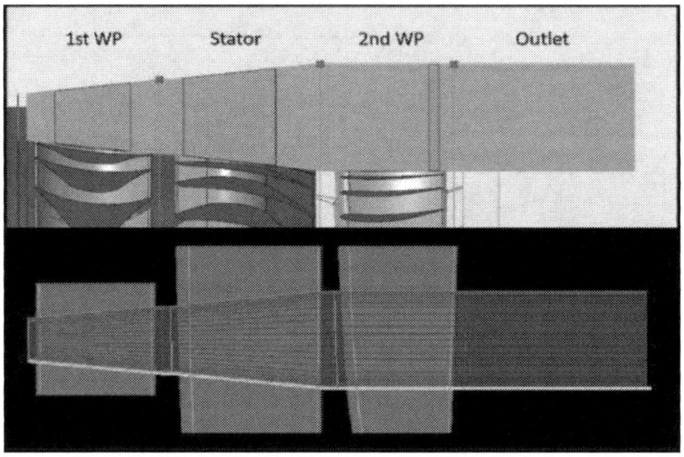

Fig. 3. Comparison of the meridional channel in CAD and CFD

Figure 3 shows the meridional flow channel of the Curtis turbine in the CAD software Creo (top) [25] and in IGG (Interactive Geometry Generator) (bottom) [22], the program to prepare the geometry and to generate the mesh for the CFD simulation. In Fig. 3 (top), the meridional channel in blue and the blades in grey can be seen. The red crosses mark the points where the channel can be adjusted. The height of the turbine outlet changes accordingly to the height at the outlet of the last rotor wheel. The inlet of the first rotor wheel is fixed due to a fixed nozzle outlet height. This simplification was necessary because the expenditure of time for manually meshing a new nozzle geometry was too high. The bottom part of Fig. 3 shows the meridional channel in blue and the blades in grey in the IGG environment. In comparison to the channel and blades at the top of the picture, only the blade heights differ because the meshing process implies prismatic blade geometries, which stand out over the hub and shroud curve of the meridional flow channel. The channel in IGG is taken directly from the CAD software via export/import. In this way, it is possible to adapt the channel in CAD and combine it with the existing blades in IGG. This implementation allows quick geometry adjustments, mesh generation and thereby a quick gain in information. In this way, it was possible, that the geometry, in this case the meridional flow channel, could be changed. The meshing process ran afterwards and the simulation was then started. After the analysis and evaluation of the results received from this CFD simulation, the authors could then adjust the geometry and start the entire process again. The CFD simulation needed about 8 h so that the overall turn-around-time for one geometry modification was about one working day (including the night for CFD calcs).

3 Results

Due to a high expenditure of time, not all of the different versions of the turbine could be analysed in detail. Because of this reason, only the most prospective version of the turbine is analysed in the following.

4 CFD Contour plots

Figure 4 shows the absolute (top) and relative (bottom) Mach number distribution for the V16 at 36,000 rpm and the design pressure ratio at 50 % blade height. The nozzles accelerate continuously as designed, with little anisotropic flow at the outlet to about Mach 2.5 to Mach 3. The flow passes through the first rotor wheel with shockwaves, probably induced by the leading edges of the rotor blades. Flow separation occurs at the suction side of some of the blades, and the flow stays attached at the pressure side. For the stator wheel, the flow enters the channel with sub- to transonic flow and accelerates steadily to about Mach 1.5 to Mach 2 in the absolute reference frame. The flow then passes through the second rotor wheel and exits the turbine through the outlet with about Ma 0.7. Surely, this flow field is representing one specific position of the frozen rotor-stator interface. The results should not be overrated, but nevertheless, the chosen approach is reliable and quick enough to be applied in a reasonable turbine design process.

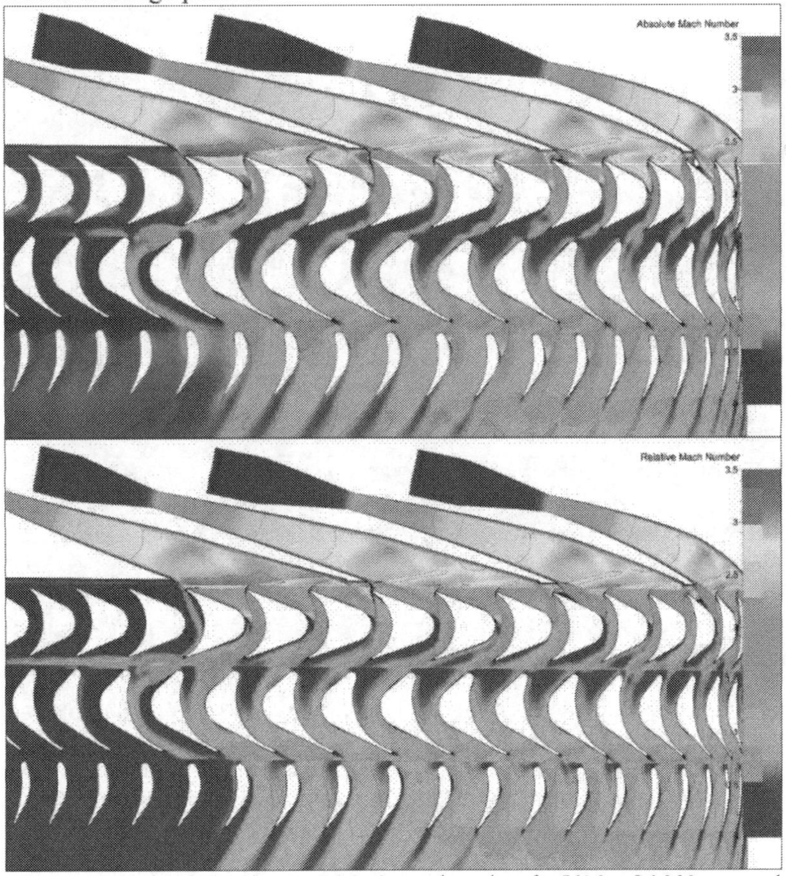

Fig. 4. Absolute (top) and Relative (bottom) Mach number plots for V16 at 36,000 rpm and design pressure ratio (87.5)

4.1 CFD Performance maps

The designed Curtis turbine will not constantly work under design conditions in the WHR plant. Condensing pressure and thereby overall turbine pressure ratio (PR) will vary, e.g. due to seasons. In order to diminish efficiency losses due to varying PR, the rotational speed of the turbine is usually adapted. Therefore, efficiency characteristics of V16 are generated by means of the described CFD approach.

The total to static isentropic efficiency as a function of the rotational speed of the V16 is shown in Figure 5. The maximum rotational speed of 36,000 rpm at the design point was chosen due to mechanical integrity limitations. Generally, the efficiency shows a comprehensible behavior over the entire rotational speed range. The efficiency of the turbine increases with rotational speeds, 63 % are achieved at the highest considered rpm of 39,000, which are not allowed in actual operation (mechanical integrity). The entire course lies on a too high efficiency level from approx. 53 % to 62 %, because not all physics are considered. Although the 360° CFD simulation takes into account steam leakages at the beginning and the end of the impinged part of the rotor wheels, unsteady effects are not considered. Furthermore, the tip gaps of the rotor wheels and the hub gap of the stator wheel are not modelled – as can be seen in Fig. 3 - due to CFD performance reasons.

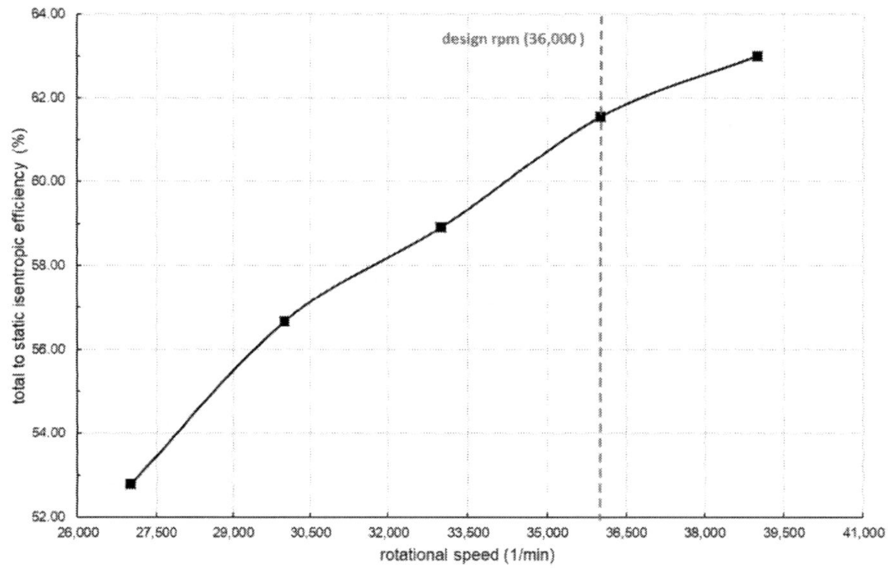

Fig. 5. Total to static isentropic efficiency as a function of the rotational speed for the V16

Figure 6 depicts the total to static isentropic efficiency as a function of the pressure ratio (PR) for V16. The pressure ratio was adjusted by changing the static pressure at the outlet of the turbine (i.e. condensing pressure, displayed at every point in the figure); the inlet pressure stayed constant at 17.5 bar. From a lower to a higher PR, the efficiency increases up to the design PR of 87.5 (17.5 absolute total inlet pressure and 0.2 bar static outlet pressure) and decreases from this point to a PR of 175. For pressure ratios lower than approx. PR = 29, the simulation did not converge because of the high unsteady flow effects and flow separations. For these high outlet pressures, the second rotor wheel even increased the total enthalpy and therefore worked as a compressor instead of a turbine. The efficiency drop for higher than designed pressure ratios can be explained by the inability of the Laval nozzles to convert the available higher enthalpy drop ($PR > PR_{design}$) into kinetic energy. As expected, the turbine efficiency shows a dependency on the pressure ratio but still performs well in the outlet pressure range between 0.15 bar and 0.3 bar. The highest efficiency at the designed PR reinforces the confidence of the authors in their design.

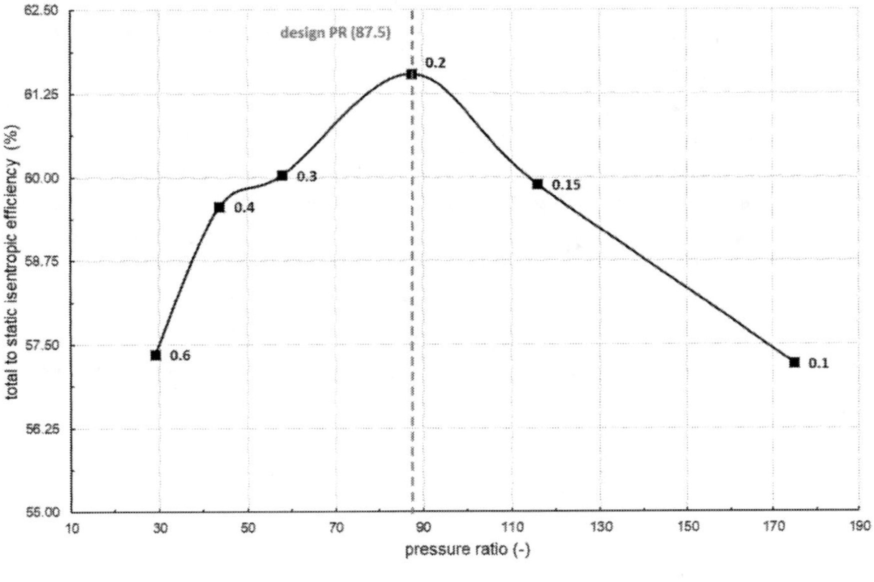

Fig. 6. Total to static isentropic efficiency as a function of the pressure ratio for the V16 with outlet pressures (bar) for every point

5 Conclusion

The physical challenges of designing a small-scale steam turbine were repeated. Thereby, it has become clear why a two wheel velocity compounded axial Curtis turbine must be chosen for the given design task. The applied design approach, consisting of an in-house developed 1D turbine design tool and a parameterized 3D CAD/CFD design, was introduced. The shortcomings of the applied 1D loss model and their consequences for the turbine geometry were addressed.

The predicted efficiency (1DTDT) of the V1 was just 53.6% compared to 57.5% in 3D CFD. By adjusting and optimizing the meridional flow channel directly in 3D space, the desired axial pressure distribution could be achieved with V16. Total to static isentropic efficiency is finally 61.5% in 3D-CFD. Thus, the parametric design, directly in 3D space, proved to be a practical and reasonable process for daily turbine development. Mach number contours at design point showed the expected pattern but furthermore, area for improvements. Finally, performance characteristics of the final design (V16) were introduced and evaluated.

The designed axial two-wheel Curtis is currently under construction. For the new design, Titanium for the rotor can be avoided and thereby material and manufacturing costs reduced. The new turbine will be installed 2022 in the WHR plant of the main sewage treatment plant in Nuremberg/Germany. There, the waste heat of two 1 MW biogas engines is converted to 40 -50 kWel by means of a micro steam Rankine cycle.

The authors gratefully acknowledge the funding of the joint project KompACT on the techno-economic optimization of an SRC plant for high temperature WHR by the Federal Ministry for Economic Affairs and Energy in the 6th Energy Research Program on the basis of the decision by the German Bundestag.

References

1. Pehnt DM, Bödeker J, Arens M, Jochem DE, Idrissova F. Die Nutzung industrieller Abwärme – technisch-wirtschaftliche Potenziale und energiepolitische Umsetzung: Bericht im Rahmen des Vorhabens „Wissenschaftliche Begleitforschung zu übergreifenden technischen, ökologischen, ökonomischen und strategischen Aspekten des nationalen Teils der Klimaschutzinitiative". Heidlberg, Karlsruhe; 2010.
2. Macchi E, Astolfi M (eds.). Organic rankine cycle (ORC) power systems: Technologies and applications. Duxford, UK, Amsterdam, Boston, Heidelberg, London, New York, Oxford, Paris, San Diego, San Francisco, Singapore, Sydney, Tokyo: Woodhead Publishing is an imprint of Elsevier; Elsevier; 2017.
3. Tanuma T. Advances in steam turbines for modern power plants. Oxford: Woodhead Publishing; 2016.
4. Raab Florian, Opferkuch Frank, Klein Harald. Dezentrale Verstromung von Abwärme: Forschungsarbeit zur Abwärmenutzung mittels Steam-Rankine-Cycle-Technlogie. In: BWK Energie.
5. Kraus MH, Deichsel M, Hirsch P, Opferkuch F, Heckel C. Hermetic 40-kW-Class Steam Turbine System for the Bottoming Cycle of Internal Combustion Engines. In: ASME Turbo Expo 2016: Turbomachinery Technical Conference and Exposition, June 13-17, 2016, Seoul, South Korea. New York, N.Y.: ASME; 2016.
6. REFPROP: NIST Reference Fluid Thermodynamic and Transport Properties Database (REFPROP). [March 30, 2021]; Available from: https://www.nist.gov/srd/refprop.
7. Seume JR, Peters M, Kunte H. Design and test of a 10kW ORC supersonic turbine generator. J. Phys.: Conf. Ser. 2017;821:12023. https://doi.org/10.1088/1742-6596/821/1/012023.
8. Klonowicz P, Witanowski Ł, Suchocki T, Jędrzejewski Ł, Lampart P. Selection of optimum degree of partial admission in a laboratory organic vapour microturbine. Energy Conversion and Management 2019;202:112189. https://doi.org/10.1016/j.enconman.2019.112189.
9. Uusitalo A, Honkatukia J, Backman J, Nyyssönen S. Experimental study on charge air heat utilization of large-scale reciprocating engines by means of Organic Rankine Cycle. Applied Thermal Engineering 2015;89:209–19. https://doi.org/10.1016/j.applthermaleng.2015.06.009.
10. Żywica G, Kaczmarczyk TZ, Breńkacz Ł, Bogulicz M, Andrearczyk A, Bagiński P. Investigation of dynamic properties of the microturbine with a maximum rotational speed of 120 krpm – predictions and experimental tests. J VIBROENG 2020;22(2):298–312. https://doi.org/10.21595/jve.2019.20816.
11. Meher-Homji CB. The Historical Evolution Of Turbomachinery 2000:281–322. https://doi.org/10.21423/R1X948.
12. Stodola A. Die Dampfturbinen: Mit einem Anhang über die Aussichten der Wärmekraftmaschinen und über die Gasturbine. 1st ed. Berlin: Springer; 2012.
13. De Laval CGP. Steam Turbine: Patent(522,066); 1889.
14. Weiß AP. Volumetric expander versus turbine - which is the better choice for small ORC plants? In: University of Liège and Ghent University, editor. ASME-ORC2015: Proceedings of the 3rd International Seminar on ORC Power Systems; 2015.

15. Harris FR. The Parsons Centenary—a Hundred Years of Steam Turbines. Proceedings of the Institution of Mechanical Engineers, Part A: Power and Process Engineering 1984;198(3):183–224. https://doi.org/10.1243/PIME_PROC_1984_198_024_02.
16. OMERSCALES EFC. The Vertical Curtis Steam Turbine. Transactions of the Newcomen Society 1990;62(1):157–8. https://doi.org/10.1179/tns.1990.008.
17. Spadacini C, Rizzi D. Radial outflow turbines for Organic Rankine Cycle expanders. In: Macchi E, Astolfi M, editors. Organic rankine cycle (ORC) power systems: Technologies and applications. Duxford, UK, Amsterdam, Boston, Heidelberg, London, New York, Oxford, Paris, San Diego, San Francisco, Singapore, Sydney, Tokyo: Woodhead Publishing is an imprint of Elsevier; Elsevier; 2017, p. 335–359.
18. Weiß AP, Popp T, Zinn G, Preißinger M, Brüggemann D. A micro-turbine-generator-construction-kit (MTG-c-kit) for small-scale waste heat recovery ORC-Plants. Energy 2019;181:51–5. https://doi.org/10.1016/j.energy.2019.05.135.
19. Weiß AP, Popp T, Müller J, Hauer J, Brüggemann D, Preißinger M. Experimental characterization and comparison of an axial and a cantilever micro-turbine for small-scale Organic Rankine Cycle. Applied Thermal Engineering 2018;140:235–44. https://doi.org/10.1016/j.applthermaleng.2018.05.033.
20. Weiß AP, Novotný V, Popp T, Zinn G, Kolovratník M. Customized Small-Scale ORC Turbogenerators- Combining a 1D-Design Tool, a Micro-Turbine-Generator-construction-kit and Potentials of 3D-Printing. In: The National Technical University of Athens, editor. ORC2019: Proceedings of the 5th International Seminar on ORC Power Systems; 2019.
21. Weiß AP, Novotný V, Popp T, Streit P, Špale J, Zinn G et al. Customized ORC micro turbo-expanders - From 1D design to modular construction kit and prospects of additive manufacturing. Energy 2020;209:118407. https://doi.org/10.1016/j.energy.2020.118407.
22. NUMECA International - FINE™/Turbo. [March 30, 2021]; Available from: https://www.numeca.com/en_eu/product/fineturbo.
23. Streit P, Popp T, Weiß AP. Simulation and analysis of the performance map of a micro-ORC-turbine - Comparison with measurements. In: 18th Conference on Power System Engineering, Thermodynamics & Fluid Flow: PSE 2019.
24. Streit P, Popp T, Winkler J, Scharf R, Weiβ AP. Numerical and experimental investigation of different technologies for adjusting the swallowing capacity of a cantilever ORC turbine. In: THERMOPHYSICAL BASIS OF ENERGY TECHNOLOGIES (TBET 2020). AIP Publishing; 2021, p. 70001.
25. PTC Creo Parametric. [March 30, 2021]; Available from: https://www.pdsvision.com/de/ptc-creo.

Analysis of the secondary circuit of a gas-cooled fast reactor

Michal Volf[1*], *Martin* Pelikán[1], and *Pavel* Žitek[1]

[1] University of West Bohemia, Faculty of Mechanical Engineering, Department of Power System Engineering, Univerzitní 22, Pilsen, Czech Republic

Abstract. The article focuses on a power conversion system for a gas-cooled fast reactor working with helium. The power conversion system, i.e., secondary and possible tertiary system of a power plant, is used to convert heat generated by nuclear fission into electrical energy. The presented research deals with the conceptual design of this system, mainly its secondary circuit, which is assumed to be a Brayton cycle. Several concepts are evaluated, including single and staged compression and possible heat regeneration. The goal of the work is to select the main parameters of such a cycle that would not only be ideal in terms of efficiency, but would also allow decay heat to be used and further converted into electricity. In this way, the secondary cycle could be used as an additional safety system for the nuclear power plant.

1 Introduction

A power conversion system is a crucial part of any power plant as it converts heat energy into broadly usable electrical energy. It goes without saying that the high efficiency of such a system is the main focus of the design; however, power plant safety is equally important, especially if the plant is a nuclear one. In nuclear power plants, the heat is generated by nuclear fission in different types of reactors and the reactor is cooled down by the secondary cycle during operation under normal operating conditions. This means that the reactor is cooled down using electricity produced by the power plant. However, in an emergency, the turbine (or secondary cycle) is shut down and an additional power supply is needed to cool down the reactor as it still produces some decay heat.

Our focus is aimed at gas-cooled fast reactors (GFR) which are classed as Gen IV reactors, and they feature a closed fuel cycle as well as enhanced safety systems. From a safety point of view, it would be beneficial if in an emergency, the turbine was not shut down, but its operation was extended so that the decay heat produced by the reactor could be further converted to electricity. This electricity could be used to power the reactor's coolant compressors and other components. Thus, the power plant would not need to rely on an external electricity supply or backup power supply to cool down the decay heat of the reactor. Such a system would thus further enhance the safety of the nuclear power plant. [1],

* Corresponding author: volfm@kke.zcu.cz

The research presented in this paper is part of a larger project aimed at the conceptual design of an innovative safety system for gas-cooled nuclear reactors. In the article we discuss possible designs of a secondary circuit for a GFR reactor working with helium. The secondary circuit is assumed to be a Brayton cycle with or without staged compression and with or without regeneration. The evaluation is done for different working fluids, namely nitrogen, carbon dioxide and helium, and the possibility of an additional tertiary circuit is discussed. [2], [3]

Bear in mind that the evaluation is done for normal operating conditions, i.e., the state in which the plant works most of the time, and thus needs be designed to have the highest feasible efficiency. The evaluation of secondary circuit behaviour in a power plant emergency (i.e. during reactor shut down) is not presented; however, the secondary circuit is designed with this requirement in mind.

2 Primary circuit

The primary circuit of the GFR for which the secondary circuit is analysed, uses helium as the working fluid. The nominal heat power is assumed to be 205 MW, mass flow rate is 85 kg/s and the output temperature is 850 °C. To cool down the reactor, it is necessary to cool the helium in the heat exchanger to 400 °C. Furthermore, it should be noted that the nominal pressure in the primary circuit is assumed to be 7 MPa and thus, for safety purposes, the pressure in the secondary circuit needs to be higher.

3 Conceptual design of power conversion system

3.1 Overall requirements

When designing the power conversion system, several aspects must be considered. From the point of utilization of decay heat, it is essential to design the secondary circuit so that it allows the turbine to run for as long as possible. However, considering the entire power plant, the cost plays a significant role and thus the highest efficiency of heat conversion into electricity is desired. Considering nuclear safety, it is obvious that the secondary circuit which should also behave as a safety system must be reliable when operating during reactor shut down. It goes without saying that making priorities between these three diverse requirements is not an easy task and requires thorough consideration. [1], [4]

For the purposes of our research, we consider two principally different approaches to the design of the power conversion system. The first assumes that as much heat as possible will be used in the secondary circuit, even at the cost of the technical complexity of such a cycle. This will lead to a higher cycle efficiency and higher power output of the secondary circuit, leaving the type of tertiary circuit for further consideration as the utilizable heat will be modest.

In the second approach, it is assumed that the simplicity of the secondary cycle is more important for power plant safety in an emergency and reactor shut down since this circuit should behave as a safety system. This means that most of the heat generated by nuclear fission will be converted to electricity in the tertiary circuit, which will likely be a Rankine cycle (i.e., a water steam cycle). This circuit is intended just to increase the overall power plant efficiency and does not play any safety role.

3.2 Technical point of view

Due to the high parameters of pressure and temperature in the primary circuit of a GFR, it is neither safe nor efficient to use a steam cycle (i.e., Rankine cycle) as the secondary circuit of the power plant. The safety risk comes from possible leakage of water into the primary circuit in the heat exchanger which cannot be ruled out, and would result in fatal consequences with severe radiation leakage. Furthermore, the output temperature from the primary circuit is very high and thus for an efficient (i.e. maximum) utilization of heat, the steam cycle would have to be designed as a supercritical one which would further decrease the safety of such a power plant. It goes without saying that this would go against the entire intention of Gen IV reactors as they are designed to excel in safety and reliability. [5]

Therefore, it follows that a three-circuit arrangement will be necessary. Both primary and secondary circuits would be designed as gas cycles, and the tertiary circuit would be either a Rankine cycle or an organic Rankine cycle, depending on the output temperature of the secondary circuit. The choice of gas and tertiary circuit will depend on the results of the optimization.

4 Secondary circuit arrangement

For the analysis of the secondary circuit, two principally different arrangements were selected. The first can be seen in Fig. 1 and represents the basic Brayton cycle. In the cycle, a compressor (single or staged) compresses the gas which then enters a heat exchanger (between the primary and secondary circuits) where it heats up by removing heat from the primary circuit. Then, hot compressed gas enters the turbine where its pressure and heat energy is transformed into the mechanical energy of a rotating shaft. After exiting the turbine, the gas is cooled down in a heat exchanger between the secondary and tertiary circuits, meaning that this heat can be further used in the tertiary circuit to generate electricity. If the parameters of this removed heat (especially its temperature) are not sufficient to further generate electrical energy, basic cooling is used instead.

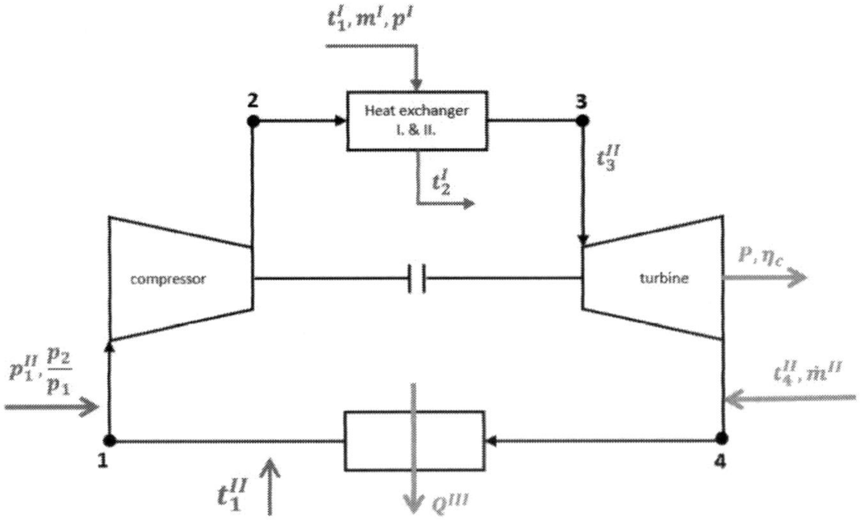

Fig. 1. Secondary circuit without regeneration

The second arrangement (see Fig. 2) adds a regenerator into the cycle. This means that the compressed gas is preheated before entering the heat exchanger between the primary and secondary cycles. For this, the heat of the gas exiting the turbine is used, which will cause a decrease in the parameters of the heat utilizable for the possible tertiary circuit. On the other hand, the advantage of the regenerator lies in the increase of the cycle efficiency.

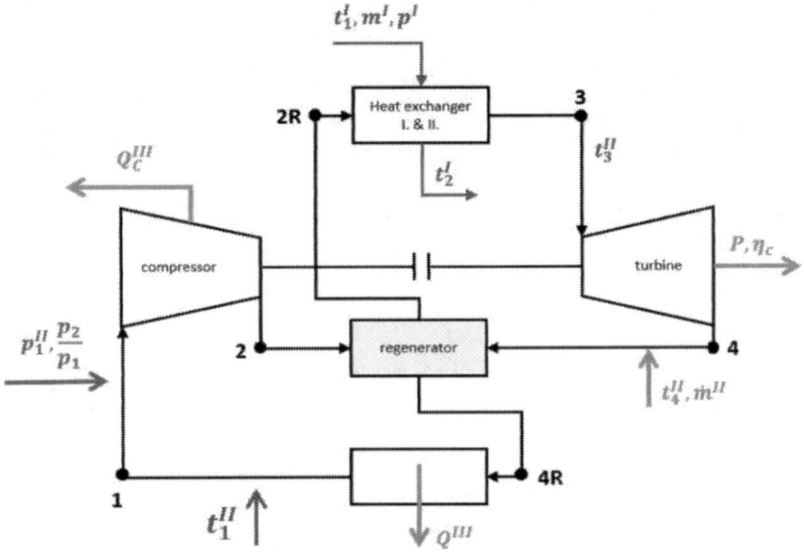

Fig. 2. Secondary circuit with heat regeneration

In both figures above, the evaluated cycle parameters are marked in green, the fixed parameters are marked in red and the varying parameters during the analysis are marked in blue.

5 Analysis of the secondary circuit

A script for the balance calculation of the secondary Brayton circuit was developed in MATLAB. It allows us to analyse the main parameters of the circuit for different working gases and various values of the boundary condition, namely the compressor pressure ratio. This in-house script was used to calculate the four variants of the Brayton cycle discussed in the preceding chapter.

For all the analysed cycles, the temperature at the compressor inlet was assumed to be 60 °C, and the lowest pressure in the circuit (i.e., the pressure at the compressor inlet) was set to 7.5 MPa so that it is not lower than the pressure in the primary circuit, which is a safety requirement. Pressure losses as well as turbine and compressor efficiencies were included. The thermodynamic efficiency of the turbine is assumed to be 90 %, and for the compressor it is assumed to be 80 %. Total pressure losses were estimated to be 600 kPa in total for the heat exchanger regardless of the working gas. We are aware that pressure losses are mainly caused by fluid friction, so for different fluids they could have different values based on viscosity and velocity. Nevertheless, most of the pressure losses occur in the heat exchangers and thus for this initial study, we assume that the heat exchanger would be designed so that the pressure loss would be similar for each fluid.

For the cycles with regeneration, it is assumed that there will be a temperature difference of at least 10 °C, meaning that the temperature at the inlet of the heat exchanger between the primary and secondary circuits (i.e., at the outlet of the regeneration heat exchanger) will not exceed 390 °C. For the cycles with staged compression, it is assumed that partially compressed gas will be cooled down in an intermediate heat exchanger to the inlet temperature, and moreover that the compression is ideally divided, meaning the pressure ratios of each stage are equal. It should be noted that only compression divided into two stages is considered.

Firstly, a study to find the optimal pressure ratio at which the highest cycle net efficiency is reached was performed. For each concept, the same varying parameter was used, meaning that the pressure ratio was changed from 1.2 up to 40. The right boundary is chosen to be this high on purpose so that the optimal pressure ratio lies in the evaluated interval, instead of hitting either the left or the right boundary. It goes without saying that for high values of pressure ratio, the pressure of the compressed gas would be unreasonably and, more importantly, unfeasibly high and thus a second study was performed. This was done assuming a fixed compressor pressure ratio of 3, which means the pressure of the compressed gas would be 22.5 MPa and thus the realization of such a circuit would still be feasible.

6 Results of the analysis

The results of the studies are summarised in Table 1 and Table 2. The column labelled "*p_2/p_1 (optimal)*" shows the results for the first analysis, i.e., the study to find the pressure ratio for which the cycle would have the highest efficiency. The column labelled "$p_2/p_1 = 3$" presents the results for the fixed pressure ratio of 3, which seems to be the highest reasonably feasible one (in terms of the maximum cycle pressure). For both analyses, the cycle net efficiency (η) is shown as well as the cycle net power (P), mass flow rate (m) and output temperature utilizable for the tertiary circuit (TIII).

Table 1. Results of the conceptual study of the secondary circuit (p_2/p_1 (optimal))

Gas	Compressor stages	Regeneration	p_2/p_1 (optimal)				
			p_2/p_1	η [%]	P [MW]	T_{III} [°C]	m [kg/s]
N_2	1°	no	7	20.6	40.0	452	358
N_2	1°	yes	4.8	23.0	44.8	423	376
N_2	2°	no	9.8	22.3	43.5	400	267
N_2	2°	yes	7.8	31.3	61.0	233	371
He	1°	no	4	18.8	36.6	432	82
He	1°	yes	3.2	21.1	41.0	406	85
He	2°	no	5.4	21.0	40.9	365	63
He	2°	yes	4.4	29.3	57.0	226	85
CO_2	1°	no	19.4	26.3	51.3	449	312
CO_2	1°	yes	10.6	29.7	57.8	411	329
CO_2	2°	no	24.4	25.1	48.9	425	207
CO_2	2°	yes	15.4	39.7	77.2	147	324

Table 2. Results of the conceptual study of the secondary circuit ($p_2/p_1 = 3$)

Gas	Compressor stages	Regeneration	$P_2/p_1 = 3$			
			η [%]	P [MW]	T_{III} [°C]	m [kg/s]
N_2	1°	no	14.4	28.1	608	274
N_2	1°	yes	20.0	38.9	433	379
N_2	2°	no	14.4	28.1	608	241
N_2	2°	yes	22.7	44.2	346	379
He	1°	no	17.1	33.4	506	70
He	1°	yes	21.0	40.9	407	85
He	2°	no	17.5	34.0	506	56
He	2°	yes	26.5	51.5	280	85
CO_2	1°	no	12.6	24.5	686	228
CO_2	1°	yes	19.4	37.8	437	352
CO_2	2°	no	12.5	24.3	686	195
CO_2	2°	yes	22.5	43.8	311	352

6.1 Analysis of the optimal pressure ratio

At first glance the results show very good efficiencies of the Brayton cycle, reaching values of over 30 % (nitrogen with staged compression, including regeneration). However, taking closer look, it can be seen that for all configurations, the optimal compressor pressure ratios are far too high and thus the maximum pressure in the cycle would reach extreme values. Assuming the feasible pressure ratio is up to the value of 3 (the corresponding maximum pressure in the cycle is 22.5 MPa), only the cycle with helium approaches this, as the optimal pressure ratio for single compression and regeneration is 3.2. Nevertheless, due to the low molar mass of helium, real-life pressure ratios for which the compressor can be designed are even lower. All in all, it can be concluded that it is not likely to use the full potential of the cycle, so the tertiary cycle is necessary to improve the overall power plant efficiency and use as much heat generated by nuclear fission as possible.

6.2 Analysis for the fixed compressor pressure ratio of 3

Looking at the efficiencies, it can be seen that the highest efficiency would be for a secondary circuit with helium and both staged compression and regeneration. However, the temperature of the residual heat for the tertiary cycle (280 °C) is not sufficient for the subsequent Rankine tertiary circuit and thus a further increase of the power plant efficiency seems to be considerably limited. From this point of view, it seems to be more convenient to use a secondary circuit working with either nitrogen or carbon dioxide, since both allow for a common Rankine tertiary cycle. Furthermore, it should be added that since helium is a very light gas, reaching a pressure ratio of 3 would require a large compressor (many axial stages if an axial one is used) and thus the real-life pressure ratio would need to be lower. It follows that for lower pressure ratios, the efficiency would further decrease (assuming all the other parameters remain the same), which makes the cycle with helium an unviable choice to proceed with.

The cycle with carbon dioxide seems to reach quite high efficiency values for both single and staged compression if regeneration is applied. The single compression would make the circuit simpler from a technical point of view which is convenient for the safety purposes discussed at the beginning of the article. Nevertheless, carbon dioxide is not chemically stable at temperatures above 633 °C [6] so using it would create further issues.

Considering all these factors, a Brayton secondary circuit with nitrogen and without staged compression seems to be the ideal choice, since it allows for a proper Rankine tertiary circuit that would significantly increase the overall power plant efficiency. Furthermore, the properties of nitrogen are very similar to the properties of air and so, unlike for helium, there would be no issues regarding chemical stability or gas leakage. Finally, this cycle configuration is also technically simple and thus it fits the safety approach in the design of the power conversion system. The T-s and h-s diagrams for this cycle (without regeneration) are shown in Fig. 3. Regeneration could be used in the secondary circuit to increase its efficiency; however, it would decrease the utilizable temperature for the tertiary circuit. Therefore, the decision whether or not to regenerate heat should be made based on an advanced analysis including the tertiary circuit, which is planned in the later stages of our project. Nevertheless, the parameters of the compressors and turbine remain the same.

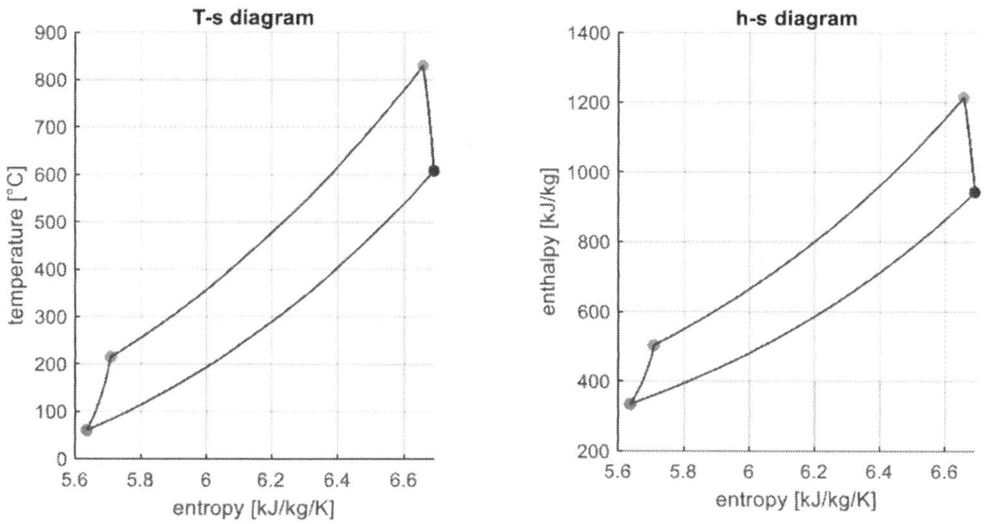

Fig. 3. T-s and h-s diagrams of the selected secondary circuit (compressor inlet: 60 °C, 7.5 MPa; compressor outlet: 215 °C, 22.5 MPa; turbine inlet: 830 °C, 21.9 MPa; turbine outlet: 608.1 °C, 8.1 MPa)

7 Conclusion

In the article, the initial study of a power conversion system for a gas-cooled fast reactor was presented. The goal of the power conversion system is not just to convert heat into electrical energy but also to behave as a safety system in an emergency and reactor shut down. Several configurations of a secondary circuit were analysed to find the ideal pressure ratio at which the net cycle efficiency would be the highest. However, based on the results, the ideal ratios were too high to be considered feasible and thus a second analysis was done assuming a fixed pressure ratio of 3. This analysis showed that the secondary circuit should use nitrogen as a working fluid and should have a single compression to achieve a technically simpler cycle. The decision of whether to utilize heat regeneration within the cycle was not made as this should be based on a subsequent larger analysis that will include the tertiary circuit.

This project is co-financed from the state budget by the Technology Agency of the Czech Republic under the THETA Programme, project number TK03030121 - Conceptual Design of an Innovative Safety System for Gas-cooled Nuclear Reactors. The presented work was also financially supported by student project SGS-2019-021 (Improving the efficiency, reliability and service life of power machines and equipment 5).

References

1. *Gas cooled fast reactor research in Europe.* Stainsby, Richard, et al. 9, s.l. : Nuclear Engineering and Design, 2011, Vol. 241. ISSN 0029-5493.
2. *Analysis of a 115MW, 3-Shaft, Helium Brayton Cycle Using Nuclear Heat Source.* Kumar, Pradeep K, Tourlidakis, A and Pilidis, P. New Orleans, Louisiana : ASME, 2001. V002T04A028.
3. Thermodynamic analysis and preliminary design of closed Brayton cycle using nitrogen as working fluid and coupled to small modular Sodium-cooled fast reactor (SM-SFR). Olumayegun, Olumide, Wang, Meihong and Kelsall, Greg. s.l. : Applied Energy, 2017, Vol. 191, pp. 436-453. ISSN 0306-2619.
4. *The Supercritical Carbon Dioxide Power Cycle: Comparison to Other Advanced Power Cycles.* Dostal, Vaclav, Hejzlar, Pavel and Driscoll, Michael J. s.l. : Nuclear Technology, 2006, pp. 283-301.
5. *Generation IV nuclear reactors: Current status and future prospects.* Locatelli, Giorgio, Mancini, Mauro and Todeschini, Nicola. s.l. : Energy Policy, 2013, Vol. 61. ISSN 0301-4215.
6. *The thermal decomposition of carbon dioxide.* M.H. Lietzke, Chantal Mullins. 1981, Journal of Inorganic and Nuclear Chemistry, pp. 1769-1771.

Assessment of condensation models for moist air transonic flow prediction

Piotr Wiśniewski[1*], *Mirosław* Majkut[1], *Sławomir* Dykas[1], *Krystian* Smołka[1], *Guojie* Zhang[2], and *Balazs* Pritz[3]

[1]Silesian University of Technology, Department of Power Engineering and Turbomachinery, Konarskiego 18, 44-100 Gliwice, Poland
[2]School of Mechanical and Power Engineering, Zhengzhou University, Zhengzhou, China
[3]Institute of Thermal Turbomachinery, Karlsruhe Institute of Technology, Karlsruhe, Germany

Abstract. The aim of this article is to thoroughly analyse the influence of condensation models on the modelling of condensation phenomena in transonic flow of moist air. The reason for the study was the fact that different condensation models are used by researchers to obtain satisfactory results of numerical modelling. The condensation models tested herein differ in the nucleation rate formula and the droplets growth equation. Four most often used condensation models were selected for detailed investigations. The results obtained from each model were compared with experiments for the nozzle flow. The main focus was on the location of the onset of the nucleation process. Moreover, the droplets growth intensity was compared and discussed. The nozzle flow CFD calculations were performed using the ANSYS Fluent commercial tool. Finally, the condensation model which is the most suitable for the moist air transonic flow was recommended.

1 Introduction

The presence of water or steam in atmospheric air exerts a significant influence on flow conditions in elements of any devices or machines, affecting their durability, accuracy and efficiency. If unfiltered or wet, atmospheric air always contains a certain amount of suspended solid particles, water droplets (fog, rain, etc.) or steam. Air cleaning or drying is not feasible in many technical applications, e.g. in turbine engines. Therefore the impact of such impurities on the flow in turbomachinery needs to be investigated.

In the case of transonic flows, based on the condensation trigger, phase change phenomena can be divided into homogeneous and heterogeneous processes. Spontaneous nucleation occurs due to the process of rapid supersaturation (or supercooling) of gas, when the appearing condensation nuclei reach the critical radius triggering the homogeneous condensation phenomenon. If air contains suspended particles, condensation may occur on them. This process is referred to as heterogeneous condensation. The tiny droplets formed due to nucleation or on the suspended particles grow further. In favourable conditions they can reach significant sizes and form so-called coarse droplets. The condensation process can

* Corresponding author: piotr.wisniewski@polsl.pl

be rapid, especially during gas expansion in the convergent-divergent nozzle (CD nozzle), where supersonic conditions are reached in the nozzle divergent part. The phase change phenomena are connected with the interphase latent heat exchange. During the condensation process, latent heat is released from condensing steam to the gaseous phase. This affects the flow structure by changing the gas pressure, temperature and, thereby, velocity. The influence of the latent heat release is of major importance during the occurrence of rapid homogeneous condensation. The so-called condensation wave then appears, followed by a sudden release of a high amount of energy. On the other hand, if air contains suspended particles, heterogeneous condensation becomes the major phase-change process, which has a more gradual character but results in the formation of droplets with a relatively large size, depending on the initial size of the foreign nuclei.

Evaporation is a process contrary to condensation. If the droplet evaporates, latent heat is absorbed from the gaseous phase in the phase-change process, leading to a local drop in the fluid temperature, pressure, and therefore – in its density, velocity, entropy, etc. If the flow is supersonic, shock waves form, and the liquid droplets interacting with the shock wave tend to evaporate. This process is particularly interesting if the liquid fraction (e.g. fog or rain) is significant because it then results in the absorption of a relatively large amount of latent heat.

Most often homogeneous and heterogeneous condensation processes appear together. Only their share in the liquid phase formation may change depending on the air parameters. The parameter usually used to define the content of steam in humid air is called relative humidity, which is the relative value of the ratio of the mass of steam in the air to the total mass of steam the air can hold. Typical humidity values range from 40% to 50%. However, due to a change in weather conditions, the relative humidity level can reach as much as 100%. If the mass of steam in atmospheric air exceeds the mass of steam the air can hold, the excess of steam condenses creating fog. It has to be emphasized that the mass of steam that the air can hold, and therefore the air relative humidity, is strictly related to the air temperature and pressure. If the air temperature rises, the steam saturation pressure increases and, as a result, air can hold more steam. Therefore, the impact of temperature is crucial if condensation occurs. Another air parameter that has a major impact on phase-change phenomena is the number and diameter of the particles suspended in air. If the number of particles is significant, heterogeneous condensation might become the major phase-change phenomenon, weakening the influence of homogeneous condensation and weakening the condensation wave. Taking account of the impact of phase-change phenomena on the most crucial parameters of the transonic moist air flow, it must be stressed that the aforementioned factors (i.e., humidity, air parameters, presence of suspended solid or liquid particles) should be taken into consideration in CFD modelling of such flows.

Internal and external flows with condensation have been investigated intensively for over 50 years and are still of great interest to researchers all over the world [1]. In the 1960's Gyarmathy proposed one of the most popular models for the droplet growth to model the steam condensing flow in the turbine. Gyarmathy's model [2]was further modified by Young [3,4] and by Fuchs and Sutigin [5]in the 1970's and provides slightly better predictions of the growth of nanosized droplets. The experimental and numerical studies regarding internal flows thorough CD nozzles conducted by Schnerr et al. [6,7,8] became the reference for researchers for the modelling of both moist air and pure steam condensing flows. Adam investigated, both experimentally and numerically, the non-stationarity of the flow, especially the impact of relative humidity on the interaction between condensation and shock waves and the oscillation frequency in a nozzle with parallel walls downstream the throat [10]. A 3D academic in-house code was developed by Dykas et al [9]. It treats the air-stream mixture as a homogeneous fluid and was validated against in-house experimental studies. The code enables robust computation of condensation in internal and external flows [12,13]. Recently, it became the basis for the model external implementation into commercial software. The functionality of the model implemented in the commercial software was extended by allowing slip velocity consideration, i.e. investigation of the difference between liquid and gaseous phase velocities, which is especially important if droplets reach a relatively large size or if the flow has a significant swirl [14,15]. Valuable techniques for condensation simulation in complex 3D structures were introduced by Yamamoto [16,17]. The numerical study of condensation is extremely demanding in terms of computing power, but the latest advancements have led to attempts to simulate the entire compressor rotor with a non-uniform circumferential distribution of wetness at the inlet [16]. The liquid water content at the jet engine inlet was studied by Moriguchi et al., who prove a significant change in air parameters at the compressor rotor outlet due to water evaporation [18]. Most numerical studies regarding condensation are performed based on the single-fluid approach, using the assumption of a continuous fluid for the mixture of air, steam and water. This approach gives reliable results if the droplet inertia and temperature between the phases are negligible. However, if the swirl of the flow of droplets is significant or if the droplets are relatively large, air and water should be considered as separate phases using the two-fluid approach. Some attempts have been made to model multi-phase humid air and wet steam flows [19,20,21]. Condensation is not only present in humid air. In fact, it is a very common phenomenon. Therefore, similar models have been adjusted to simulate condensation in pure steam [22,23,24,25,26,27] or in other gases, such as CO2 [28].

In this paper, the CFD computations are performed using the ANSYS Fluent commercial tool. The tool was extended by a condensation model using User Defined Functions (UDF's) [29]. The authors consider the influence of different condensation models on the condensation phenomena in the convergent-divergent nozzle.

2 Numerical and mathematical models

In this paper, the condensing flow is modelled numerically using the single-fluid approach, whereby the flow governing equations are stated for the mixture of air, steam and water. The slip velocity between the phases is neglected. All fluid properties, e.g. viscosity, thermal conductivity etc., are computed for mixture.

The RANS model for the compressible flow and the transport equations:

$$\frac{\partial}{\partial t}\rho + \frac{\partial}{\partial x_j}\rho \mathbf{v}_j = 0 \tag{1}$$

$$\frac{\partial}{\partial t}\rho \mathbf{v}_i + \frac{\partial}{\partial x_{ij}}\left(\rho \mathbf{v}_i \mathbf{v}_{ij} + \delta_{ij} p - \boldsymbol{\tau}_{ij}\right) = 0 \tag{2}$$

$$\frac{\partial}{\partial t}\rho E + \frac{\partial}{\partial x_j}\left(\mathbf{v}_j \rho E + \mathbf{v}_j p\right) = \frac{\partial}{\partial x_j}\left(k_{\text{eff}}\frac{\partial}{\partial x_j}T + \boldsymbol{\tau}_{ij}\mathbf{v}_j\right) + S_E \tag{3}$$

where ρ is density, t is time, $\mathbf{v}$ is the velocity vector, δ is Kronocker delta, p is pressure, $\boldsymbol{\tau}$ is stress tensor, E is energy, k_{eff} is effective conductivity, T is temperature and S_E is source term due to condensation and it has following form:

$$S_E = \left(S_{y_{hom}} + S_{y_{het}}\right) \cdot L \tag{4}$$

where L is the latent heat.

The k-ω Shear-Stress-Transport (SST) model was adopted as the Eddy-Viscosity Turbulence Model [30]. The thorough description of the turbulence model can be found in ANSYS Fluent Theory Guide [31].

Additional transport equations were required to model the phase change duto homogeneous and heterogeneous condensation. Partial differential equations (PDEs) were included that describes the number of droplets per kilogram (n), created due to the nucleation process, and the liquid mass fraction (y), due to the homogeneous and heterogeneous condensation. These additional transport equations were introduced using the ANSYS Fluent UDFs. In total, there are three additional PDs, which in case of 2D flow, together with the RANS model, gives a system of 9 PDEs to solve. The number of solid particles in the heterogeneous process is constant and based on the literature and our previous research [32,14,9]. The additional transport equations for the liquid phase take the following form:

$$\frac{\partial \rho y_{hom}}{\partial t} + \frac{\partial}{\partial x_j}\left(\rho \mathbf{v}_j y_{hom}\right) = S_{y_{hom}} \quad (5)$$

$$\frac{\partial \rho n_{hom}}{\partial t} + \frac{\partial}{\partial x_j}\left(\rho \mathbf{v}_j n_{hom}\right) = S_{n_{hom}} \quad (6)$$

$$\frac{\partial \rho y_{het}}{\partial t} + \frac{\partial}{\partial x_j}\left(\rho \mathbf{v}_j y_{het}\right) = S_{y_{het}} \quad (7)$$

where S represents the source term: for eq. 5 – the source of mass of the condensed liquid due to homogenous condensation, for eq. 6 – the source of droplets per kilogram of air generated due to the nucleation process, for eq. 7 – the source of mass of water resulting from the heterogeneous condensation process. The source terms are defined by the following formulae:

$$S_{y_{hom}} = \frac{4}{3}\pi\rho\rho_l r^{*3} J_{hom} + 4\pi\rho\rho_l n_{hom} r_{hom}^2 \frac{\mathrm{d}r_{hom}}{\mathrm{d}t} \quad (8)$$

$$S_{n_{hom}} = \rho J_{hom} \quad (9)$$

$$S_{y_{het}} = 4\pi\rho\rho_l n_{het} r_{het}^2 \frac{\mathrm{d}r_{het}}{\mathrm{d}t} \quad (10)$$

where J is the nucleation rate presenting the number of nuclei with critical radius r^* formed in one kilogram of moist air in one second; r is the droplet radius, which is calculated based on the water mass fraction and number of droplets.

The homogeneous condensation process is triggered by spontaneous nucleation, whereby nuclei form in the fluid as the result of rapid supercooling, reaching the critical radius. In our model, the critical radius was based on the Kelvin equation:

$$r^* = \frac{2\sigma}{\rho_l R_v T_v \ln(S)} \quad (11)$$

where σ is the steam surface tension, R_v is the individual gas constant, T_v is the steam temperature and $S = p_v/p_s$ is supersaturation, the ratio of partial steam pressure to saturation pressure. The classical nucleation theory was implemented to describe the formation of condensation nuclei in one kilogram of moist air within one second. The nucleation rate corrected by Kantrowitz [33] is defined as:

$$J_{hom} = C\sqrt{\frac{2\sigma}{\pi m_v^3}\frac{\rho_v}{\rho_l}}\mathrm{e}^{\left(-\frac{4\pi r^{*2}\sigma}{3kT_v}\right)} \tag{12}$$

$$C = \frac{q_c}{1 + \frac{2(\gamma-1)}{(\gamma+1)}\frac{(h_v - h_l)}{R_vT_v}\left(\frac{(h_v - h_l)}{R_vT_v} - \frac{1}{2}\right)} \tag{13}$$

where γ is the ratio of specific heat capacities, m_v is the mass of one molecule, ρ_l is the liquid density at temperature T_v, k is the Boltzmann constant and q_c represents the condensation coefficient, which is equal to one.

The so-called non-isothermal Kantrowitz correction C is based on the assumption that all growing droplets have the same excess temperature. In the presence of a large volume of the gaseous phase, like in the case of moist air, the isothermal assumption is justified because collisions of condensing molecules with inert gas molecules are much more common than those with condensing droplets. Finally, it reduces nucleation by a factor of 50–100.

According to the gas dynamics theory, the frequency of the collision of steam molecules with condensation nuclei is related to the average free path of the steam molecules and the radius of the water droplets. The larger the Knudsen number (Kn), the smaller the diameter of the droplet. The equation for calculating Kn is:

$$Kn = \frac{l_S}{2r} = \frac{\mu_l\frac{\sqrt{9/8\pi RT}}{p}}{2r} \tag{14}$$

where ls is the free mean path of the steam molecule, μ is the dynamic viscosity of moist air, T is temperature, R is the gas constant; p is the pressure of moist air, r is the droplet radius.

When $Kn < 0.01$, the droplet growth process can be described as a continuous flow region. When $Kn > 4.5$, the droplet growth is governed by the free molecular flow. When $0.01 \leq Kn \leq 4.5$, the transition region is the case.

The most popular and the one most often used by researchers is the continuous droplet growth model developed by Gyarmathy [2], which is applied for a bigger size of droplets:

$$\frac{\mathrm{d}r}{\mathrm{d}t} = \frac{1}{\rho_l}\cdot\frac{\left(1-\frac{r^*}{r}\right)}{r(1+3.18Kn)}\cdot\left(\frac{\lambda_vR_vT_v{}^2}{(h_v-h_l)^2}\right)\cdot\ln(S) \tag{15}$$

where λ_v is thermal conductivity of vapour.

The correction for this model was proposed in [3] and it is known as the Fuchs-Sutugin model:

$$\frac{\mathrm{d}r}{\mathrm{d}t} = \frac{1}{\rho_l}\cdot\frac{(1+2Kn)\left(1-\frac{r^*}{r}\right)}{r(1+3.42Kn+5.32Kn^2)}\cdot\left(\frac{\lambda_vR_vT_v{}^2}{(h_v-h_l)^2}\right)\cdot\ln(S) \tag{16}$$

Another modification of Gyarmathy's model was proposed by Young, who, drawing on the experience from the nozzle flow modelling, introduced the following correction [3,4]:

$$\frac{\mathrm{d}r}{\mathrm{d}t} = \frac{1}{\rho_l} \cdot \frac{1}{r\left(1 + 3.78(1-v)\frac{Kn}{Pr}\right)} \left(1 - \frac{r^*}{r}\right) \cdot \left(\frac{\lambda_v R_v T_v{}^2}{(h_v - h_l)^2}\right) \cdot \ln(S) \tag{17}$$

where:

$$v = \frac{R_v T_v}{(h_v - h_l)} \left(\alpha - 0.5 - \frac{2 - q_c}{2q_c} \cdot \left(\frac{\gamma + 1}{2\gamma}\right)\left(\frac{c_p T}{(h_v - h_l)}\right)\right) \tag{18}$$

coefficients α and q_c were set to 1 in the calculations presented herein.

In the continuous droplet growth models, the temperature of the droplets (T_l) can be determined by the capillarity effect using the expression given by Gyarmathy [2]:The correction for this model was proposed in [3] and it is known as the Fuchs-Sutugin model:

$$T_l = T_s - (T_s - T_v)\frac{r^*}{r} v \tag{19}$$

where T_s is the steam saturation temperature for given pressure. Here, steam temperature equals the temperature of air.

If the water droplets are very small, for which Kn>4.5, the growth of the droplets should be governed by considering the molecular and macroscopic transport process. The droplet growth equation for the free-molecular regime is given by the kinetic Hertz-Knudsen model [6,7,8]:

$$\frac{\mathrm{d}r}{\mathrm{d}t} = \frac{\alpha_c}{\rho_l} \cdot \frac{p_v - p_s}{\sqrt{2\pi R_v T}} \tag{20}$$

In eq.20 α_c is a condensation coefficient describing how many steam molecules colliding with the water droplet settle on its surface. The value of 1 for this coefficient means that all steam molecules are captured by the water droplet surface. If this value equals 0, none of the steam molecules are captured by the water droplet surface, all are deflected. For the Hertz-Knudsen droplet growth model, the droplets, steam and air have the same temperature: $T_v = T_l = T$.

3 Numerical study

The geometry of the circular nozzle presented below (Fig. 1) was applied to investigate moist air condensing flows using different droplet growth models. This nozzle was previously used in our numerical as well as experimental research [13,15]. The 2D mesh presented on Figure 1 was created and it has ~21 000 elements, the boundary layer with growth rate equal to 1.2 was introduced to obtain $y^+ = 1$. As a convergence criterion of the iteration process, it was assumed that the mass flow rate difference between inlet and outlet was less than 0.01%.

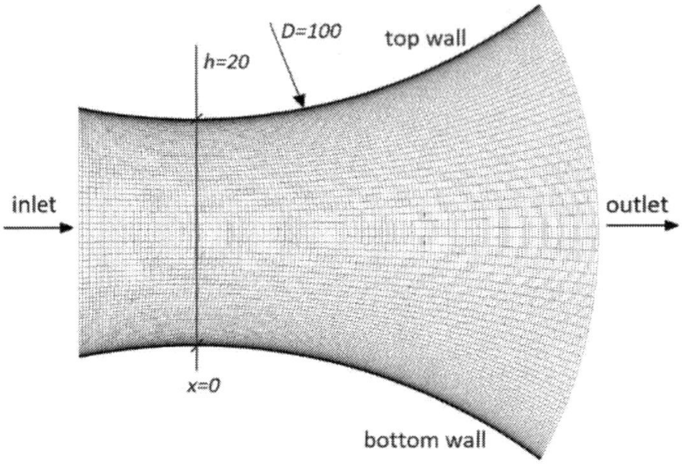

Fig. 1. Mesh of the applied circular nozzle

Bearing in mind that the homogeneous condensation process consists of nucleation and droplet growth phenomena, four different droplet growth models will be used below keeping the same nucleation model. The computational results will be compared with the experimentally obtained static pressure distribution along the central axis of the nozzle. The static pressure distribution reveals the location and intensity of the condensation process. The aim is to check which droplet growth model captures the experimental results in the best way.

Table 1 gathers the boundary conditions for the two cases corresponding to the experimental measuring campaigns [13] .

Table 1. Boundary conditions

	Case 1	Case 2
Relative humidity, %	43	68
Inlet total pressure, kPa	98.8	98.8
Inlet total temperature, K	297.15	297.15
Outlet	Supersonic	Supersonic

Generally, two types of the droplet growth model are considered: continuous and kinetic. The Knudsen number decides which model should be used (cf. eq. 14). Therefore, the steam molecule free mean path l_s should be determined first to find the size of the droplets that should be expected choosing one of the two types of the droplet growth model under consideration (Fig.2).

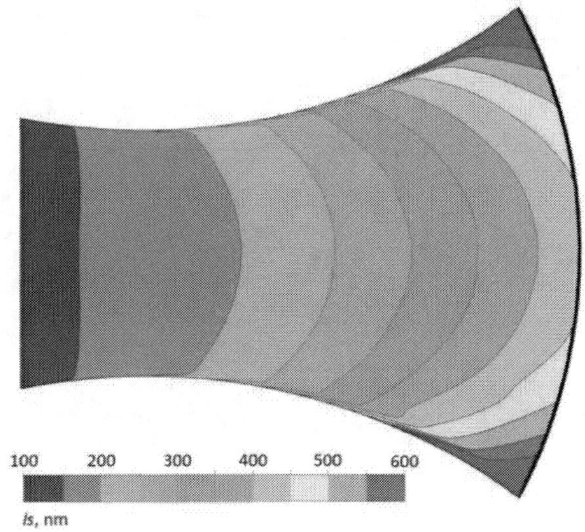

Fig. 2. Distribution of the free mean path of steam molecules

Analysing the values of the free mean path of the steam molecule (Fig. 2), it can be noticed that for the continuous droplet growth model the expected size of the droplet is bigger than $1 \cdot 10^{-5}$ m, which is much bigger than the size of droplets which are usually formed due to the nucleation process. In the case of moist air condensing flows, the so-called coarse droplets arise only when heterogeneous condensation takes place on the water droplets or solid particles already existing in the flow.

3.1 Comparison of droplet growth models

Due to the fact that in transonic condensing flows the range of the Knudsen number values can be wide, the use of the kinetic Hetrz-Knudsen (HK) droplet growth model seems to be a simplification. However, the majority of continuous droplet growth models already include a correction with respect to the Knudsen number, such as the Gyarmathy (GY), the Fuchs-Sutugin (FS), or the Young (YO) correction. Nevertheless, these corrections are rarely used in moist air condensing flow modelling. Fig. 3 presents a comparison of the moist air condensing flow modelling in a circular nozzle by means of four different droplet growth models. The simulation results were compared with the experimentally obtained static pressure distribution along the nozzle midline. The calculations results obtained from simulations using the Hertz-Knudsen (HK) model, with condensation coefficient α_c=1, and the continuous model with the Fuchs-Sutugin (FS) correction are similar and close to the experimental data. On the other hand, the continuous models with the Gyarmaty (FS) and the Young (YO) correction differ from the experiment significantly for both cases. For Case 1, with the lower relative humidity value, the HK and the FS droplet growth models are very close to each other, even in terms of the droplet size.

Using the continuous droplet growth model, information is obtained about the temperature difference between the droplet and steam $\Delta T = T_\mathrm{l} - T_\mathrm{v}$. In the case of the kinetic model it is assumed that this difference is zero. Fig. 4 presents the distribution of the rise in the droplet temperature due to condensation. The highest value of this temperature difference is observed for the model with the Fuchs-Sutugin correction.

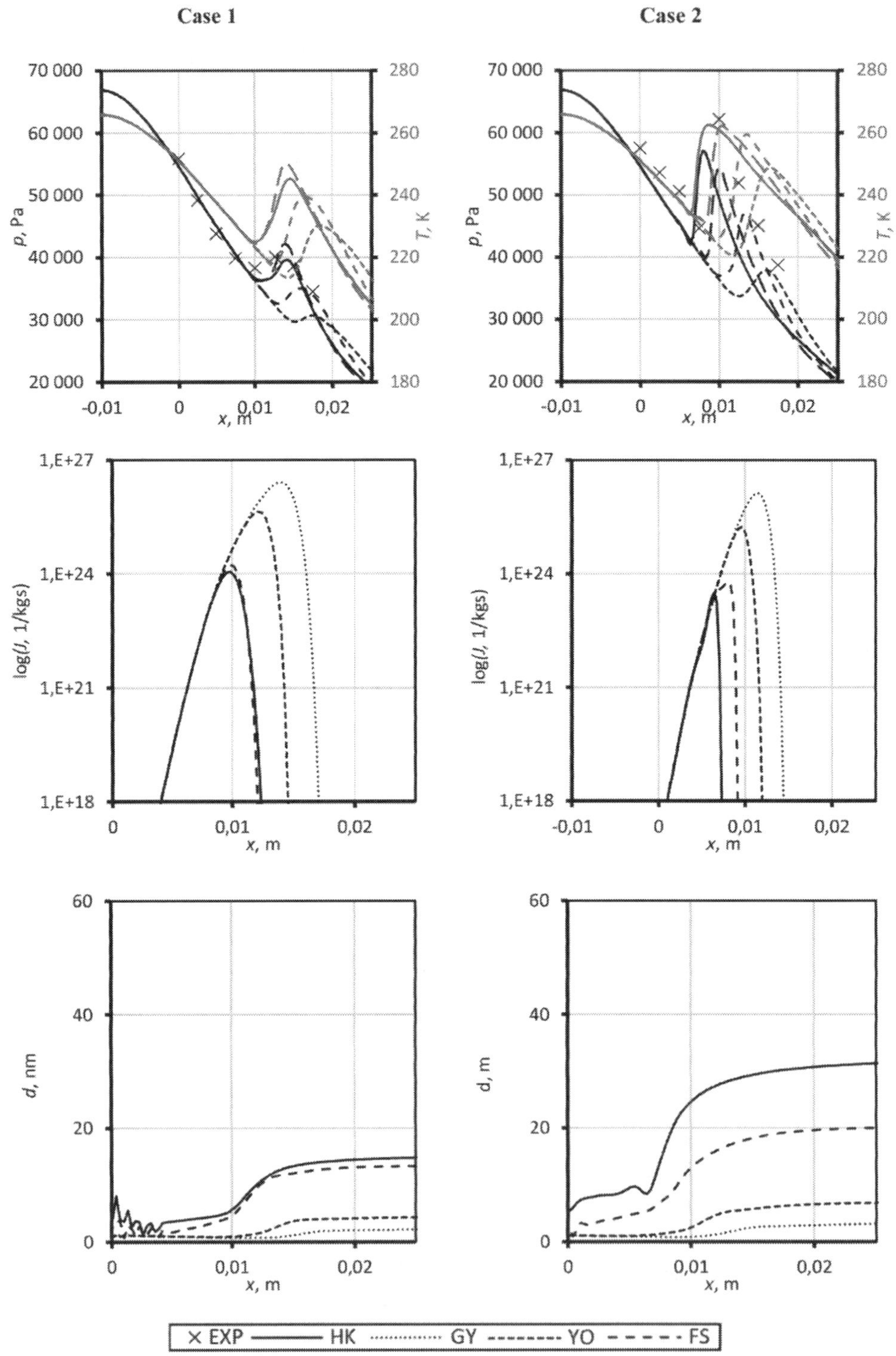

Fig. 3. Comparison of numerical results with different droplet growth models: static pressure and static temperature; nucleation rate and droplets size distributions along the nozzle axis

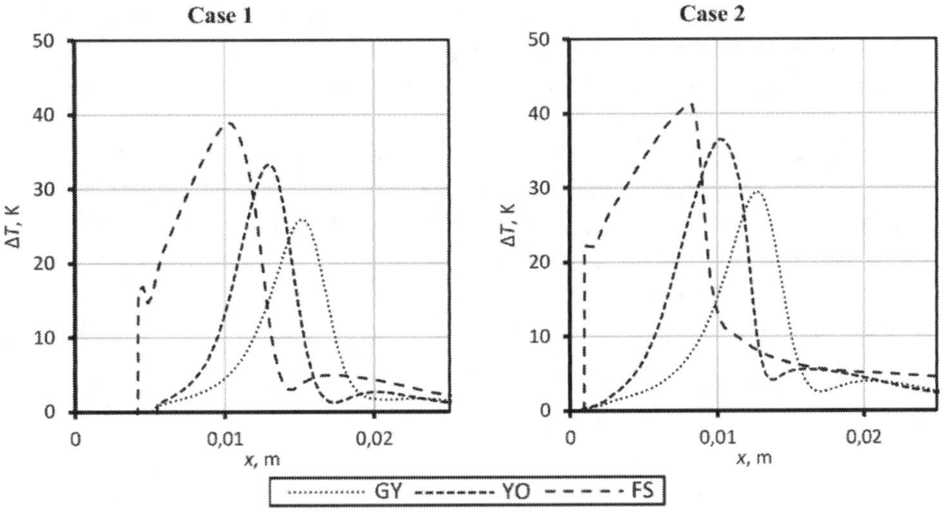

Fig. 4. Comparison of temperature difference $\Delta T = T_l - T_v$; distributions along the nozzle midline calculated using different continuous droplet growth models

4 Conclusions

The present numerical study investigates the suitability of different droplet growth models for the moist air condensing flow. The obtained results lead to the following conclusions:

- For moist air condensing flows where only homogeneous condensation is considered, the Hertz-Knudsen droplet growth model is recommended.
- The Fuchs-Sutugin correction for the continuous droplet growth model gives results close to the experiment, providing additional information on the temperature difference between the droplet and steam.
- Due to the wide range of the Knudsen number values, a hybrid droplet growth model is recommended, especially when heterogeneous condensation is considered as well.

Summing up, the paper shows that each selection of the condensation model should not be arbitrary and it requires a validation process first, for the experimental data for moist air flow, not steam flow. The study presented herein should make clear that the choice of droplet growth model depends on the flow conditions, what seems to be trivial, but in the case of heterogeneous condensation, it may be crucial.

Our future research in this field will concentrate on the application of the hybrid droplet growth model for the condensing flow modelling in more complex geometries.

This research was co-funded by the Silesian University of Technology within Initiative of Excellence - Research University (grant 08/050/SDU20/10-2703) and by the European Union through the European Social Fund (grant POWR.03.05.00-00-Z305).

References

1. Rhee J, Im J, Kim J, Song SJ. Humidity effects on the aerodynamic performance of a transonic compressor cascade. International Journal of Heat and Mass Transfer. (2019): p. 743-751.
2. Gyarmathy G. Grundlagen einer Theorie der Nassdamfturbine Juris-Verlag; (1962).

3. Young JB. Spontaneous Condensation of Steam in Supersonic Nozzles: N81-13307. , Whittle Laboratory, University of Cambridge; (1980).
4. Young JB. Two-dimensional, nonequilibrium, wet-steam calculations for nozzles and turbine cascades. Journal of Turbomachinery. (1992): p. 569-579.
5. Fuchs N, Sutugin A. Topics in Current Aerosol Research New York; (1971).
6. Schneer GH, Mundinger G. Similarity, drag, and lift in transonic flow with given internal heat addition. European Journal of Mechanics - B/Fluids. (1993): p. 597-611.
7. Schneer GH, Dohrmann U. Drag and lift in non-adiabatic transonic flow. AIAA Journal. (1994): p. 101-107.
8. Schneer GH, Dohrmann U. Transonic flow around airfoils with relaxation and energy supply by homogeneous condensation. AIAA Journal. (1990): p. 1187-1193.
9. Dykas S. Badania przepływów transonicznych z kondensacją pary wodnej Gliwice: Wydawnictwo Politechniki Śląskiej; (2006).
10. Adam S. In Adam S. Numerische und experimentelle Untersuchung instationarer Dusenstromungen mit Energiezufuhr durch homogene Kondensation. Karlsruhe; (1996).
11. Goodheart KA, Dykas S, Schneer GH. Numerical modelling of heterogeneous/homogeneous condensation on the ONERA M6 wing. In Proceedings of the 12th International Conference on Fluid Flow Technologies; (2003); Budapest. p. 335-342.
12. Dykas S, Majkut M, Smołka K, Strozik M. Comprehensive investigations into thermal and flow phenomena occurring in the atmospheric air two-phase flow through nozzles. International Journal of Heat and Mass Transfer. (2017) November: p. 1072-1085.
13. Dykas S, Majkut M, Smołka K. Influence of Air Humidity on Transonic Flows with Weak Shock Waves. Journal of Thermal Science. (2019).
14. Wiśniewski P, Dykas S, Yamamoto S. Importance of Air Humidity and Contaminations in the Internal and External Transonic Flows. Energies. (2020) June.
15. Wiśniewski P, Dykas S, Yamamoto S, Pritz B. Numerical approaches for moist air condensing flows modelling in the. International Journal of Heat and Mass Transfer. (2020);(162).
16. Yamamoto S, Hagari H, Murayama M. Numerical simulation of condensation around the 3-D wing. Transactions of the Japan Society for Aeronautical and Space Sciences. (2000): p. 182-189.
17. Yamamoto S. Computation of practical flow problems with release of latent heat. Energy. (2005): p. 197-208.
18. Moriguchi S, Endo T, Miyazawa H, Furusawa T, Yamamoto S. Numerical Simulation of Unsteady Moist-air Flows through Whole-annulus Rotor Blade Rows in Transonic Compressor. In Proceedings of the ASME-JSME-KSME 2019 8th Joint Fluids Engineering Conference AJKFluids2019 ; (2019); San Francisco.
19. Zhang G, Zhang X, Wang F, Dingbiao W, Zunlong J. The relationship between the nucleation process and boundary conditions on non-equilibrium condensing flow based on the modified model. International Journal of Multiphase Flow. (2019) May: p. 180-191.
20. Karabelas SJ, Markatos NC. Water vapor condensation in forced convection flow over an airfoil. Aerospace Science and Technology. (2008): p. 150-158.

21. Edathol J, Brezgin D, Aronson K, Dong Kim H. Prediction of non-equilibrium homogeneous condensation in supersonic nozzle flows using Eulerian-Eulerian models. International Journal of Heat and Mass Transfer. (2020).
22. Zhang G, Zhang X, Wang F, Wang D, Jin Z, Zhou Z. Design and optimization of novel dehumidification strategies based on modified nucleation model in three-dimensional cascade. Energy. (2019) November.
23. Zhang G, Zhang X, Wang F, Wang D, Jin Z, Zhou Z. Numerical investigation of novel dehumidification strategies in nuclear plant steam turbine based on the modified nucleation model. International Journal of Multiphase Flow. (2019) November.
24. Zhang G, Wang F, Wang D, Wu T, Qin X, Jin Z. Numerical study of the dehumidification structure optimization based on the modified model. Energy Conversion and Management. (2019) February: p. 159-177.
25. Zhang G, Dykas S. Optimization of the primary nozzle based on a modified condensation model in a steam ejector. Applied Thermal Engineering. (2020).
26. Zhang G, Dykas S, Majkut M, Smołka K, Cai X. Experimental and numerical research on the effect of the inlet steam superheat degree on the spontaneous condensation in the IWSEP nozzle. International Journal of Heat and Mass Transfer. (2021).
27. Zhang G, Dykas S, Li P, Li H, Wang J. Accurate condensing steam flow modeling in the ejector of the solar-driven refrigeration system. Energy. (2020).
28. Sun W, Cao X, Yang W, Jin X. Numerical simulation of CO2 condensation process from CH4-CO2 binary gas mixture in supersonic nozzles. Separation and Purification Technology. (2017): p. 238-249.
29. ANSYS. [Online]. [cited 2020 01 20. Available from: https://www.ansys.com/.
30. Menter FR. Two-equation Eddy-Viscosity Turbulence Models for Engineering Applications. AIAA Journal. (1994): p. 1598-1605.
31. ANSYS Fluent Theory Guide, 2021R1. [Online]. [cited 2021 July 1.
32. Woś A. Meteorologia dla geografów [Meteorology for geographers] Warszawa: Wydawnictwo Naukowe PWN; (2002)
33. . Kantrowitz A. Nucleation in very rapid vapor expansions. J. Chem. Phys.`. (1951): p. 1097-1100.
34. Kermani MJ, Gerber AG. A general formula for the evaluation of thermodynamic and aerodynamic losses in nucleating steam flow. International Journal of Heat and Mass Transfer. (2003).
35. Wróblewski W, Dykas S, Gardzilewicz A, Kolvratnik M. Numerical and Experimental Investigations of Steam Condensation in LP Part of a Large Power Turbine. Journal of Fluids Engineering. (2009).

The structure of turbulent flow behind the NACA 0012 airfoil at high angles of attack and low Reynolds number

Vitalii Yanovych[1, 2*], *Daniel* Duda[1], *Václav* Uruba[1, 2], and *Pavlo* Kosiak[1]

[1]Department of Power System Engineering, Faculty of Mechanical Engineering, University of West Bohemia in Pilsen, Univerzitní 22, 306 14 Pilsen, Czech Republic.
[2]Department of Fluid Dynamics, Institute of Thermomechanics, Czech Academy of Sciences, Dolejškova 5, 182 00, Prague, Czech Republic.

Abstract. This paper shows the results of a study of the turbulent structure behind the NACA 0012 airfoil. During the experiment, the flow velocity was 20 m·s^{-1}. That corresponds the Reynolds number $1.3\cdot10^5$. The point behind the trailing edge was chosen for experimental studies. Measurements were performed at six angles of attack $\alpha = 0°$, 15°, 30°, 45°, 60°, 75° and various cases of positioning the measuring section. Namely at a constant cross-section and a constant distance behind the airfoil. The NetScanner pressure system and hot-wire technique were used for experimental studies. The obtained data allowed us to investigate the wake topology. The average velocity and standard deviation distributions are clearly grouped depending on the angle of attack. Thus, flowing around the airfoil is better up to $\alpha \leq 15°$. Distributions by power density and dissipation spectra also have a similar grouping tendency. Finally, we investigated the turbulent structure according to the research program. We found that at $\alpha \geq 45°$, depending on the measurement case, there is a clear difference in the distribution of standard deviation, Eulerian length scale, Taylor micro-scale, and Reynolds number based on the Taylor micro-scale. The obtained values at the constant distance, in contrast to the constant cross-section, are reduced.

1 Introduction

The study of the flow structure around an airfoil is a common task. Because even the smallest changes in their geometry play a key role in the overall evolution of the streamlined flow. In recent years, NACA family profiles have become widespread. They have a great variety and are actively used for wind energy, aviation, and energy systems.

It should be noted that the NACA 0012 profile deserves special attention among many profiles. This is because it has a relatively simple shape geometry and high aerodynamic characteristics. Which makes it a benchmark not only for practical uses but also for a variety of experimental and analytical studies. That is why many papers are devoted to the study of the flow structure around this aerodynamic profile. For example, the distribution of pressure

* Corresponding author: yanovych@kke.zcu.cz

profile and wake profile behind airfoil is very well described in [1]. Based on the obtained experimental data, the authors found the coefficients of lift, drag, and torque twist. In addition, they found that the use of a gurney flap can increase the lift rate by 25%. Similar studies are also shown in the papers [2-4]. Where the authors focused on the study of aerodynamic coefficients and features of the wake topology behind airfoil. The investigations were performed at angles of attack up to 40°. While the paper [5] shown the results of the study of the flow evolution around the airfoil up to 90°. But with a relatively high degree of turbulence intensity of the overall flow.

Frequently NACA 0012 airfoil is used as a basis for analyzing the shape effect of the trailing edge on the structure of the turbulent flow. For example, [6] shows a study of the effect of vortex instability on the aerodynamic characteristics of the airfoil with multiscale modifications of its trailing edge. The authors concluded that large multi-scale trailing edges at some distance from each other break down the vortex shedding mechanism. That improves the aerodynamic characteristics of the airfoil and reduces the amount of noise generated. It should be noted that the latter is a key issue in the development of modern wind turbines. This phenomenon is well described in the paper [7]. Where the authors conducted a comparative analysis of experimental studies of noise generated by stationary and rotating blades of the wind turbine rotor.

Nowadays, computational fluid dynamics (CFD) is used in many cases to better understanding flow dynamics. The NACA 0012 airfoil is usually used to compare the aerodynamics of the streamlined body or to test the flow simulation algorithm around them. For example, the authors [8] based on numerical simulations showed a comparative assessment of aerodynamic characteristics between S809 and NACA 0012. Whereas the paper [9] presents the results of the numerical investigation of the separation mechanisms and prediction of the turbulence transition in a separate shear layer behind symmetric airfoil.

Generally, much of the papers about NACA 0012 airfoil show the study of aerodynamic coefficients and flow topology at angles of attack that do not exceed critical values. However, studying the structure of the flow around the airfoil at a high angle of attack is an important task also. That is especially true for aviation. For example, at high and low altitudes, where even small maneuvers may require high angles of attack. In the first case due to the low density of air in the upper atmosphere. In the second at relatively low flight speed. In addition, when landing, all aircraft use flaps at a high angle of attack. Thus, for military aircraft, a high angle of attack can significantly increase their maneuverability. Accordingly, the F-18 HARV, Su-27M, or Su-35 aircraft can use angles of 70°, 120°, and 180°, respectively, for their maneuvers.

Many researchers are using hot-wire to study the structure of turbulence around an airfoil. Basically, they perform measurements at some cross-sections behind a bluff body. Even at different angles of attack, the distance remains the same. However, as the angle of attack increases, the distance between the trailing edge and the measuring section growing also. As a result, the structure of the turbulent flow, on the way to the measuring point evolve more. The main idea of this paper is to investigate this evolve behind the symmetric profile of NACA 0012.

2 Experimental facilities

The experiment was conducted in a low-speed open-type wind tunnel at the Department of Energy Systems Engineering, West Bohemian University [10, 11]. The internal dimensions of the width, height, length of the test section were 0.3 m, 0.2 m and 0.75 m, respectively.

To study the structure of turbulent flow behind an asymmetrical airfoil, the profile NACA 0012 was chosen. It is a typical outboard airfoil for utility-scale wind turbines. The original dimensions of the airfoil were the chord length c=100 mm, the span $L{\cdot}c^{-1} \approx 1$ and the thickness $b{\cdot}c^{-1} \approx 0.12$. At these parameters, the blockage ratio, relevant to the test section, was a proximally 2% at an attack angle α=0°. The airfoil was manufactured of plastic using 3D printing technology. We used a PRUSA i3 printer and PLA thermoplastic filament materials for this purpose. After that by P2500 sandpaper, we reduced the surface roughness of airfoil to 5 μm.

To change the angle of attack, the airfoil was attached to the shaft of the stepper motor. Which was placed on the cover of the test section. Thus, the aerodynamic center of the airfoil was at $\Delta z{\cdot}c^{-1}{\approx}1.53$ and $\Delta x{\cdot}c^{-1} \approx 2.23$. Where Δz and Δx are the distance to the wall and the inlet of the test section, respectively. The scheme of the experimental investigation is shown in Fig. 1. Measurements were performed at six positions of angles from α=0° to α=75° in steps of 15°. Accordingly, at α = 0° the airfoil chord has the same direction as the general flow. The Reynolds number was based on the length of the airfoil chord and was equal to $1.3{\cdot}10^5$. That corresponds to the flow velocity of 20 $m{\cdot}c^{-1}$.

The 9116 series NetScanner pressure system and the five-hole Pitot tube were used to estimate the wake topology depends on the angle of attack. The measurements took place in the plane behind the airfoil at $l_1{\cdot}c^{-1} \approx 0.28$ at step $w_z{\cdot}c^{-1} \approx 0.05$.

The 55P15 hot-wire probe was used to study the *u*-component of turbulent fluctuation behind airfoil. The position of the measuring point according to the experimental procedure was changing by the traverse system. The hot-wire signal was sampled at 75 kHz. While the measurement time at one point was 10 s. We also used filtering for the received signal: for the High Pass - 10 Hz and the Low Pass - 30 kHz.

Of particular interest in these studies was the structure of turbulence directly behind the trailing edge. That is why in the range $r{\cdot}c^{-1} \approx 0.28$ the distance between the measuring points was smaller $k_z{\cdot}c^{-1} \approx 0.02$ than outside $k_z{\cdot}c^{-1} {\approx}0.05$. In the general case, the measuring cross-section from $z{\cdot}c^{-1}{\approx}$-1.2 to 0.8 was located at a height $y{\cdot}c^{-1}{\approx}1.8$. The latter value also corresponds to the center of span airfoil. While the location of the cross-section was different. In the first case, the position remains unchanged relative to the coordinates of the test section. Thus, at a growing angle of attack, the distance from the measuring cross-section to the airfoil increased. Whereas in the second case, it was contrary. The distance to the trailing edge always stayed the same $l_2{\cdot}c^{-1}{\approx}0.15$.

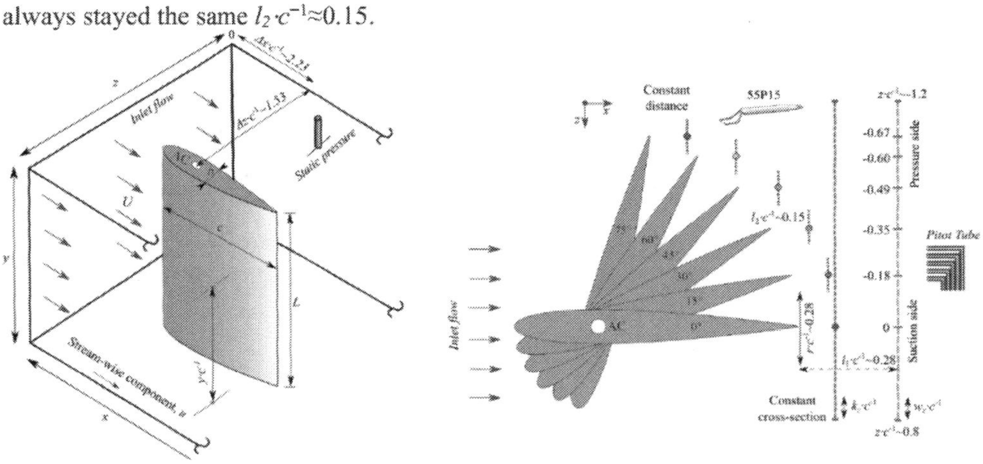

Fig. 1. Scheme of the experimental investigation. The left shows the positioning of the airfoil in the test section. The arrows in blue indicate the direction of airflow U. Where $z{\cdot}c^{-1}$ and $x{\cdot}c^{-1}$ respectively are the width and length of the measuring fields. The $k_z{\cdot}c^{-1}$, $w_z{\cdot}c^{-1}$, $y{\cdot}c^{-1}$ and $r{\cdot}c^{-1}$ are the steps between measuring points for hot-wire and NetScanner pressure system, the height of the location and area of the trailing edger, respectively. Whereas $\Delta z{\cdot}c^{-1}$ and $\Delta x{\cdot}c^{-1}$ are the coordinates of the aerodynamic center (AC). Blue and red dashed lines are a set of measuring cross section for hot-wire and pressure system, respectively.

3 Results and discussion

3.1 Wake topology

A series of experimental studies were performed using the 9116 series NetScanner system. As a result, we obtained the surfaces of the flow velocity distribution behind the airfoil at a different angle of attack (Fig. 2). This made it possible to evaluate the general characteristic of the wake topology. It was found that with increasing the angle of attack from 0° to 75°, the width wake increases by 8.5 times. While the wake depth begins to increase sharply after α=15° and has the maximum value $U{\cdot}u^{-1}\approx$-0.57 at α=45°. The physically negative value of this parameter means the presence of backflow. Also, it should be noted that exactly at α=15° we can clearly observe the upper and downer wingtip vortices. Further increase in the angle of attack leads to the disappearance of this feature. This is due to the increase in the area of the velocity deficit behind the airfoil. At the same time, outside this area, the flow velocity also increases. Thus, at α=75° the normalized velocity is equal to $U{\cdot}u^{-1}\approx$1.4. This pattern is related to the blockage ratio of the airfoil at different α.

To further study the structure of the turbulent flow by hot-wire, we chose the point M. Which was placed exactly behind trailing edge. As noted above, the experiments were performed for two cases: at constant cross-section and constant distance.

The experimental results are presented as a series of graphs. They show the distribution of the stream-wise velocity u and standard deviation σ_u normalized by average flow velocity U depending on the angle of the attack (Fig. 3 a, b, d, e). For convenience, the resulting curves are located relative to the position of the trailing edge at α=0°.

The obtained distributions for the two measurement cases at different angles of attack have similar behavior. The obtained curves clearly show the developed of the instability region behind the airfoil. Moreover, relative to the trailing edge, this region is expanding in direction to the suction side. While at the boundaries of the wake the values of $u{\cdot}U^{-1}$ and $\sigma_u{\cdot}U^{-1}$ is growing. But unlike $u{\cdot}U^{-1}$ the distribution of $\sigma_u{\cdot}U^{-1}$ has two local peaks. The largest of them is shifted to the pressure side. It is because at positive angle α, the vortices which passing from the suction side interact with the vortices from the pressure side. This leads to the emergence of a dominant vortex in the general flow. Whereas at α=0°, as a result of delaying boundary layer separation, this feature disappears. In general, sharp changes in the wake topology are observed after $\alpha\geq$30°. While at α=0° and α=15° the width and depth of the wake remain the smallest.

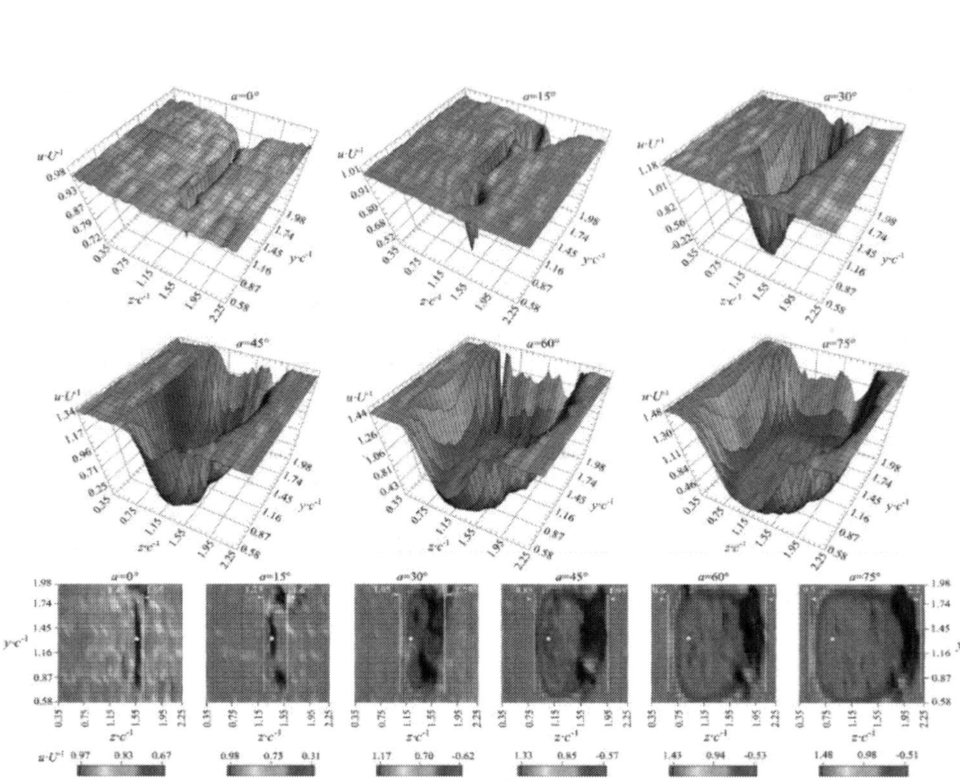

Fig. 2. The distributions of normalized velocity $u \cdot U^{-1}$ depending on the angle α at $l_1 \cdot c^{-1} \approx 0.28$ and Re $\approx 1.3 \cdot 10^5$. The upper part shows the flow velocity distribution surface. Whereas below there are its projections on the coordinates x and y. The white lines outline the velocity deficit area. The point M is the position of the measuring behind the trailing edge.

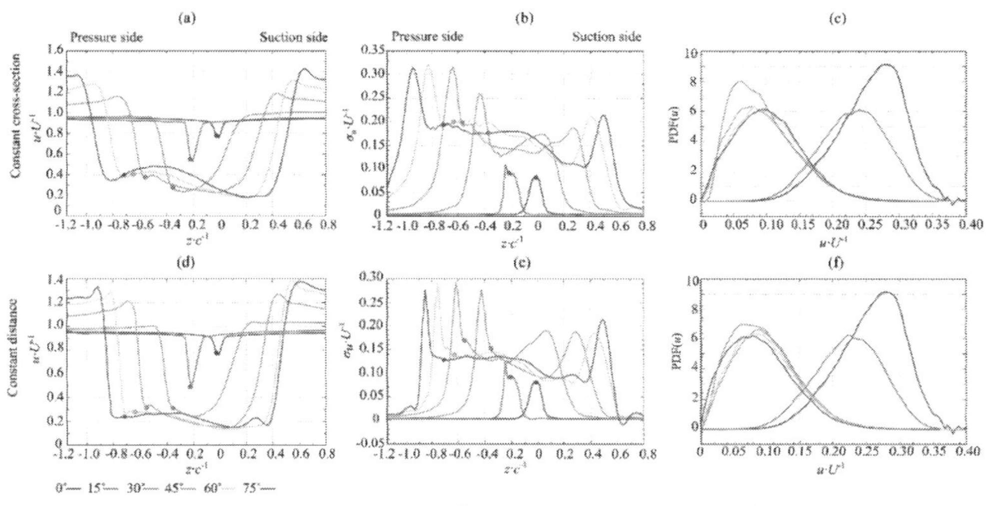

Fig. 3. (a, b, d, e) The normalized velocity $u \cdot U^{-1}$ and standard deviation $\sigma_u \cdot U^{-1}$ distribution depending on angle of attack at $l_1 \cdot c^{-1} \approx 0.15$ and Re $\approx 1.3 \cdot 10^5$. (c, f) The right two graphs represent probability density functions of the velocity fluctuations PDF(u) at the measuring point M. The upper (a, b, c) and lower (d, e, f) graphs show the flow characteristics measured at constant cross-section and constant distance, respectively. Colored lines and dots respectively reflect the distribution of flow characteristics and the position of the measuring point depending on variation α.

The probability density function (PDF(u)) was used to describe the stream-wise fluctuation at the measurement points M. As generally expected for turbulence, we found that fluctuations are well described by a Gaussian distribution. In addition, it is clearly observed that for both measuring cases the average velocity at $\alpha \leq 15°$ is approximately 5 times higher than at $\alpha \geq 30°$.

Thus, we can conclude that the flowing around airfoil is better up to $\alpha \leq 15°$. Further increase in the angle of attack leads to the development of an instability region there produces active perturbations. This phenomenon has a significant impact on the location of the point where boundary layer separation begins. Many publications are devoted to this issue [12-14]. The researchers note the significant impact of flow separation on the lift and drag characteristics of the airfoil. That is why each airfoil has an optimal angle of attack. For symmetrical airfoil in most cases, it is up to 15°. Where the ratio of lift and drag coefficients will be the highest.

3.2 The power and dissipation spectra

Power spectral density was used to determine the distribution of flux energy intensity between different scale components at measuring point M. Graphical interpretation of normalized Eulerian power spectral densities of velocity fluctuations depending on the angle α at different cases of measuring are shown in Fig. 4 (a, b). For representing the Kolmogorov dimensionless coordinates were used. The obtained spectrum was normalized by $(\epsilon\nu^5)^{1/4}$ and the wave number $k \cdot \eta$. Where η, ϵ and ν is dissipation scale, dissipation rate and kinematic viscosity of air $1.5 \cdot 10^{-5}$ [m^2c^{-1}], respectively. The different flow velocity at the measurement point leads to some shift of the obtained curves relative to the wavenumber.

As is known, according to Kolmogorov's law, the spectral density of turbulence energy decreases with increasing wave number. As a result, the resulting curve has a slope $f = -5/3$ [15, 16]. This pattern can be clearly observed for the point *M* regardless of the angle of attack. It should be noted that the behavior of the obtained curves for the two measuring cases is almost the same. The main difference is observed in the inertial subrange at $\alpha = 30°$. Physically, it characterizes the range where turbulence kinetic energy is transferred from larger to smaller scales without loss. This is due to the distance between the measuring point and the trailing edge as the angle of attack increases.

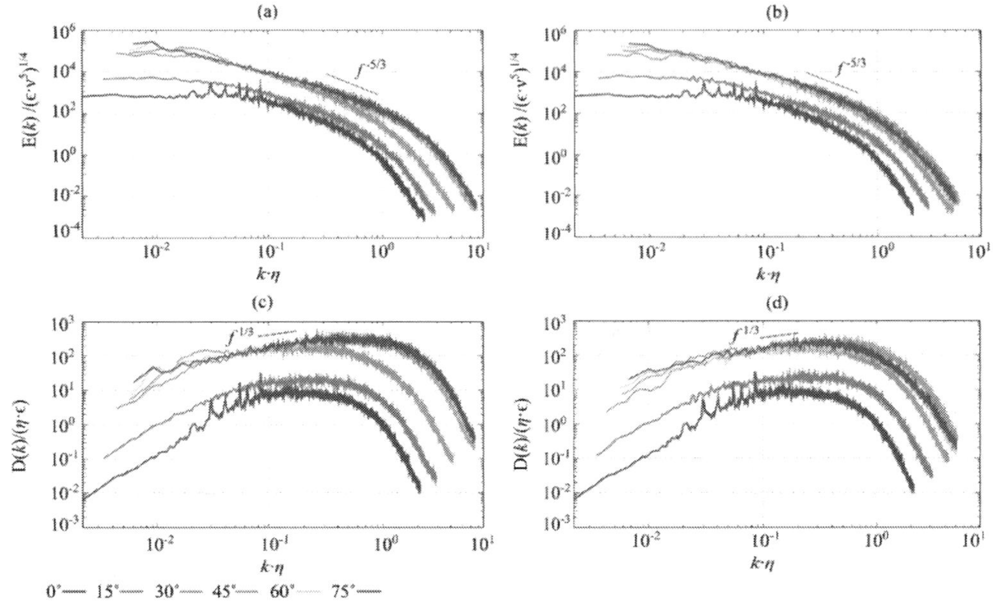

Fig. 4. Normalized Eulerian power *E(k)* and dissipation *D(k)* spectral of stream-wise fluctuations at point M depending on the angle of attack and case of measurement. The right (a, c) and left (b, d) graphs represent the flow characteristics at constant cross-section and at constant distance, respectively. The experimental Reynolds numbers was $1.3 \cdot 10^5$. Colored lines reflect the value of the angle α.

In both cases of measurement, the received curves can be divided into two groups. Thus, before α=15° and after, there is a clear difference in the energy characteristics of the flow fluctuations. Moreover, for high angles of attack at $\alpha \geq 30°$, it is almost the same.

In addition, the dissipation spectrum for the measuring points M was found (Fig. 4 c, d). The dissipation phenomenon occurs at high wavenumbers where the smallest vortices dissipate turbulent energy due to viscosity. The dissipation spectra can be estimated as:

$$D(k) = 2\nu k^2 E(k), \tag{1}$$

where *E(k)* is Eulerian power spectral densities [m^2c^{-1}]. While the $k = f \cdot \bar{u}^{-1}$ is spatial wavenumber, where $\bar{u}$ is the average flow velocity at the measuring location.

For non-dimensional coordinates, we used normalization in the form $D\,(k)/\eta \cdot \varepsilon$ and for wave number $k \cdot \eta$. The measuring point *M* has a classical form of energy dissipation distribution. The obtained curves in the dissipation interval have a slope *f = 1/3*. As in the energy spectrum, depending on the angle of attack, they have similar distribution groups. Moreover, the maximum value of dissipation is observed at $\alpha \geq 30°$. This pattern clearly correlates with the obtained distributions of the flow velocity behind the airfoil (see Fig. 3 a, d).

3.3 The structure of the turbulent flow at the point

To estimate the structure of turbulent flow (stream-wise component) depending on positioning of the measuring cross-section and angle of attack, we chose the point *M*. For this purpose, we used a methodology that is well described in the paper [17]. The authors used correlation time analysis to evaluations the mean temporal and spatial scales of turbulent flows. Where T_E and L_E characterize the average duration and lifetime of vortices, respectively. While the second-order Eulerian longitudinal structure-function was used to estimations the dissipation rate ϵ.

Using the obtained values of ϵ, the dissipation scale η and the dissipation time τ_η are estimated from classical relations:

$$\eta = (\nu^3/\epsilon)^{1/4} \quad (2)$$

$$\tau_\eta = (\nu/\epsilon)^{1/2} \quad (3)$$

Also we found the Reynolds number based on the Taylor micro-scale:

$$R_\lambda = (\sigma_u \lambda)/\nu \quad (4)$$

where $\lambda = \sigma_u\sqrt{15\,\nu/\epsilon}$ is the Taylor micro-scale of the flow. While $\sigma_u = \sqrt{N^{-1}\sum_{i=0}^{N-1}(u_i - \bar{u})^2}$ is standard deviation of the streamwise velocity. Accordingly N and u_i is the total number of points in the data set and instantaneous velocity at the measuring location, respectively.

The calculated parameters of the turbulent flow at the point M depending on the angle α shown in Fig. 5.

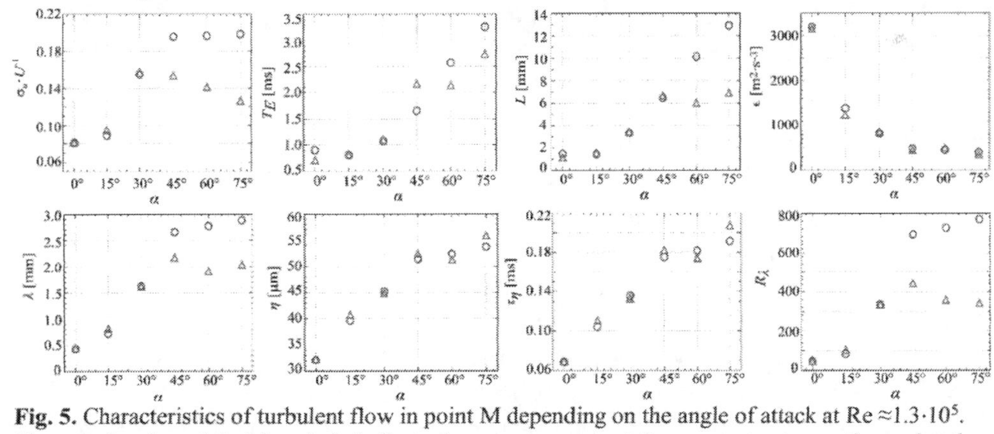

Fig. 5. Characteristics of turbulent flow in point M depending on the angle of attack at Re ≈1.3·10^5. The red triangle and the blue circle indicate the measurements at the constant cross-section and at the constant distance, respectively.

The obtained distributions of T_E, ϵ, η, and τ_η, regardless of the measuring case, have similar dynamics. Accordingly, their maximum and minimum value at α=75° placed around $\epsilon \approx$ 450 m^2·s^{-3}, η≈55 μm and $\tau_\eta \approx$ 0.2 ms. Whereas at $\alpha \geq$45°, we can see some differences in the distributions of $\sigma_u \cdot U^{-1}$, L_E, λ, and R_λ. The obtained values at the constant distance, in contrast to the constant cross-section, are reduced. For example, in the first and second measuring cases, the value of Taylor microscale Reynolds number at α=75° is $R_\lambda \approx$ 800 and $R_\lambda \approx$ 320, respectively. A similar trend is also observed for σ_u·U^{-1}.

It should be noted that the maximum value of the lift coefficient at low Reynolds number for NACA 0012 is observed at α≈45° [18]. Thus, it is possible to make assumptions about some relationship between the structure of the turbulent flow and the lift coefficient.

4 Conclusion

The wake topology and structure of the turbulent flow behind NACA 0012 airfoil was investigated. Accordingly, the 9116 series NetScanner system and hot-wire anemometry with probe 55P15 were used for this purpose. The experimental studies are conducted at low Reynolds numbers $Re \approx 1.3 \cdot 10^5$ and angles of attack up to 75° with a step of 15°. Two positions of measuring sections were used in the study. In the first case, with a constant cross-section, the position of which does not change at variations of the angle of attack. Whereas in the second, the distance from the trailing edge to the measuring cross-section was constant. Based on the obtained experimental data, several graphical distributions were obtained to the flow velocity and standard deviation. This allowed us to clearly evaluate the wake topology behind airfoil. We found that regardless of the measurement case, the obtained curves are grouped depending on the angle of attack. Thus, we can conclude that the flowing around the airfoil is better up to $\alpha \leq 15°$. A similar trend is observed in the case of analysis of power and dissipation spectra.

To assess the structure of the turbulent flow, the measuring point behind the trailing edge was chosen. Depending on the measuring cases the obtained data have some differences in the distribution of integral length scale, Taylor micro-scale and Reynolds number based on the Taylor micro-scale. The obtained values at $\alpha \geq 45°$ and the constant distance, in contrast to the constant cross-section, are reduced.

In general, the analysis showed a certain relationship between the parameters of turbulent flow and the aerodynamic characteristics of the airfoil. The obtained data clearly indicate the position of the optimal angle of attack $\alpha \approx 15°$ for the NACA 0012 airfoil.

This work was supported by the project of Technology Agency of the Czech Republic TACR No. TH02020057 "Program Epsilon" and the Internal Doctoral Educational Grant programme (IDEG).

References

1. Y. Li, J. Wang and P. Zhang, Effects of Gurney Flaps on a NACA0012, *Airfoil. Flow, Turbulence and Combustion* **68**, *27–39* (2002).
2. A. Masayuki, T. Nonomura, H. Aono, A. Oyama, K. Fujii, H. Nagai and K. Asai, Computational and experimental analysis of a high-performance airfoil under low-Reynolds-number flow condition, *Journal of Aircraft* **51**, 1864–1872 (2014).
3. F. S. Hover, Ø. Haugsdal, M. S and M. Triantafyllou, Effect of angle of attack profiles in flapping foil propulsion, Journal of Fluids and Structures 19: 1, 37–47 (2004).
4. M. Banafi and M. Banafi, An Experimental research on wake behind airfoil NACA 0012 under various angles, International Journal of Theoretical and Applied Mechanics 2, 164–170 (2017).
5. Ph. Devinant, T. Laverne, J. Hureau, Experimental study of wind-turbine airfoil aerodynamics in high turbulence, Journal of Wind Engineering and Industrial Aerodynamics 90, 689–707 (2000).
6. J. Nedić and J. C. Vassilicos, Vortex Shedding and Aerodynamic Performance of Airfoil with Multiscale Trailing-Edge Modifications, AIAA Journal 53:11, 3240–3250 (2015).
7. B. Zajamšek, C. Doolan and Z. Prime, "Experimental measurements of rotating and stationary wind turbine rotor blade noiseh", 19th Australasian Fluid Mechanics Conference, AFMC (2014).

8. E. Douvi and D. Margaris, "Aerodynamic characteristics of s809 vs. NACA 0012 airfoil for wind turbine applications", 5th International Conference from Scientific Computing to Computational Engineering, (2012)
9. I. Rodríguez, O. Lehmkuhl, R. Borrell, A. Oliva, Direct numerical simulation of a NACA 0012 in full stall, International Journal of Heat and Fluid Flow 43, 194–203 (2013).
10. V. Yanovych, D. Duda, V. Horáček and V. Uruba, "Research of a wind tunnel parameters by means of cross-section analysis of air flow profiles", AIP Conference Proceedings 2189, (2019).
11. V. Yanovych, D. Duda, V. Horáček and V. Uruba, "Creation of recombination corrective algorithm for research of a wind tunnel parameters", AIP Conference Proceedings 2118, (2019).
12. M. Phillip and T. Kunihiko. "Separation control on NACA 0012 airfoil using momentum and wall-normal vorticity injection", 32nd AIAA Applied Aerodynamics Conference AIAA 2014-2685, (2014).
13. Md. S. Islam, Sh. M. Hakim,M. Alil, and Md. Q. Islam, "Numerical investigation on boundary layer control through moving surface in NACA 0012 airfoil", AIP Conference Proceedings 1851, 020111 (2017).
14. K. Yousefi, R. Saleh, Three-dimensional suction flow control and suction jet length optimization of NACA 0012 wing, Meccanica 50, 1481–1494 (2015).
15. A.N. Kolmogorov, V. Levin, J.C.R. Hunt, O.M. Phillips, D. Williams, Dissipation of energy in the locally isotropic turbulence. *Proceedings of the Royal Society of London. Series A: Mathematical and Physical Sciences* **434**, 15–17 (1991).
16. J. L. Lumley, G. Newman, The return to isotropy of homogeneous turbulence. *J. Fluid Meach.* **82**, 161–178 (1977).
17. V. Yanovych, D. Duda, V. Uruba, Structure turbulent flow behind a square cylinder with an angle of incidence, *European Journal of Mechanics, B/Fluids* **85**, 110–123 (2021).
18. R. E. Sheldahl and P. C. Klimas, "Aerodynamic characteristics of seven symmetrical airfoil sections through 180-degree angle of attack for use in aerodynamic analysis of vertical axis wind turbines", Technical Report, United States, 1981.

Theoretical foundations of increasing the efficiency of wide-angle diffusers and practical results of new solutions to this problem

Arkadiy Zaryankin[1], *Mikhail* Cherkasov[1*], and *Ivan* Lavyrev[1]

[1]National Research University "Moscow Power Engineering Institute, Institute of power engineering and mechanic, Department of steam and gas turbines, Krasnokazarmennaya 14, 111250 Moscow, Russian Federation

Abstract. It is shown theoretically that the efficiency of wide-angle diffusers mainly depends on the method of supplying the working medium to the diffuser channel and on the nature of the interaction of this medium with the subsequent streamlined surface of the diffuser. The first condition is due to those changes in the acting force factors within the wall (boundary layer), which occur during the transition from the inlet confuser channel to the subsequent diffuser, and the second follows from the need to ensure an uninterrupted flow of the working medium moving against the increasing (in the direction of travel) static pressure. The above computational studies of wide-angle diffusers have shown that, subject to theoretically justified conditions, it is possible to ensure uninterrupted flow in flat and conical diffusers with an increase in the opening angle α of their flow path to 20 ° and with the same degrees of expansion ratio n to obtain pressure recovery coefficients commensurate with those in diffusers with an angle α = 7 °.

1 Introduction

Conical and annular diffusers with a straight axis are structurally the simplest, but very effective, devices that convert the kinetic energy of moving liquid and gaseous media into potential energy, determined by the value of static pressure in these media.

However, the final effect of such energy conversion depends on a number of additional factors, among which, naturally, the first place is occupied by the geometric parameters of the diffusers. With regard to conical and flat diffusers, such parameters are (see Fig. 1): α - opening angle of the flow path; $n = \frac{F_2}{F_1}$ - the expansion ratio of the channel, where F_1 is the area of the inlet section, and F_2 is the area of the outlet section); $\bar{L} = \frac{L}{D_1}$ – relative length (for flat diffusers $\bar{L} = \frac{L}{H_1}$), where D_1 and D_2 are the diameter of the inlet and outlet sections of the conical diffuser and H_1 and H_2 - are the heights of the inlet and outlet sections.

All three parameters are related by obvious relationships:

* Corresponding author: mikhail.cherkasov95@yandex.ru

$$\sqrt{n} = 1 + \frac{L}{D_1} tg \frac{\alpha}{2} \text{ for conical diffusers}$$

$$n = 1 + \frac{L}{h} tg \frac{\alpha}{2} \text{ for flat diffusers.}$$

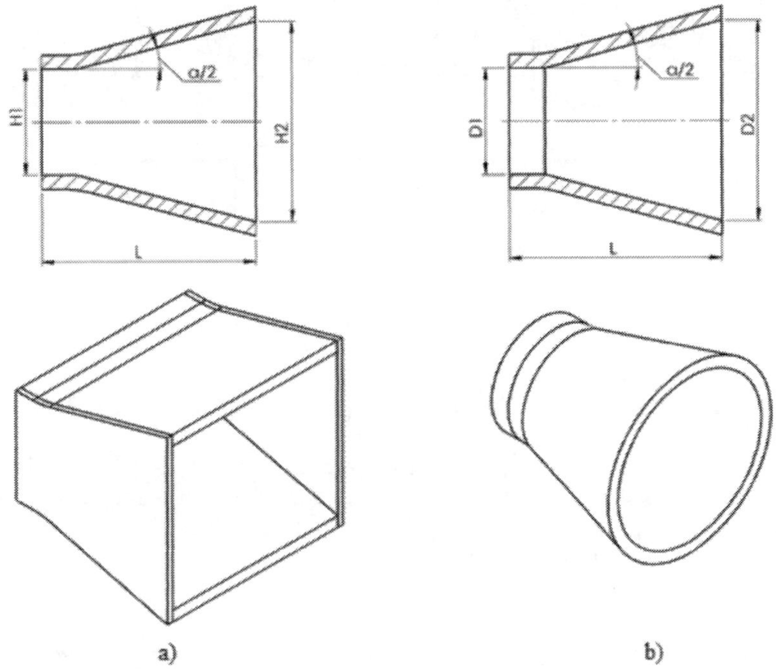

Fig. 1. Types of wide angle diffusers a) - flat diffuser; b) - conical diffuser

From the point of view of obtaining the maximum possible pressure increase in diffusers, the main parameter is the expansion ratio n, an increase in which theoretically leads to an incorrect increase in their efficiency.

However, with a fixed degree of expansion ratio n, the degree of pressure recovery (increase in potential energy) in diffusers is decisively influenced by the conditions for the formation of the velocity field when the flow enters the devices under consideration.

In flat and conical diffusers, these conditions largely depend on the opening angle of their flow path.

To illustrate the noted influence on Fig. 2 shows how the coefficient of total energy losses in conical diffusers changes from an increase in the expansion ratio at different values of the opening angle α [1].

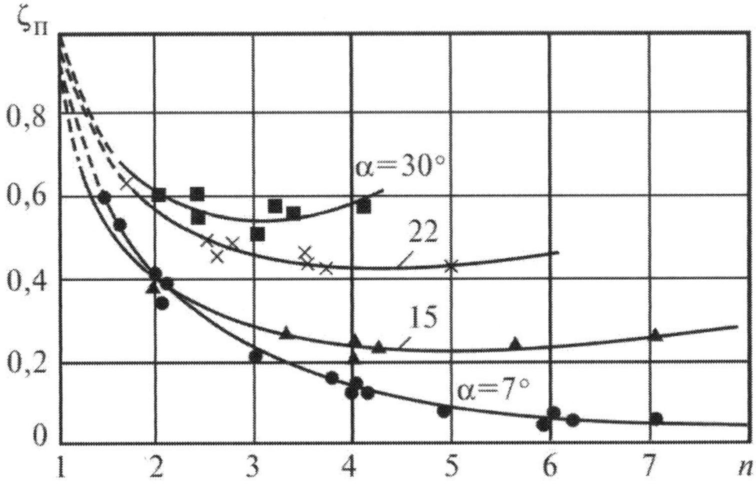

Fig. 2. Influence of the degree of expansion ratio of the conical diffuser n on the coefficient of total energy loss at α = const [1]

It is clearly seen here that only at the angle α = 7 ° there is a continuous decrease in the indicated coefficient.

For all values of the angle α exceeding 10 °, the dependence ζp = f (n, α) has clearly defined values of the coefficient ζp, at which this value reaches its minimum value. A further increase in the angle α leads to an increase in losses and, accordingly, to a decrease in the efficiency of the diffusers, due to the emerging separation of the flow from the walls of the diffuser channel.

Thus, the problem of increasing the efficiency of diffusers is reduced to solving theoretical problems associated, firstly, with the physical picture of the flow during the transition from the converging flow in the inlet part of the diffuser to the subsequent diffuser flow, and, secondly, to find a simple and effective method of elimination, or at least would be to the displacement of the section of separation of the flow from the streamlined surfaces in the direction of the outlet section of the diffuser.

Consider the theoretical and practical ways of solving the indicated tasks.

2 Theoretical foundations for the practical implementation of optimal conditions for the inlet of working media in flat and conical diffusers

The flow path of classical diffusers shown in Fig. 1 includes an inlet section designed to form a relatively uniform inlet velocity profile at the entrance to the diffuser, the magnitude of the irregularity of which significantly affects the nature of the flow in the subsequent diffuser channel. While maintaining a continuous flow and a fixed back pressure behind the diffuser P2 with an increase in the degree of expansion ratio of the channel n, the pressure P1 in the inlet section continuously decreases, which entails a continuous increase in the pressure recovery factor

$$\xi\ (\xi = (P_2 - P_1)/\left(\frac{\rho c_1^2}{2}\right)).$$

However, as already indicated in the introduction, in practice such a process is realized only at small opening angles of the flow path of the diffusers, not exceeding 10 °.

With an increase in the angle α over 10 °, in the absence of additional measures of external influence on the nature of the flow in the diffuser channel, flow separation from the walls occurs (almost always one-sided). Downstream of the section where flow separation occurred, a pressure close to the pressure P_2 behind the diffuser is established, and, accordingly, further pressure recovery does not occur, but an increase in energy losses takes place.

According to the data shown in Fig. 2, with an increase in the angle α, the section where the flow separation occurs (the minimum of the total loss coefficient in Fig. 2) shifts towards the inlet section of the diffuser and at α = 30 ° the flow separation occurs practically in the inlet section, and the change in the total loss coefficient ζp with an increase in the expansion ratio n corresponds to a similar dependence in a channel with a sudden expansion ratio of its transverse area.

The experimental data presented clearly indicate that it is the conditions of the flow entering the diffuser that determine the entire subsequent process of converting the kinetic energy of the flow into potential energy determined by the increase in static pressure when the flow of working media moves to the outlet section of the diffuser. In order to understand the physical picture of the change in the initial uniform velocity field when it approaches the inlet section of the diffuser, we present a standard inlet section and point out the reasons that occur during the transition from confuser to diffuser flow (see Fig. 3).

Purely structurally, such a transition most often occurs as a result of a simple attachment of the diffuser itself to the inlet part of the supply confuser. As a result, an angular bend of the streamlined surface occurs in the inlet section 1-1 (see Fig. 3). When flowing around such a surface devoid of an ideal fluid of friction forces, any of its angular kinks is a singular point where the velocity breaks.

In a real liquid, such a velocity discontinuity, of course, is absent, but the indicated angular bend is a source of strong disturbance leading to a local increase in velocity from the side of the confuser supply of the liquid to the diffuser and its no less intense local decrease at the inlet section of the diffuser.

As a result, at the inlet section of the diffuser, the nature of the change in the longitudinal velocity near the streamlined surface, as shown in Fig. 3 turns out to be significantly different from the average flow rate.

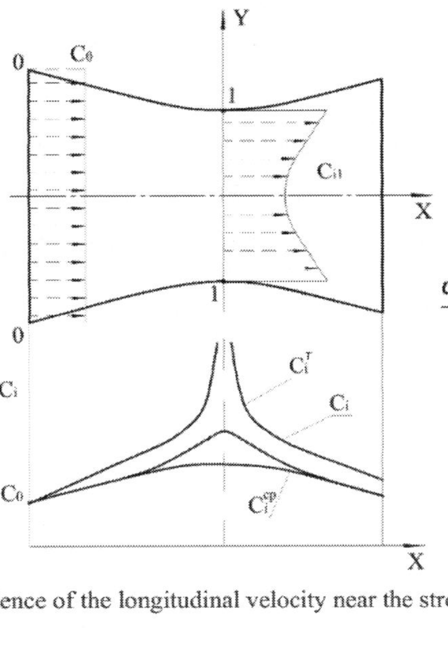

$$\left.\frac{dC_i^{cp}}{dX}\right|_{X=X_1} < \left.\frac{dC_i}{dX}\right|_{X=X_1}$$

Fig. 3. Dependence of the longitudinal velocity near the streamlined surface of the diffuser

A sharp increase in the local velocity in the near-wall region of the inlet section 1-1 from the side of the confuser channel with its subsequent no less sharp deceleration at the inlet part of the diffuser itself changes the entire velocity field in section 1-1 (see Fig. 3) and has a very strong effect on on the subsequent flow of the working medium in the subsequent channel.

The velocity field in the inlet section 1-1 of the diffuser takes the form of an "inverted" parabola in the meridial section with apex on the longitudinal axis of the channel under consideration (see Fig. 3).

With such a velocity field, there is a continuous increase in the transverse velocity gradient $\frac{dC_i}{dy}$ from the center of the channel to the periphery. An inevitable consequence of such a velocity profile is an increase in the indicated direction and shear stresses τ proportional to the transverse velocity gradient squared in a turbulent flow regime.

Thus, if, with a uniform (gradientless) velocity field in the transverse direction, the viscosity is limited by the boundary layer region, then in the presence of a flat conical or annular diffuser at the entrance to a flat conical or annular diffuser, the effect of the viscous properties of real liquid and gaseous media covers the entire cross section of the channel. In this case, in accordance with the energy equation at the inlet section of the diffuser in its cross sections, the pressure field also turned out to be nonuniform.

In addition, the fact of a sharp change in the local gradients of velocities and pressures in the region of the angular bend of the streamlined surface is very important (see Fig. 3). When approaching the corner point, the negative value of the local longitudinal pressure gradient near the wall sharply increases, followed by an equally sharp increase in the positive pressure gradient immediately after the flow inlet into the diffuser.

Accordingly, the local positive pressure gradient at the entrance to the diffuser turns out to be significantly higher than the average static value of this value, which significantly increases the probability of flow separation immediately at the entrance to the diffuser.

All the negative factors noted increase both with an increase in speed and with an increase in the opening of the diffuser, which contributes to an earlier separation of the flow from the streamlined surfaces.

It should also be noted that the velocity strongly varying at the inlet section of the diffusers, increasing from the wall to the center of the channel, excludes the possibility of determining the average pressure P1 in the inlet section 1-1 (see Fig. 3), since the local pressure measured on the wall of the inlet section , turns out to be significantly lower than the average flow pressure in this section.

In view of the foregoing, an obvious solution to reduce the unevenness of the velocity profile when the flow enters the diffuser is the elimination of the angular bend of the surface in this section by smooth radial conjugation of the converging inlet section with the subsequent diffuser.

The larger the junction radius, the smaller the difference between the local pressure on the wall and the pressure averaged over the flow rate in the inlet section of the diffuser, and the smaller the inlet positive pressure gradient. Accordingly, one should expect a lower probability of flow separation directly at the flow entry into the diffuser.

However, as studies in [2] have shown, this conclusion turns out to be true only partially.

Smooth flow entry into the diffuser really eliminates flow separation, but at diffuser opening angles above 10 °, local flow separation in the inlet section nevertheless occurs. In fig. 4 shows a diagram of a setup for studying diffusers, where pressure oscillograms were taken in a flat diffuser with an opening angle $\alpha = 15$ ° along its longitudinal axis using low-inertia pressure sensors [2].

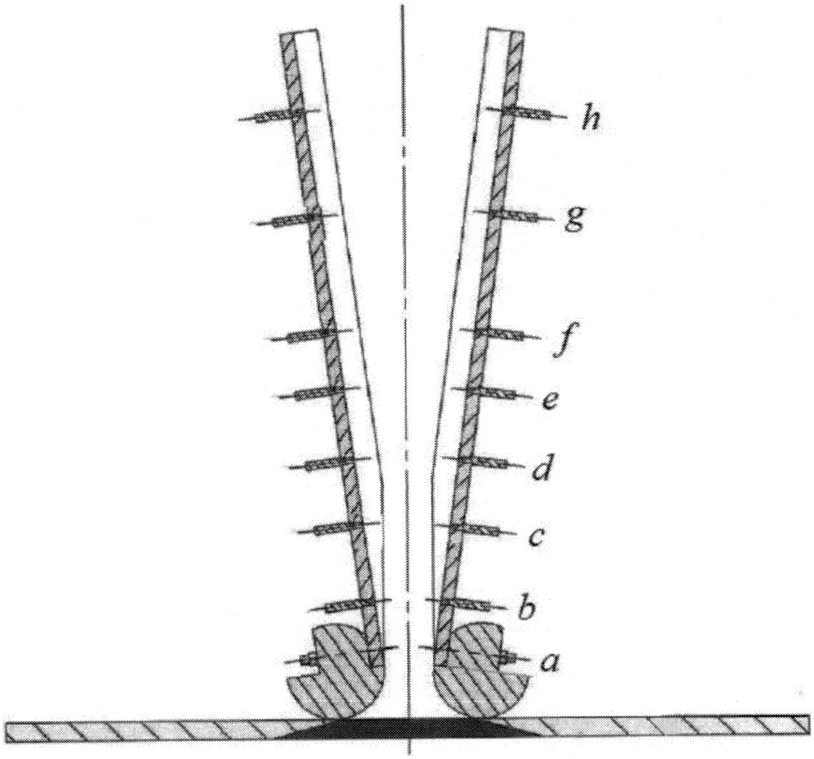

Fig. 4. Installation diagram of pressure sensors in a flat diffuser channel $\alpha = 15°$ [2]

The results of these measurements, shown in Fig. 5 clearly show the presence of maximum amplitudes of pulsations just after the inlet section of the diffuser, followed by a decrease in pressure fluctuations towards the outlet section.

This result requires a deeper analysis, which is given below. Its essence is based on L. Prandtl's averaged equation for a flat boundary layer:

$$\rho\left(u\frac{\partial u}{\partial x} + \vartheta\frac{\partial u}{\partial y}\right) = -\frac{dP}{dx} + \frac{d\tau}{dy} \qquad (1)$$

Where ρ - is the flux density; u, ϑ - velocity components in the Cartesian coordinate system; P- is the static pressure; $\tau = \tau_p + \tau_t$ is the shear stress in the boundary layer, equal to the sum of the shear stress caused by molecular friction (τ_p) and the shear stress (τ_t) caused by turbulence within the boundary layer.

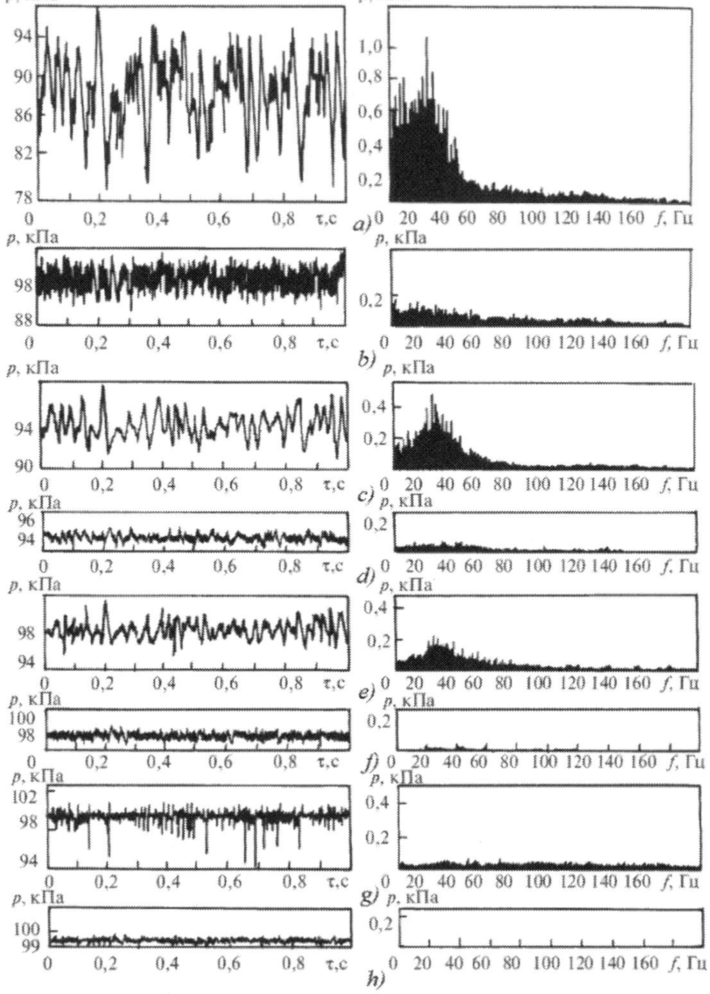

Fig. 5. Oscillogram of pressure pulsations on the wall of a flat diffuser [2]

For further analysis, this fact is very important that directly on the streamlined surface, where $u = \vartheta = 0$, from equation (1) it follows:

$$\frac{dP}{dx} = \frac{d\tau}{dy} \qquad (2)$$

That is, the sign of the transverse stress gradient in the boundary layer on a solid surface is always equal to the sign of the longitudinal pressure gradient.

Further, it should be noted that equations (1) is the balance equation of the force factors acting within the boundary layer.

Indeed, when multiplying all the terms of equation (1) by the elementary volume of liquid $dV = dx \cdot dy \cdot dz$, it turns out that the force due to the acting longitudinal pressure gradient $dF_p = dP \cdot dy \cdot dz$ is equalized by the inertial forces $dF_{ин} = \rho \cdot dx \cdot dy \cdot dz \left(u\frac{\partial u}{\partial x} + \vartheta\frac{\partial u}{\partial y}\right)$ and forces due to the gradient of the friction stress $dF_\tau = d\tau \cdot dx \cdot dz$.

As a result, the balance equation of the force factors acting within the boundary layer takes the following form:

$$dF_p = dF_\tau = dF_{in} \tag{3}$$

In this record, on the left, there is an external force action on a moving fluid, determined in this case by the geometry of the channel, and on the right side is the response of the flow to an external action that balances the magnitude of the specified action. As long as the above ratio of the force factors is fulfilled, the continuous flow in any channel is also maintained.

It is also important to note one more circumstance, which is that according to the second equation of L. Prandtl for the boundary layer, the transverse pressure gradient within it does not change in the direction of the y-axis. That is, $\frac{dP}{dy}$ in a fixed section of the channel remains constant and, therefore, in any cross section $\frac{dP}{dy} = 0$, and, therefore,

$$dF_p = const = dF_\tau - dF_{in} \tag{4}$$

From relation (4) follows a direct connection between the nature of the change in the velocity profile within the boundary layer with the profile of the change in the friction stress in this layer.

Based on the boundary conditions on the streamlined surface, where $\frac{dP}{dx} = \frac{d\tau}{dy}$, and on the outer boundary of the boundary layer, $\tau = 0$, we construct a qualitative picture of the change in shear stresses during the transition from confuser to diffuser flow (see Fig. 6).

Using the ones shown in Fig. 6 diagrams of the distribution of shear stresses, we will construct the same diagrams for the transverse gradients of friction stresses (see Fig. 7).

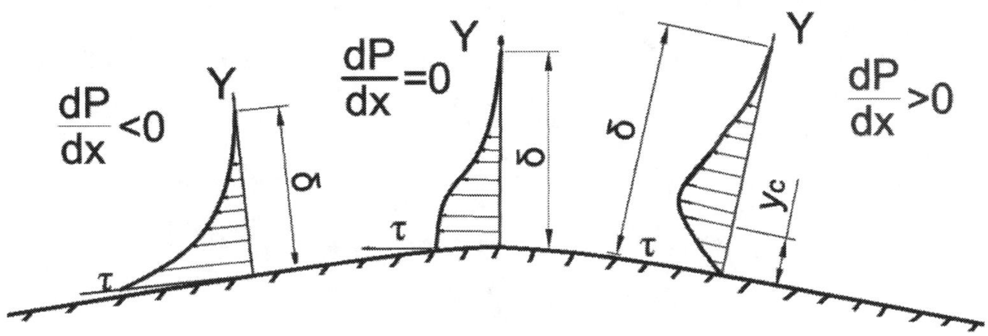

Fig. 6. Diagrams of changes in shear stresses during the transition from converging to diffuser flow

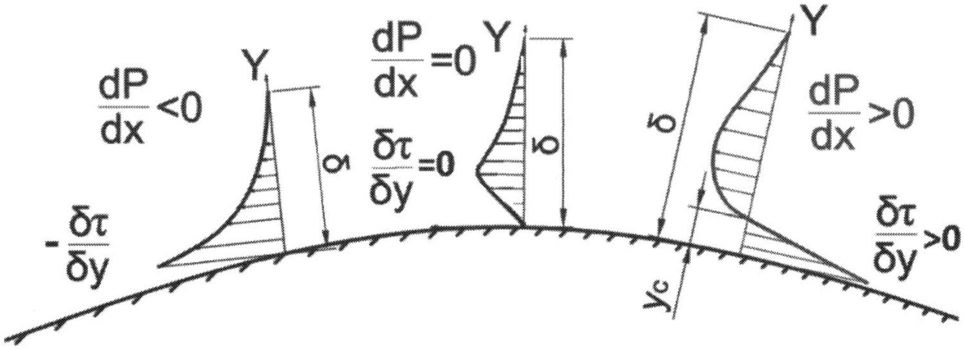

Fig. 7. Diagrams for transverse gradients of friction stresses

The presented distributions of transverse gradients of friction stresses $\frac{d\tau}{dy} = f(y)$ during the transition from confusor flow to diffuser flow show that at a relatively small linear distance, a fundamental change in the distribution pattern of the considered gradients of friction stresses occurs.

If in the confuser flow region (the inlet part of the fluid supply to the diffuser) the shear stresses act throughout the entire thickness of the boundary layer in the direction opposite to the fluid motion, then immediately after the flow enters the diffuser, an almost critical restructuring of the distribution of transverse gradients of tangential stresses and, consequently, forces occurs. caused by these stresses.

Near the streamlined surface at a certain distance yc from the wall, the transverse gradients of tangential stresses change their sign and begin to act in the direction of fluid movement, providing the possibility of fluid movement in the near-wall region against the increasing pressure in the diffuser channel.

In a real flow with inertia, such a restructuring of the acting forces at small linear distances cannot occur, which violates the above balance equation (4), and in practice leads to a local separation of the flow when the liquid enters the diffuser channel even with smooth coupling of the converging channel with diffuser channel (see Fig. 5).

The considered picture of the redistribution of force factors acting within the boundary layer during the transition from confuser to diffuser flow, for the first time theoretically substantiated the need to introduce a cylindrical section in diffusers after the inlet confuser, which then smoothly mates with the diffuser [3].

Note that constructively such a solution was used in the diffuser seats of all new control valves considered in [3].

Our laboratory studies of flat diffusers have shown that the changes made to the scheme of organizing the supply of working media to such diffusers really make it possible to eliminate the local flow separation at the entrance to the diffuser and sharply reduce the amplitude of pressure pulsations in this area. To illustrate what has been said in Fig. 8 shows two oscillograms of pressure pulsations at the inlet to a flat diffuser.

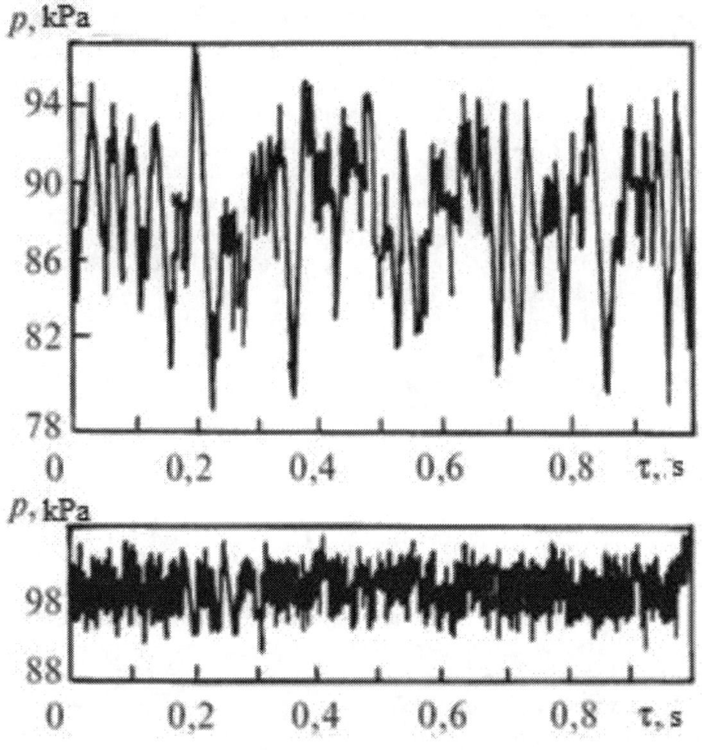

Fig. 8. Oscillograms of pressure pulsations at the inlet to a flat diffuser [2]

The first of them refers to the standard system for supplying air to the diffuser, and the second in the presence of a plane-parallel channel between the inlet confuser and the subsequent diffuser.

However, such structural changes, having reduced the level of dynamic loads on the walls of the diffusers, do not appreciably change the nature of the flow in their subsequent sections and, accordingly, do not lead to an expansion ratio of the maximum permissible expansion ratio angles α, which ensure uninterrupted flow in the considered devices.

To solve this main problem, a variety of geometric and aerodynamic methods of external influence on liquid and gaseous media moving in diffusers are used [1].

Among them, the most promising method for maintaining a continuous flow in wide-angle diffusers is a method based on changing the conditions of interaction of moving liquid and gaseous media with streamlined surfaces.

So in works [4, 5], by passing in flat and conical diffusers from smooth streamlined surfaces to longitudinally ribbed surfaces, it was possible to increase the limiting values of the expansion ratio angle of the flow parts of these diffusers up to 15 ° with a simultaneous decrease in several times (at an angle of 15 ° in 3, 5 times) vibration accelerations on their walls [2, 5].

Continuing this direction, let us consider the theoretical prerequisites for those changes in the shape of streamlined surfaces that can dramatically change the nature of the flow in wide-angle diffusers.

3 Theoretical substantiation of the advisability of using slotted surfaces in the direction of flow of streamlined surfaces

The basis for solving this problem is the classical logarithmic velocity profile, since it was he who formed the idea of the structure of the turbulent boundary layer.

It is this structure that until now underlies almost all works that consider the physical picture of the interaction of all liquid media with streamlined surfaces.

Consider the extent to which this structure corresponds to reality.

To answer this question, let us return to the origins, on the basis of which the structural diagram of the turbulent boundary layer was formed.

It is based on L. Prandtl's semi-empirical theory of turbulence, which made it possible to structurally establish the relationship between turbulent tangential stresses τ with the averaged longitudinal velocity $\bar{u}$ (we will omit the line in what follows).

$$\tau = \rho l^2 \left(\frac{du}{dy}\right)^2 \tag{5}$$

Where ρ - is the density of the moving liquid medium; l is the length of the mixing path, at a distance of which turbulent layers moving chaotically in a turbulent flow retain their momentum.

The above relation was taken as a differential equation that determines the velocity profile in a turbulent boundary layer. When integrating it, two conditions $l = \chi y$ and $\tau = \text{const}$ were accepted, which are satisfied only in the immediate vicinity of the streamlined surface.

Nevertheless, the subsequent integration of equation (5) was carried out without specifying the limit of integration. As a result of such integration, a well-known equation was obtained that relates the longitudinal velocity u in a cylindrical pipe with the transverse coordinate y.

$$\frac{u}{\vartheta_*} = \frac{1}{\chi} lny + C \tag{6}$$

Where $\vartheta_* = \sqrt{\frac{\tau}{\rho}}$, having the dimension of speed, is called "dynamic speed", and χ is some constant determined experimentally.

As always, the constant of integration C is determined from the boundary conditions of the problem being solved.

With regard to the velocity profile in a tube with a developed turbulent flow or in the boundary layer, the velocity u on the streamlined surface, due to the "sticking" hypothesis, should be equal to zero. For equation (6), this boundary condition, which is central in fluid dynamics, turned out to be impracticable, since at y = 0 the velocity u tends to infinity.

Hence, it quite naturally follows that the real velocity profile in a turbulent flow regime cannot be described by the obtained relation (6).

However, in this case, the hypothesis came to the rescue that turbulence cannot exist in the near-wall layer and a viscous laminar sublayer remains here, where the friction stresses are determined by Newton's formula ($\tau = \mu \frac{du}{dy}$) [6, 7].

In this regard, the constant of integration C in formula (6) should be determined not from the condition on the streamlined surface, but at a certain distance from it, equal to the thickness of the hypothetical laminar sublayer y_l at a velocity $u = u_l$.

As a result, using the experimental data of Nikuradze downstream in pipes, the now classical logarithmic velocity profile was obtained:

$$\frac{u}{\vartheta_*} = 2{,}5 ln \frac{\vartheta_* y}{\nu} + 5{,}5 \quad (7)$$

In fluid dynamics, this is apparently a natural case when the structure of the flow of a liquid or gas in the boundary layer was obtained not as a result of a direct physical experiment, but as a result of fitting this structure to dependence (6).

The thus obtained two-layer structure of the turbulent boundary layer turned out to be very convenient for the physical explanation of the independence of the drag coefficient on the Reynolds number when the fluid moves in rough pipes [6, 7].

At first glance, all textbooks on fluid dynamics give quite logical considerations about the real existence of a laminar (viscous) sublayer, and only A.P. Melnikov in 1956 [8] and L.D. Landau with E.M. Lifshitz in 1966 [9] asserted the turbulent nature of the flow in the entire cross section of the turbulent boundary layer.

This statement was substantiated in more detail in [10], where it was shown that Eq. (6) does not satisfy any of the three boundary conditions at all.

To prove this, we represent the logarithmic velocity profile (6) in the usual fixed coordinates $\frac{u}{u_t} = f\left(\frac{y}{\delta}\right)$:

$$\frac{u}{u_t} = \frac{\vartheta_*}{\chi u_t}\left(ln\frac{y}{\delta} + ln\frac{\vartheta_*\delta}{\nu} + C\right) \quad (8)$$

At the outer boundary of the boundary layer at y = δ, the relative velocity $\frac{u}{u_t}$ should be equal to 1, a $\frac{d\left(\frac{u}{u_t}\right)}{dy} = 0$ (naturally, the boundary conditions follow from the definition of the boundary layer). If we judge by equation (8) at $\frac{y}{\delta} = 1$, that is, on the outer boundary of the boundary layer:

$$\frac{u}{u_t} = \frac{\vartheta_*}{\chi u_t}\left(ln\frac{\vartheta_*\delta}{\nu} + C\right) > 1$$

$$\frac{d}{dX}\left(\frac{u}{u_t}\right) = \frac{\vartheta_*}{\chi u_t} > 0$$

The results obtained convincingly show that the logarithmic velocity profile does not satisfy any of the three natural boundary conditions, and it cannot be used in the presented form (8) even when creating a hypothesis of a laminar sublayer.

As for the "brilliant" coincidence of Nikuradze's experiments with the logarithmic profile in handicap (8), it was shown in [9] that when using "floating" coordinates that change with a change in the Reynolds number the dimensions of the quadrants where the corresponding velocity profile is located, then in this case the coinciding experimental points generally turn out to be in different sections of the compared velocity profiles.

The situation changes fundamentally if the integration constant is determined not from the condition on the wall, but from the condition on the outer boundary of the boundary layer, where at y = δ; u = ut. In this case, using relation (8), we obtain:

$$\mathrm{C} = \frac{u_t}{\vartheta_*} - ln\frac{\vartheta_*\delta}{\nu} \rightarrow \frac{u}{u_t} = 1 + \frac{\vartheta_*}{u_t} ln\left(\frac{y}{\delta}\right) \quad (9)$$

The logarithmic velocity profile in the form (9) practically satisfies all three boundary conditions without the need to introduce the concept of a laminar sublayer, since the ratio $\frac{u}{u_t} = 0$ at a distance from the wall $\frac{y}{\delta} = e^{-\frac{u_t}{\vartheta_*}}$

Since $\frac{u_t}{\vartheta_*}$ in Nikuradze's experiments varies within $10 \div 35$, then $\frac{y}{\delta} < 0{,}01$ mm.

The analysis shows that it is not a viscous sublayer that actually contacts the streamlined surface, but a turbulent flow with a sufficiently high friction stress, which makes it possible to overcome large positive longitudinal pressure gradients acting in the diffuser.

In support of what has been said, attention should be paid to the fact that a necessary condition for the onset of turbulence is the presence of a shear flow [11], the maximum value of which occurs precisely near the streamlined surface.

Accordingly, it is this region that is not a region of transition from a viscous flow to a subsequent turbulent one within the boundary layer, but a region of intense turbulence generation followed by its attenuation in the direction normal to the streamlined surface [1].

However, in spite of the above considerations about the two-layer structure of the turbulence of the velocity profile, the very idea of such a structure can turn out to be very fruitful if a stable laminar layer is created artificially between the turbulent flow and the streamlined surface, ensuring the contact of the turbulent flow not with a solid surface, but with an artificially created near the wall. a laminar interlayer of a moving liquid or gaseous medium.

Technically, this problem can be solved by passing from smooth streamlined surfaces to slit surfaces in the direction of the flow.

A variant of a flat diffuser with such walls is shown in Fig. 9. Here the streamlined surfaces are made with very narrow longitudinal slots, the width of which does not exceed b = 2 mm at the channel depth h = 7–10 mm and the thickness of the partitions between the slots a = 0.8–1 mm.

With this design of a streamlined surface in rectangular slots with a large relative depth $\frac{h}{b} \approx 5 \div 6$, due to the large influence of viscosity and small dimensions of the slots, the transition to a turbulent flow cannot occur. As a result, the turbulent flow external to the thin edges will no longer come into contact with a solid surface, where, according to the adhesion hypothesis, the velocity vanishes, but with a laminar flow, the velocity of which on the plane of contact with the external turbulent flow will no longer be zero, but much less than the external velocity flow.

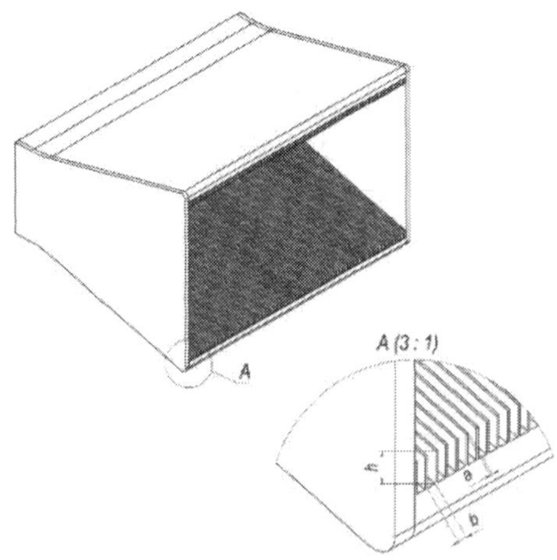

Fig. 9. Flat diffuser with ribbed surfaces

Such a measure will sharply reduce the transverse velocity gradient at the outer boundary of the ribbed surface and cause additional turbulization of the flow in the region of two flows with different velocities, thereby ensuring the possibility of maintaining an unseparated flow in diffusers with large degrees of expansion ratio n, expansion ratio angles α exceeding 10 ÷ 15 °.

4 Conclusions

1. As a result of the analysis, the deformation of the velocity profile during the supply of working media to the inlet section of the diffuser showed that this deformation primarily limits the opening angles of the flow paths of flat and conical diffusers, exceeding which causes a local separation of the flow immediately behind the inlet section.
2. It is shown that in order to prevent local flow separation when the flow enters the diffuser between the supplied confuser and the subsequent diffuser, it is necessary to introduce a plane-parallel and cylindrical section, which provides the possibility of restructuring the direction of the acting transverse gradients of friction stresses in the near-wall region from negative to positive values.
3. The analysis of the logarithmic velocity profile in a turbulent boundary layer does not correspond to any of the three natural boundary conditions arising from the definition of the boundary layer, and its two-layer structure, which asserts the existence of a viscous (laminar) sublayer near the walls, is only a consequence of the desire to impart a logarithmic velocity profile physical meaning.
4. Using the hypothesis of a laminar sublayer, it is proposed to artificially create a stable laminar flow near the wall by passing to streamlined slit surfaces, and thus ensuring contact of the main stream not with a solid wall, but with a laminar flow in the interslit space.

References

1. Zaryankin A.E. Mechanics of incompressible and compressible fluids. Publishing House MPEI 2014.
2. Zaryankin A.E., Gribin V.G., Paramonov A.N., Fichoryak O.M., Noskov V.V. The influence of the opening angle of diffusers on their vibration state and some ways to reduce these vibrations, Teploenergetika 2012, no. 9 p. 27-36.
3. Zaryankin A.E., Simonov B.P. Steam turbine control and one hundred-porn-control valves. Moscow: MPEI Publishing House, 2005.
4. Grigoriev E.Yu. Development and research of ways to reduce vibration of annular diffusers of gas turbines: dis ... cand. tech. Sciences: 05.04.12 / Grigoriev E.Yu. –M .: 2014. - 259 p.
5. Noskov V.V. Experimental study of flow modes on the vibrodynamic state of diffuser elements of the flow parts of turbomachines: dis ... cand. tech. Sciences: 05.04.12 / Noskov V.V. –M .: 2010. - 230 p.
6. Loytsyansky L.G. Mechanics of liquids and gases. Publishing house "Science" 1970.
7. Schlichting G. Theory of the boundary layer. Publishing house "Science" 1969.
8. Melnikov A.P. Fundamentals of Theoretical Aerodynamics. LKVVIA Publishing House 1953.
9. Landau L.D., Lifshits E.M. Continuum mechanics. SHTTL. M. 1954 - 532 p.

10. Zaryankin A.E. To the question of a logarithmic velocity profile correspondence with the experimental
11. Schubauer G.B., Chen K.M. Turbulent flow and heat transfer / Per. from English // ed. Lin-Jia-Jio. Moscow: Foreign Literature Publishing House. 1963 .- 563 p.

AUTHOR INDEX

EDP Sciences – Web of Conferences
17, Avenue du Hoggar
Parc d'Activité de Courtabouf
BP 112
F-91944 Les Ulis Cedex A
France

ISBN 978-1-7138-3801-2

CURRAN ASSOCIATES INC.
proceedings.com